Environmental
Pollution
and Control

Environmental Pollution
and Control

FOURTH EDITION

J. Jeffrey Peirce
Duke University

Ruth F. Weiner
Sandia National Laboratories

P. Aarne Vesilind
Duke University

Butterworth-Heinemann
An Imprint of Elsevier

Boston   Oxford   Johannesburg   Melbourne   New Delhi   Singapore

Butterworth–Heinemann
An Imprint of Elsevier

 Recognizing the importance of preserving what has been written, Elsevier prints its books on acid-free paper whenever possible.

 Elsevier supports the efforts of American Forests and the Global ReLeaf program in its campaign for the betterment of trees, forests, and our environment.

Library of Congress Cataloging-in-Publication Data

Peirce, J. Jeffrey.
   Environmental pollution and control. — 4th ed. / J. Jeffrey
Peirce, Ruth F. Weiner, P. Aarne Vesilind.
     p. cm.
   Rev. ed. of: Environmental pollution and control / P. Aarne
Vesilind. 3rd ed. ©1990.
   Includes bibliographical references (p.   ) and index.
    ISBN-13: 978-0-7506-9899-3   ISBN-10: 0-7506-9899-3 (alk. paper)

    1. Environmental engineering. I. Weiner, Ruth F. II. Vesilind,
P. Aarne. III. Vesilind, P. Aarne. Environmental pollution and
control. IV. Title.
TD145.V43 1997
628—DC21   97-20292
                                              CIP

British Library Cataloguing-in-Publication Data
A catalogue record for this book is available from the British Library.

The publisher offers special discounts on bulk orders of this book.
For information, please contact:
Manager of Special Sales
Butterworth-Heinemann
225 Wildwood Avenue
Woburn, MA 01801-2041
Tel: 781-904-2500
Fax: 781-904-2620
For information on all Butterworth-Heinemann publications
available, contact our World Wide Web home page at:
http://www.bh.com

Transferred to Digital Printing 2009

To

Elizabeth Davis Rasnic, Shayn, and Leyf
Lisa, Annie, Sarah, and Rachel
Pamela, Steve, and Lauren

Contents

Preface

Since this book was first published in 1972, several generations of students have become environmentally aware and conscious of their responsibilities to planet earth. Many of these environmental pioneers are now teaching in colleges and universities, and have students with the same sense of dedication and resolve that they themselves brought to the discipline. In those days, it was sometimes difficult to explain what environmental science or engineering was, and why the development of these fields was so important to the future of the earth and to human civilization. Today there is no question that the human species has the capability of destroying its home and that we have taken major steps toward doing exactly that.

And yet, while much has changed in a generation, much has not. We still have air pollution; we still contaminate our water supplies; we still dispose of hazardous materials improperly; we still destroy natural habitats as if no other species mattered. And, worst of all, we still populate the earth at an alarming rate. The need for this book, and for the college and university courses that use it as a text, continues; it is perhaps more acute now than it was several decades ago.

Although the battle to preserve the environment is still raging, some of the rules have changed. Now we must take into account risk to humans and be able to manipulate concepts of risk management. With increasing population, and fewer alternatives to waste disposal, this problem has intensified. Environmental laws have changed and will no doubt continue to evolve. The economic cost of preservation and environmental restoration continues to increase. Attitudes toward the environment are often couched in what has become known as the environmental ethic. Finally, the environmental movement has become politically powerful, and environmentalism sometimes can be made to serve a political agenda.

In revising this book, we incorporate the evolving nature of environmental sciences and engineering by adding chapters as necessary and eliminating material that is less germane to today's students. We have nevertheless maintained the essential feature of this book—the packaging of the more important aspects of environmental engineering science and technology in an organized manner and the presentation of this mainly technical material to a nonengineering audience.

This book has been used as a text in courses that require no prerequisites, although a high school knowledge of chemistry is important. A knowledge of college-level algebra is also useful, but calculus is not required for an understanding of the technical and scientific concepts.

We do not intend this book to be scientifically and technically complete. In fact, many complex environmental problems have been simplified to the threshold of pain for many engineers and scientists. Our objective, however, is not to impress nontechnical students with the rigors and complexities of pollution control technology but rather to make some of the language and ideas of environmental engineering and science more understandable.

J. Jeffrey Peirce
Ruth F. Weiner
P. Aarne Vesilind

Chapter 1

Pollution and Environmental Ethics

*"If seven maids with seven mops
Swept it for half a year,
Do you suppose," the Walrus said,
"That they could get it clear?"
"I doubt it," said the Carpenter,
And shed a bitter tear.*
—Lewis Carroll

Could the Walrus and the Carpenter have been talking about our earth? And is the situation really this grim? Is it time to start shedding bitter tears, or is there something we can do to control environmental pollution?

The objective of this book is to at least begin to answer these questions. As the title suggests, this book focuses first on the problems of environmental pollution, but then concentrates on methods of control—what we humans can do to prevent and control the pollution of our planet.

We define *environmental pollution* as the contamination of air, water, or food in such a manner as to cause real or potential harm to human health or well-being, or to damage or harm nonhuman nature without justification. The question of when harm to nonhuman nature is justified is a sticky one and is addressed below in the discussion on ethics.

In this first chapter we begin by asking why we seem to have such problems with environmental pollution. Where do these problems originate, and what or who is to blame for what many consider to be the sorry state of the world? Next we discuss our environmental problems within the framework of ethics. We begin by showing how the most basic concepts of environmental pollution that reflect public health concerns are really ethical issues. We then discuss how these ethics have been used to extend the concerns with pollution beyond public health to include the despoliation of the planet, including the extinction of species and destruction of places. All of these problems are still within the context of harm to humans. Finally, we discuss issues that have nothing to do with public health or human well-being, but nevertheless are important to us in terms of environmental quality.

FIGURE 1–1. Human excreta disposal, from an old woodcut. [Source: Reyburn, W., *Flushed with Pride*, London: McDonald (1969).]

THE ROOTS OF OUR ENVIRONMENTAL PROBLEMS

Much of the history of Western civilizations has been characterized as exploitation, destruction, and noncaring for the environment. Why are we such a destructive species? Various arguments have been advanced to explain the roots of our environmentally destructive tendencies, including our religions, our social and economic structure, and our acceptance of technology.

Religion. In the first chapter of Genesis, people are commanded by God to subdue nature, to procreate, and to have dominion over all living things. This anthropocentric view of nature runs through the Judaeo-Christian doctrine, placing humans at the pinnacle of development and encouraging humans to use nature as we see fit.

In his essay, "The Historical Roots of Our Ecological Crisis," Lynn White argues that those who embrace the Judaeo-Christian religions are taught to treat nature as an enemy and that natural resources are to be used to meet the goals of human survival and propagation. From this dogma (so goes the argument) have developed technology and capitalistic economy, and, ultimately, environmental degradation.

Because the Judaeo-Christian traditions are most prominent in the United States, we often forget that this is not a majority religious tradition in the world. Billions of people embrace very different deities and dogmas, and yet they also live in capitalistic economies with perhaps even greater destruction of environmental quality. So it cannot be just the Judaeo-Christian religions that are to blame.

Remember also that Christianity and Islam both developed at a time when there were a number of competing religions from which to choose. For many, the Christian ideas and ethics derived from the Judaic traditions seemed to fit most comfortably with their existing ethics and value systems, while others chose Islam over other religions. It seems quite obvious that Christianity was not the *reason* for the development of science, capitalism, and democracy, but simply provided an ethical environment in which they flourished (at least in Europe). It seems farfetched, therefore, to blame our environmental problems on our religions.

Social and Economic Structures. Perhaps it is our social structures that are responsible for environmental degradation. Garrett Hardin's "The Tragedy of the Commons" illustrates this proposition with the following story:[1]

> A village has a common green for the grazing of cattle, and the green is surrounded by farmhouses. Initially, each farmer has one cow, and the green can easily support the herd. Each farmer realizes, however, that if he or she gets another cow, the *cost* of the additional cow to the farmer is negligible because the cost of maintaining the green is shared, but the *profits* are the farmer's alone. So one farmer gets more cows and reaps more profits, until the common green can no longer support anyone's cows, and the system collapses.

Hardin presents this as a parable for overpopulation of the earth and consequent resource depletion. The social structure in the parable is capitalism—the individual ownership of wealth—and the use of that wealth to serve selfish interests. Does that mean that noncapitalist economies (the totally and partially planned economies) do a better job of environmental protection, natural resource preservation, and population control?

The collapse of the Soviet Union in 1991 afforded the world a glimpse of the almost total absence of environmental protection in the most prominent socialist nation in the developed world. Environmental devastation in the Commonwealth of Independent States (the former USSR) is substantially more serious than in the West. In the highly structured and centrally controlled communist

---

[1]Hardin, G., "The Tragedy of the Commons," *Science* 162 (1968): 1243.

system, *production* was the single goal and environmental degradation became unimportant. Also, there was no such thing as "public opinion," of course, and hence nobody spoke up for the environment. When production in a centrally controlled economy is the goal, all life, including human life, is cheap and expendable.[2]

Some less industrialized societies, such as some Native American tribes, the Finno-Ugric people of northern Europe, and the Pennsylvania Amish in the United States, have developed a quasi–steady-state condition. These sociopolitical systems incorporate animistic religion, holding that nature contains spirits that are powerful, sometimes friendly, and with whom bargains can be struck. The old Estonians and Finns, for example, explained to the spirit of the tree why cutting it down was necessary.[3] As another example, Estonians began the wheat harvest by putting aside a shaft of wheat for the field mice. This mouse-shaft (*hiirevihk*) did not appear to have religious significance; it was explained as a means of assuring the mice of their share of the harvest.[4]

These societies were not all environmentally stable, however, nor did they deliberately act to protect their environment. Those that are still in existence coexist with the industrialized societies that have not achieved a steady state, use the products and marketing mechanisms of those states, and lose their young people to societies where there is wider opportunity. Society is the reflection of the needs and aspirations of the people who establish and maintain it. Re-establishment of a nonindustrialized society would be doomed to failure, because such societies have already demonstrated that they do not meet people's needs.

The democratic societies of the developed world have in fact moved consciously toward environmental and resource protection more rapidly than either totally planned economies or the less developed nations. The United States has the oldest national park system in the world, and pollution control in the United States predates that of other developed nations, even Canada, by about 15 years.

So much for blaming capitalism.

Science and Technology. Perhaps the problem is with science and technology. It has become fashionable to blame environmental ills on increased knowledge of nature (science) and the ability to put that knowledge to work (engineering). During the industrial revolution the Luddite movement in England violently resisted the change from cottage industries to centralized factories; in the 1970s a pseudo-Luddite "back-to-nature" movement purported to reject technology altogether. However, the adherents of this movement made considerable use of the fruits of the technology they eschewed, like used vans and buses, synthetic fabrics, and, for that matter, jobs and money.

---

[2]Solzhenytsin, A., *The Gulag Archipelago,* New York: Bantam Books (1982).
[3]Paulsen, I., *The Old Estonian Folk Religion,* Bloomington, IN: Indiana University Press (1971).
[4]According to F. Oinas of the University of Indiana.

People who blame science and technology for environmental problems forget that those who alerted us early to the environmental crisis, like Rachel Carson in *Silent Spring*,[5] Aldo Leopold in *A Sand County Almanac*,[6] and Barry Commoner in *The Closing Circle*,[7] were *scientists*, sounding the environmental alarm *as a result of scientific observation*. Had we not observed and been able to quantify phenomena like species endangerment and destruction, the effect of herbicides and pesticides on wildlife, the destruction of the stratospheric ozone layer, and fish kills due to water pollution, we would not even have realized what was happening to the world. Our very knowledge of nature is precisely what alerted us to the threats posed by environmental degradation.

If knowledge is value-free, is technology to blame? If so, less technologically advanced societies must have fewer environmental problems. But they do not. The Maori in New Zealand exterminated the moa, a large flightless bird; there is considerable overgrazing in Africa and on the tribal reservations in the American Southwest; the ancient Greeks and Phoenicians destroyed forests and created deserts by diverting water. Modern technology, however, not only provides water and air treatment systems, but continues to develop ways in which to use a dwindling natural resource base more conservatively. For example, efficiency of thermal electric generation has doubled since World War II, food preservation techniques stretch the world's food supply, and modern communications frequently obviate the need for energy-consuming travel, and computer use has markedly decreased the use of paper.

If technology is not to blame, does it have the "wrong" values, or is it value-free? Is knowledge itself, without an application, right or wrong, ethical or unethical? J. Robert Oppenheimer faced this precise dilemma in his lack of enthusiasm about developing a nuclear fusion bomb.[8] Oppenheimer considered such a weapon evil in itself. Edward Teller, usually credited with its development, considered the H-bomb itself neither good nor evil, but wished to keep it out of the hands of those with evil intent (or what he perceived to be evil intent). The developers of the atomic bomb, although defending the position that the bomb itself was value-free, nonetheless enthusiastically promoted the peaceful uses of atomic energy as a balance to their development of a weapon of destruction. The ethics of technology is so closely entwined with the ethics of the uses of that technology that the question of inherent ethical value is moot. On balance, technology can be used to both good and evil ends, depending on the ethics of the users.

Assessment of the ethics of the use of any technology depends on our knowledge and understanding of that technology. For example, at this writing, scientists are investigating whether or not proximity to the electric and magnetic

---

[5]Carson, Rachel, *Silent Spring*.

[6]Leopold, Aldo, *A Sand County Almanac*, New York: Oxford University Press (1949).

[7]Commoner, Barry, *The Closing Circle*.

[8]Newhouse, J., *War and Peace in the Nuclear Age*, New York: Alfred A. Knopf (1988).

fields associated with electric power transmission increases cancer risk. Clearly, the ethics associated with transmission line location depends on the outcome of these investigations. Acceptance or rejection of any technology on ethical grounds must depend on an understanding of that technology. The weakness of the Luddite argument lay in the Luddites' ignorance of what they were fighting.

We seem to be left with little to blame for environmental pollution and destruction except ourselves. That is, if we are to reverse the trend in environmental degradation, we need to change the way we live, the way we treat each other and our nonhuman environment. Such ideas can be connected by what has become known as *environmental ethics*. Environmental ethics is a complex term and requires some explanation. First, we need to understand what we mean by *ethics* and what justification we have for wishing that everyone be ethical.

ETHICS

Ethics is the systematic analysis of morality. Morality, in turn, is the perceptions we have of what is right and wrong, good or bad, or just or unjust. We all live by various moral values such as truth and honesty. Some, for example, find it very easy to tell lies, while others will almost always tell the truth. If all life situations required nothing more than deciding when to tell the truth or when to lie, there would be no need for ethics. Very often, however, we find ourselves in situations when some of our moral values conflict. Do we tell our friend the truth, and risk hurting his feelings, or do we lie and be disloyal? How do we decide what to do? Ethics makes it possible to analyze such moral conflicts, and people whose actions are governed by reflective ethical reasoning, taking into account moral values, are said to be ethical people.

We generally agree among ourselves to be ethical (that is, to use reflective and rational analysis of how we ought to treat each other) because to do so results in a better world. If we did not bother with morality and ethics, the world would be a sorry place, indeed. Imagine living in an environment where nobody could be trusted, where everything could be stolen, and where physically hurting each other at every opportunity would be normal. While some societies on this globe might indeed be like that, we must agree that we would not want to live under such conditions. So we agree to get along and treat each other with fairness, justice, and caring, and to make laws to govern those issues of greatest import and concern.

The most important point relative to the discussion that follows is that ethics only makes sense if we assume reciprocity—the ability of others to make rational ethical decisions. You don't lie to your friend, for example, because you don't want him or her to lie to you. To start lying to each other would destroy the caring and trust you both value. Truth-telling therefore makes sense because of the social contract we have with others, and we expect others to participate. If they do not, we do not associate with them, or if the breach of the contract is great enough, we send them to jail and remove them from society.

Environmental ethics is a subcategory of ethics. Its definition can be approached from three historical perspectives: environmental ethics as public health, environmental ethics as conservation and preservation, and environmental ethics as caring for nonhumans.

ENVIRONMENTAL ETHICS AS PUBLIC HEALTH

During the middle of the nineteenth century, medical knowledge was still comparatively primitive, and public health measures were inadequate and often counter-productive. The germ theory of disease was not as yet appreciated, and great epidemics swept periodically over the major cities of the world. Some intuitive public health measures did, however, have a positive effect. Removal of corpses during epidemics and appeals for cleanliness undoubtedly helped the public health.

We in modern-day America have difficulty imagining what it must have been like in cities and farms not too many years ago.

Life in cities during the Middle Ages, and through the Industrial Revolution, was difficult, sad, and usually short. In 1842, the *Report from the Poor Law Commissioners on an Inquiry into the Sanitary Conditions of the Labouring Population of Great Britain* described the sanitary conditions in this manner:

> Many dwellings of the poor are arranged around narrow courts having no other opening to the main street than a narrow covered passage. In these courts there are several occupants, each of whom accumulated a heap. In some cases, each of these heaps is piled up separately in the court, with a general receptacle in the middle for drainage. In others, a plot is dug in the middle of the court for the general use of all the occupants. In some the whole courts up to the very doors of the houses were covered with filth.

The 1850s witnessed what is now called the "Great Sanitary Awakening." Led by tireless public health advocates like Sir Edwin Chadwick in England and Ludwig Semmelweiss in Austria, proper and effective measures began to evolve. John Snow's classic epidemiological study of the 1849 cholera epidemic in London stands as a seminal investigation of a public health problem. By using a map of the area and thereon identifying the residences of those who contracted the disease, Snow was able to pinpoint the source of the epidemic as the water from a public pump on Broad Street. Removal of the handle from the Broad Street pump eliminated the source of the cholera pathogen, and the epidemic subsided.[9] Ever since, waterborne diseases have become one of the major concerns of the public health. The reduction of such diseases by providing safe and pleasing water to the public has been one of the dramatic successes of the public health profession.

---

[9] Interestingly, it wasn't until 1884 that Robert Koch proved that *Vibrio comma* was the microorganism responsible for the cholera.

Public health has historically been associated with the supply of water to human communities. Permanent settlements and the development of agricultural skills were among the first human activities to create a cooperative social fabric. As farming efficiency increased, a division of labor became possible and communities began to build public and private structures. Water supply and wastewater drainage were among the public facilities that became necessary for human survival in communities, and the availability of water has always been a critical component of civilizations.[10] Some ancient cities developed intricate and amazingly effective systems, even by modern engineering standards. Ancient Rome, for example, had water supplied by nine different aqueducts up to 80 km (50 mi) long, with cross-sections from 2 to 15 m (7 ft to 50 ft). The purpose of the aqueducts was to carry spring water, which even the Romans knew was better to drink than Tiber River water.

As cities grew, the demand for water increased dramatically. During the eighteenth and nineteenth centuries, the poorer residents of European cities lived under abominable conditions, with water supplies that were grossly polluted, expensive, or nonexistent. In London, the water supply was controlled by nine different private companies, and water was sold to the public. People who could not afford to pay often begged for or stole their water. During epidemics, the privation was so great that many drank water from furrows and depressions in plowed fields. Droughts caused water supplies to be curtailed, and great crowds formed to wait their turn at the public pumps.

In the New World, the first public water supply system consisted of wooden pipes, bored and charred, with metal rings shrunk on the ends to prevent splitting. The first such pipes were installed in 1652, and the first citywide system was constructed in Winston-Salem, North Carolina, in 1776. The first American water works was built in the Moravian settlement of Bethlehem, Pennsylvania. A wooden water wheel, driven by the flow of Monocacy Creek, powered wooden pumps that lifted spring water to a hilltop wooden reservoir from which it was distributed by gravity. One of the first major water supply undertakings in the United States was the Croton Aqueduct, started in 1835 and completed six years later, that brought clear water to Manhattan Island, which had an inadequate supply of groundwater.·

Although municipal water systems might have provided adequate quantities of water, the water quality was often suspect. As one writer described it, tongue firmly in cheek:[11]

> The appearance and quality of the public water supply were such that the poor used it for soup, the middle class dyed their clothes in it, and the very rich used it for top-dressing their lawns. Those who drank it filtered it through a ladder, disinfected it with chloride of lime, then lifted out the dangerous germs which survived and killed them with a club in the back yard.

---

[10]A fascinating account of the importance of water supply to a community through the ages may be found in James Michener's novel *The Source.*

   ith, G., *Plague on Us,* Oxford: Oxford University Press (1941).

Water filtration became commonplace toward the middle of the nineteenth century with the first successful water supply filter constructed in Parsley, Scotland, in 1804. Many less successful attempts at filtration followed, a notable one being the New Orleans system for filtering water from the Mississippi River. In this case the water proved to be so muddy that the filters clogged too fast for the system to be workable. The problem with muddy water was not alleviated until aluminum sulfate (alum) began to be used as a pretreatment to filtration in 1885. Disinfection of water with chlorine began in Belgium in 1902 and in America, in Jersey City, New Jersey, in 1908. Between 1900 and 1920 deaths from infectious disease dropped dramatically, owing in part to the effect of cleaner water supplies.

Human waste disposal in early cities was both a nuisance and a serious health problem. Often the method of disposal consisted of nothing more than flinging the contents of chamberpots out the window. Around 1550, King Henri II repeatedly tried to get the Parliament of Paris to build sewers, but neither the king nor Parliament proposed to pay for them. The famous Paris sewer system was finally built under Napoleon III, in the nineteenth century.

Stormwater was considered the main drainage problem, and it was in fact illegal in many cities to discharge wastes into the ditches and storm sewers. Eventually, as water supplies developed,[12] the storm sewers were used for both sanitary waste and stormwater. Such *combined sewers* exist in some of our major cities even today.

The first system for urban drainage in America was constructed in Boston around 1700. There was surprising resistance to the construction of sewers for waste disposal. Most American cities had cesspools or vaults, even at the end of the nineteenth century, and the most economical means of waste disposal was to pump these out at regular intervals and cart the waste to a disposal site outside the town. Engineers argued that although sanitary sewer construction was capital intensive, sewers provided the best means of wastewater disposal in the long run. Their argument prevailed, and there was a remarkable period of sewer construction between 1890 and 1900.

The first separate sewerage systems in America were built in the 1880s in Memphis, Tennessee, and Pullman, Illinois. The Memphis system was a complete failure because it consisted of small-diameter pipes, intended to be flushed periodically. No manholes were constructed, and because the small pipes clogged, cleanout became a major problem. The system was later removed and larger pipes, with manholes, were installed.[13]

Wastewater treatment first consisted only of screening for removal of the large floatables to protect sewage pumps. Screens had to be cleaned manually, and wastes were buried or incinerated. The first complete treatment systems

---

[12]In 1844, to limit the quantity of wastewater discharge, the city of Boston passed an ordinance prohibiting the taking of baths without doctor's orders.

[13]American Public Works Association, *History of Public Works in the United States, 1776–1976,* Chicago: American Public Works Association (1976).

were operational by the turn of the century, with land spraying of the effluent being a popular method of final wastewater disposal.

The quest for public health also drives the concern with the extinction of species. Not too many years ago the public would have agreed with a paper mill executive when he said, "It probably won't hurt mankind a hell of a whole lot in the long run if a whooping crane doesn't quite make it."[14] The opposing view, that preservation of species and species diversity is at least as important as economic development, is now recognized as having significant public health import. Once a species is extinct, its unique chemical components will no longer be available to us for making medicines or other products. Because of this concern, the extinction of species has been codified as the federal Endangered Species Act and numerous state laws. Note that the driving force in these laws is not the value of the species itself but its potential value to human beings.

In summary, the first form of the environmental ethic makes the destruction of resources and despoliation of our environment unethical because doing so might cause other humans to suffer from diseases. Our unwillingness to clean up after ourselves is unethical because such actions could make other people sick or prevent them from being cured of disease. Because ethics involves a social contract, the rationale for the *environmental* ethic in this case is that we do not want to hurt other people by polluting the environment.

ENVIRONMENTAL ETHICS AS CONSERVATION AND PRESERVATION

A second form of the environmental ethic recognizes that nonhuman nature has value to humans above and beyond our concern for public health. We realize that the destruction or despoliation of the environment would be taking something from others—not much different from stealing. A river, for example, has value to others as a place to fish, and contaminating it takes something from those people. Cutting down old growth forests prevents us and our progeny from enjoying such wilderness, and such actions are therefore unethical.

The concept that nature has value is a fairly modern one. Until the mid-nineteenth century, nature was thought of as something to fight against—to destroy or be destroyed by. The value in nature was first expressed by several farsighted writers, most notably Ralph Waldo Emerson. He argued that nature had *instrumental value* to people, in terms of material wealth, recreation potential, and aesthetic beauty. Instrumental value can usually be translated into economic terms, and the resulting environmental ethic (from this argument) requires us to respect that value and not to destroy what others may need or enjoy. The concern of Theodore Roosevelt and Gifford Pinchot about the destruction of American forests[15] was not because they believed that somehow the

---

[14]Fallows, J.M., *The Water Lords,* New York: Bantam Books (1971).

Nash, R., *The American Environment,* New York: Prentice-Hall (1976).

forests had a right to survive but because they felt that these resources should be conserved and managed for the benefit of all. Such an environmental ethic can be thought of as *conservation environmental ethics* because its main aim is to conserve the resources for our eventual long-term benefit.

A modified form of the conservation environmental ethic evolved during this time, championed by John Muir, the founder of The Sierra Club and an advocate for the preservation of wilderness. This *preservation environmental ethic* held that some areas should be left alone and not developed or spoiled because of their beauty or significance to people. Muir often clashed with Pinchot and the other conservationists because Muir wanted to preserve wilderness while Pinchot wanted to use it wisely. Often this distinction can be fuzzy. When President Theodore Roosevelt, for example, speaking of the Grand Canyon of the Colorado, said, "Leave it as it is. The ages have been at work on it and man can only mar it,"[16] he was being both a conservationist and a preservationist.

The condition of our rivers and lakes has been one of the more visible aspects of environmental pollution. Not too many years ago, the great rivers in urbanized areas were in effect open sewers that emptied into the nearest watercourse, without any treatment. As a result, many lakes and rivers became grossly polluted and, as an 1885 Boston Board of Health report put it, "larger territories are at once, and frequently, enveloped in an atmosphere of stench so strong as to arouse the sleeping, terrify the weak and nauseate and exasperate everybody."

The condition of the rivers in England was notorious. The River Cam, for example, like the Thames, was for many years grossly polluted. There is a tale of Queen Victoria visiting Trinity College at Cambridge and saying to the Master as she looked over the bridge abutment: "What are all those pieces of paper floating down the river?" To which, with great presence of mind, he replied: "Those, ma'am, are notices that bathing is forbidden."[17]

While people today are still worried about the effect of pollution on their health, most are also adamantly opposed to the despoliation of the environment, for purely aesthetic reasons. We simply do not like to see our planet contaminated and spoiled. Nor do we want to see species or places destroyed without justification, and we argue for both conservation and preservation because we believe that nonhuman nature has value to us and its destruction makes the lives of our children poorer.

Thus the environmental ethic of conservation and preservation places value on nature because we want it conserved (so it can continue to provide us with resources) and preserved (so it can continue to be enjoyed by us). Environmental pollution is bad either because such pollution can be a public health concern or because such pollution can be a public nuisance, cost us money, or prevent us from enjoying nature. In the first case we want our water, air, food, and our

---

[16]Leydet, F., *Time and the River Flowing,* San Francisco: Sierra Club Books (1964).
[17]Raverat, G., *Period Piece,* as quoted in Reyburn, W., *Flushed with Pride,* London: McDonald (1969).

living place not to be polluted because we do not want to get ill. In the second case we do not want to have pollution because it decreases the quality of our lives. We also do not want to destroy species because, in the first instance, these species may be useful to us in what they can provide to keep us alive longer or because, in the second sense, we enjoy having these species as our co-inhibitors. These two views represent what has become known as an *anthropocentric environmental ethic,* that is, people centered. We do not want to cause pollution or destroy things because of the value these may have to humans, in terms of either public health or quality of life.

There is, however, a second kind of environmental ethic, one that recognizes all of the above concerns but also places a value on the environment, including animals, plants, and places. That is an *intrinsic value,* a value of and by itself, independent of what value we might place on it. Such an environmental ethic can be thought of as the ethics of simply caring for nonhuman nature.

ENVIRONMENTAL ETHICS AS CARING FOR NONHUMAN NATURE

Rene Dubos wrote in "A Theology of the Earth" that everything has its place and reason for being, and that all things, be they people, animals, trees, or rocks, are deserving of consideration.[18] This is a truly revolutionary idea. Why indeed do animals, trees, or rocks deserve moral consideration and moral protection? Why should we extend the environmental ethic to cover the nonhuman world? Based on the rationalization for ethics, there cannot be a very strong argument for such an extension of the moral community. Because there is no reciprocity (so goes the argument), there can be no ethics. Our caring for nonhuman nature, then, cannot ever be rationally argued and defended.

This leads to the temptation to give up the search for a rational environmental ethic and recognize that the scholarly field of ethics cannot ever provide us with the answers we seek. Using ethics to try to understand our attitudes and to provide guidance for our actions toward the nonhuman world is simply asking too much of it. Ethics was never intended to be used in this way, and we should not be disappointed that it fails to perform.

And yet we clearly do care for the nonhuman world. We would condemn anyone who wantonly destroyed natural places or who tortured animals. Why is it that we feel this way?

One possibility is that our attitudes toward other species and nonhuman nature in general is spiritual. Spiritual feelings toward nature are not new, of course, and we might have much to learn from the religions of our forebears. Many ancient religions, including Native American, are animistic, recognizing the existence of spirits within nature. These spirits do not take human form, as in the Greek, Roman, or Judaic religions. They are simply within the tree, the

---

[18]Dubos, R., "A Theology of the Earth," in Barbour, I.G., ed., *Western Man and Environmental Ethics,* Reading, MA: Addison-Wesley Publishing Co. (1973).

brook, or the sky. It is possible to commune with these spirits, to talk to them, to feel close to them. In many animistic religions any natural entity, such as a tree, has its own spirit, and one has to take these spirits seriously. If a tree is to be cut down in order to build a house, this action has to be explained to the spirit before cutting begins. If the tree is to be used for a beneficial purpose, the spirit has no objections and actually moves with the tree to the house (which assumes its own spirit). Each piece of furniture, each tool, has its own spirit, and as a result all of these objects deserve respect and consideration.

Such a *spiritual environmental ethic* based on respect does not prevent us from using the resources of the world for legitimate benefit. All life has to kill other life to survive. Woodpeckers poke holes in trees, whales eat algae, and parasitic bacteria use their host for reproduction, just as people kill chickens, harvest corn, or drain wetlands. Life as it has been designed requires us to kill other life in order to survive, and to use resources for our benefit. In doing so, however, we are required by the spiritual environmental ethic to be aware, grateful, and careful with what we kill, what we use, and what we damage.

A spiritual environmental ethic is a new paradigm for our environmental morality. We recognize that the approach provided by classical ethics does not provide the basis for explaining our attitudes toward nature. The sooner we realize this the sooner will we be able to formulate cogent, useful, and defensible arguments for doing the right thing for our environment.

APPLICATION AND DEVELOPMENT OF THE ENVIRONMENTAL ETHIC

As long as we use the anthropocentric environmental ethic in making decisions concerning the environment, conflicts between people can be resolved in the time-honored fashion with compromise, understanding, and mutual interest. But what if the concerns include that of nature, in the form of the spiritual environmental ethic? How can resolution of disagreements between people be achieved?

Because the spiritual environmental ethic is not based on reciprocity, conflicts can occur when the rights of nonhuman nature come into conflict with humans. Such situations occur when preserving or protecting the environment results in financial loss to humans, often couched in terms of "environment vs. jobs." When asked, "Should trees have rights?" most people answer "Of course." But when asked "Would you agree to lose your job (or someone else's) in order to save some trees?" the answer changes.

Under the Endangered Species Act old growth forest habitat for the Northern spotted owl does indeed take precedence and has resulted in the removal of many acres from potential logging. The town of Hoquiam, Washington, for example, lost its only industry, a mill that employed 600 workers, most of whom had lived all their lives in Hoquiam and were the grandchildren of the original mill workers. Environmental ethics come into real conflict with our moral responsibilities not to hurt other people. Do we have a right to hold that the preservation of nonhuman nature is more important than the welfare of humans? Although people and nonhuman life may have equal rights to co-exist, situations can occur when *people* must choose between human and nonhuman well-being. Granted, one can make such a choice for oneself, but has one the right to make the choice for another? One can even try to make such a choice democratically in one country, but can one country choose for another?

Confronting this dilemma is a step in the maturation of the environmental ethic. In the best of all worlds, we will confront it honestly and find a solution that is a compromise: The preservation may not be perfect or total, but complete destruction will not be tolerated. Or as Aldo Leopold, famous for his seminal *A Sand County Almanac*, said, "The first rule of intelligent tinkering is to save all the pieces."

CONCLUSION

Historically, ethics is not concerned with the natural environment. Instead, it is an attempt to answer either of two questions: "How ought I to treat others?" and "What actions that affect others are morally right?" Nature is not included in these arguments, except as some part of nature that "belongs" to some human being. That is, polluting a stream is morally wrong only if this action diminishes someone's enjoyment of the stream or its utility. The stream, or its inhabitants, has no value and no ethical standing. Today, however, we realize that we owe responsibility to nature for its own sake, and not just because it might have instrumental value. But how to do this? Scientists and engineers look at the world objectively with technical tools, and often these technical tools are inappropriate for solving value problems. We are ill-equipped to make decisions where the value of nature or other species, or even future generations of humans, is concerned.

In response to these difficult, value-laden questions, a new form of applied ethics, *environmental ethics,* has evolved and attempts to address issues on human interactions with nonhuman nature.[19] Environmental ethics helps us develop an ethical attitude towards nonhuman nature, not unlike the role of classical ethics in helping us develop a moral stance toward our fellow human beings. Such an attitude helps us make decisions where value questions come into play.

In the book, we weave together aspects of environmental engineering and environmental ethics. We believe that a basic understanding of environmental sciences and engineering is impossible without such a broad perspective. However, we also caution the reader that it is impossible to introduce these topics in any depth in this introductory textbook, and that the student should seek out more advanced courses that address all of these components of environmental pollution and control.

---

[19]See, for example, *Environmental Ethics,* a quarterly professional journal.

Chapter 2
Environmental Risk Analysis

Risk analysis allows us to estimate impacts on the environment and on human health when we have not measured or cannot measure or directly observe those impacts. It also lets us compare these impacts. In this chapter, we introduce the concept of risk analysis and risk management. The former is the measurement and comparison of various forms of risk; the latter involves the techniques used to reduce these risks.

RISK

Most pollution control and environmental laws were enacted in the early 1970s in order to protect public health and welfare.[1] In these laws and throughout this text, a substance is considered a pollutant if it has been perceived to have an adverse effect on human health. In recent years, increasing numbers of substances appear to pose such threats; the Clean Air Act listed seven hazardous substances between 1970 and 1989, and now lists approximately 300! The environmental engineer thus has an additional job: to help determine the comparative risks from various environmental pollutants and, further, to determine which risks are most important to decrease or eliminate.

Adverse effects on human health are sometimes difficult to identify and to determine. Even when such an adverse effect has been identified, it is still difficult to recognize those components of the individual's environment that are associated with it. Risk analysts refer to these components as *risk factors*. In general, a risk factor should meet the following conditions:

- Exposure to the risk factor precedes appearance of the adverse effect.
- The risk factor and the adverse effect are consistently associated. That is, the adverse effect is not usually observed in the absence of the risk factor.
- The more of the risk factor there is, or the greater its intensity, the greater the adverse effect, although the functional relationship need not be linear or monotonic.

---

[1]See Chapters 11, 17, and 22 for the details of land, water, and air laws and regulations in the United States.

- The occurrence or magnitude of the adverse effect is statistically significantly greater in the presence of the risk factor than in its absence.

Identification of a risk factor for a particular adverse effect may be made with confidence only if the relationship is consonant with, and does not contradict, existing knowledge of the cellular and organismic mechanisms producing the adverse effect.

Identification of the risk factor is more difficult than identification of an adverse effect. For example, we are now certain that cigarette smoke is unhealthy, both to the smoker—primary smoke risk—and to those around the smoker—secondary smoke risk. Specifically, lung cancer, chronic obstructive pulmonary disease, and heart disease occur much more frequently among habitual smokers than among nonsmokers or even in the whole population including smokers. The increased frequency of occurrence of these diseases is statistically significant. Cigarette smoke is thus a risk factor for these diseases; smokers and people exposed to secondhand smoke are at increased risk for them.

Notice, however, that we do not say that cigarette smoking *causes* lung cancer, chronic obstructive pulmonary disease, or heart disease, because we have not identified the actual causes, or etiology, of any of them. How, then, has cigarette smoking been identified as a risk factor if it cannot be identified as the cause? This observation about cigarette smoke was not made, and indeed could not be made, until the middle of the twentieth century, when the lifespan in at least the developed countries of the world was long enough to observe the diseases that had been correlated with exposure to cigarette smoke. In the first half of the twentieth century, infectious diseases were a primary cause of death. With the advent of antibiotics and the ability to treat such diseases, the lifespan in the developed nations of the world lengthened, and cancer and heart disease became the leading causes of death. From the early 1960s, when the average lifespan in the United States was about 70, lifelong habitual cigarette smokers were observed to die from lung cancer at ages between 55 and 65. This observation, which associated early death with cigarette smoke, identified cigarette smoke as a risk factor.

ASSESSMENT OF RISK

Risk assessment is a system of analysis that includes four tasks:

1. Identification of a substance (a toxicant) that may have adverse health effects
2. Scenarios for exposure to the toxicant
3. Characterization of health effects
4. An estimate of the probability (risk) of occurrence of these health effects

The decision that the concentration of a certain toxicant in air, water, or food is acceptable is based on a risk assessment.

Toxicants are usually identified when an associated adverse health effect is noticed. In most cases, the first intimation that a substance is toxic is its association with an unusual number of deaths. Mortality risk, or risk of death, is easier to determine for populations, especially in the developed countries, than morbidity risk (risk of illness) because all deaths and their apparent causes are reported on death certificates, while recording of disease incidence, which began in the relatively recent past, is done only for a very few diseases. Death certificate data may be misleading: An individual who suffers from high blood pressure but is killed in an automobile accident becomes an accident statistic rather than a cardiovascular disease statistic. In addition, occupational mortality risks are well documented only for men; until the present generation, too few women worked outside the home all their lives to form a good statistical base.

These particular uncertainties may be overcome in assessing risk from a particular cause or exposure to a toxic substance by isolating the influence of that particular cause. Such isolation requires studying two populations whose environment is virtually identical except that the risk factor in question is present in the environment of one population but not in that of the other. Such a study is called a *cohort study* and may be used to determine morbidity as well as mortality risk. One cohort study, for example, showed that residents of copper smelting communities, who were exposed to airborne arsenic, had a higher incidence of a certain type of lung cancer than residents of similar industrial communities where there was no airborne arsenic.

Retrospective cohort studies are almost impossible to perform because of uncertainties on data, habits, other exposures, and the like. Cohorts must be well matched in size, age distribution, life-style, and other environmental exposures, and they must be large enough for an effect to be distinguishable from the deaths or illnesses that occur anyway.

DOSE-RESPONSE EVALUATION

Dose-response evaluation is required both in determining exposure scenarios for the pollutant in question and in characterizing a health effect. The response of an organism to a pollutant always depends in some way on the amount or dose of pollutant to the organism. The magnitude of the dose, in turn, depends on the exposure pathway. The same substance may have a different effect depending on whether it is inhaled, ingested, or absorbed through the skin, or whether the exposure is external. The exposure pathway determines the biochemistry of the pollutant in the organism. In general, the human body detoxifies an ingested pollutant more efficiently than it does an inhaled pollutant.

The relationship between the dose of a pollutant and the organism's response can be expressed in a dose-response curve, as shown in Figure 2–1. The figure shows four basic types of dose-response curve possible for a dose of a specific pollutant and the respective responses. For example, such a curve may

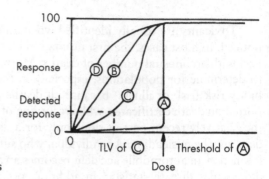

FIGURE 2–1
Possible dose-response curves

be drawn for various concentrations of carbon monoxide (the dose), plotted against the associated blood concentrations of carboxylated hemoglobin (the response). Some characteristic features of the dose-response relationship are:

1. *Threshold.* The existence of a threshold in health effects of pollutants has been debated for many years. A threshold dose is the lowest dose at which there is an observable effect. Curve A in Figure 2–1 illustrates a threshold response: There is no observed effect until a particular concentration is reached. This concentration is designated as the threshold. Curve B shows a linear response with no threshold; that is, the intensity of the effect is directly proportional to the pollutant dose, and an effect is observed for any detectable concentration of the pollutant in question. Curve C, sometimes called *sublinear,* is a sigmoidal dose-response curve, characteristic of many pollutant dose-response relationships. Although Curve C has no clearly defined threshold, the lowest dose at which a response can be detected is called the *threshold limit value* (TLV). Occupational exposure guidelines are frequently set at the TLV. Curve D displays a *supralinear* dose-response relationship, which is found when low doses of a pollutant appear to provoke a disproportionately large response.

2. *Total body burden.* An organism, or a person, can be exposed simultaneously to several different sources of a given pollutant. For example, we may inhale about 50 μg/day of lead from the ambient air and ingest about 300 μg/day in food and water. The concentration of lead in the body is thus the sum of what is inhaled and ingested and what remains in the body from prior exposure, less what has been eliminated from the body. This sum is the total body burden of the pollutant.

3. *Physiological half-life.* The physiological half-life of a pollutant in an organism is the time needed for the organism to eliminate half of the internal concentration of the pollutant, through metabolism or other normal physiological functions.

4. *Bioaccumulation and bioconcentration.* Bioaccumulation occurs when a substance is concentrated in one organ or type of tissue of an organism. Iodine,

for example, bioaccumulates in the thyroid gland. The *organ dose* of a pollutant can thus be considerably greater than what the total body burden would predict. Bioconcentration occurs with movement up the food chain. A study of the Lake Michigan ecosystem found the following bioconcentration of DDT:[2]

0.014 ppm (wet weight) in bottom sediments

0.41 ppm in bottom-feeding crustacea

3 to 6 ppm in fish

2400 ppm in fish-eating birds

Pollution control criteria for which an engineer designs must take both bioconcentration and bioaccumulation into account.

5. *Exposure time* and *time vs. dosage.* Most pollutants need time to react; the exposure time is thus as important as the level of exposure. An illustration of the interdependence of dose and exposure time is given for CO exposure in Chapter 18, in Figure 18–7. Because of the time-response interaction, ambient air quality standards are set at maximum allowable concentrations for a given time, as discussed in Chapter 21.

6. *Synergism.* Synergism occurs when two or more substances enhance each other's effects, and when the resulting effect of the combination on the organism is greater than the additive effects of the substances separately. For example, black lung disease in miners occurs much more often in miners who smoke than in those who do not. Coal miners who do not smoke rarely get black lung disease, and smokers who are not coal miners never do. The synergistic effect of breathing coal dust *and* smoking puts miners at high risk. The opposite of synergism is *antagonism*, a phenomenon that occurs when two substances counteract each other's effects.

7. LC_{50} and LD_{50}. Dose-response relationships for human health are usually determined from health data or epidemiological studies. Human volunteers obviously cannot be subjected to pollutant doses that produce major or lasting health effects, let alone fatal doses. Toxicity can be determined, however, by subjecting nonhuman organisms to increasing doses of a pollutant until the organism dies. The LD_{50} is the dose that is lethal for 50% of the experimental animals used; LC_{50} refers to lethal concentration rather than lethal dose. LD_{50} values are most useful in comparing toxicities, as for pesticides and agricultural chemicals; no direct extrapolation is possible, either to humans or to any species other than the one used for the LD_{50} determination. LD_{50} can sometimes be determined retrospectively when a large population has been exposed accidentally, as in the accident at the Chernobyl nuclear reactor.

---

[2]Hickey, J.J., et al., "Concentration of DDT in Lake Michigan," *Journal of Applied Ecology* 3 (1966): 141.

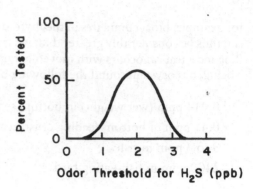

FIGURE 2–2
Distribution of odor thresholds
in a population

POPULATION RESPONSES

Individual responses to a particular pollutant may differ widely; dose-response relationships differ from one individual to another. In particular, thresholds differ; threshold values in a population, however, generally follow a Gaussian distribution. Figure 2–2 shows the distribution of odor thresholds for hydrogen sulfide in a typical population for example.

Individual responses and thresholds also depend on age, sex, and general state of physical and emotional health. Healthy young adults are on the whole less sensitive to pollutants than are elderly people, those who are chronically or acutely ill, and children. In theory, allowable releases of pollutants are restricted to amounts that ensure protection of the health of the entire population, including its most sensitive members. In many cases, however, such protection would mean zero release.

The levels of release actually allowed take technical and economic control feasibility into account, but even so are set below threshold level for 95% or more of the U.S. population. For nonthreshold pollutants, however, no such determination can be made. In these instances, there is no release level for which protection can be ensured for everyone, so a comparative risk analysis is necessary. Carcinogens are all considered to be in this category of nonthreshold pollutants.

EXPOSURE AND LATENCY

Characterization of some health risks can take a very long time. Most cancers grow very slowly and are detectable (expressed) many years, or even decades, after exposure to the potentially responsible carcinogen. The length of time between exposure to a risk factor and expression of the adverse effect is called the *latency period*. Cancers in adults have apparent latency periods of between 10 and 40 years. Relating a cancer to a particular exposure is fraught with inherent inaccuracy. Many carcinogenic effects are not identifiable in the lifetime of a single individual. In a few instances, a particular cancer is found only on exposure to a particular agent (e.g., a certain type of hemangioma is found only on exposure to vinyl chloride monomer), but for most cases the connection between exposure and effect is far from clear. Many carcinogens are identified through animal studies, but one cannot always extrapolate from animal to human results. The U.S. Environmental Protection Agency (EPA) classifies known animal carcinogens, for which there is inadequate evidence for human carcinogenicity, as probable human carcinogens.

There is a growing tendency to regulate any substance for which there is any evidence, even inconclusive, of adverse health effects. This is considered a conservative assumption, but it may not be valid in all cases. Such a conservative posture toward regulation and control is the result of the cumulative uncertainty surrounding the epidemiology of pollutants. The cost of such control has recently been determined to be far greater than the cost of treating or mitigating the effect.[3] For example, vinyl chloride emission control is estimated to cost 1.6 million dollars per year of life saved, while leukemia treatment by bone marrow transplant costs $12,000 per year of life saved.

EXPRESSION OF RISK

In order to use risks in determining pollution standards, as EPA does, it is necessary to develop quantitative expressions for risk. The quantitative expressions reflect both the proportionality of the risk factor to the adverse effect and the statistical significance of the effect.

Risk is defined as the product of probability and consequence, and is expressed as the probability or frequency of occurrence of an undesirable event. It is important to note that *both* probability and consequence must play a role in risk assessment. Arguments over pollution control often concentrate on consequence alone; members of the public fear a consequence (like the Bhopal isocyanate release) irrespective of its remote likelihood or low frequency of occurrence. However, pollution control decisions, like other risk-based decisions, cannot be made on the basis of consequence alone. If we were to determine actions only with regard to their consequences, we would never travel by bicycle, automobile, or airplane, would never start a campfire or burn wood in a stove or fireplace, and never eat solid food, because death (and a very unpleasant death) is one possible consequence of all of these actions. We actually take relative risk, and thereby relative probability of harm, into account in all such decisions. For example, if 10% of the students in a course were randomly given an F, the "risk" of getting an F would be 0.1 per total number of grades assigned. The probability is 0.1 and the consequence is F. An expression of risk incorporates both the probability and some measure of consequence. In discussing

[3]Tengs, T.O., et al., "Five Hundred Life Saving Interventions and Their Cost Effectiveness," *Risk Analysis* 15 (1995): 369–389.

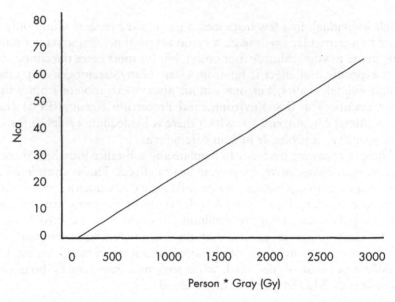

FIGURE 2–3. Sample population-response curve

human health or environmental risk, the consequences are adverse health effects or adverse effects on some species of plant or animal.

Challenges to the linear nonthreshold theory of carcinogenesis have been raised recently, particularly with respect to the effects of ionizing radiation. Bond et al.[4] re-examined data from atom bomb survivors and observed the dose-response shown in Figure 2–3.

Several studies of the Marshall Islanders, who were exposed during the atmospheric nuclear tests in the 1950s, indicate that there may be a threshold of radiation exposure for a population before any excess cancers are seen in that population,[5] and a similar phenomenon has been observed in analysis of cancer incidence in radium dial painters. Recent Russian data for a population accidentally exposed to plutonium-contaminated water from 1949 to 1956 also show the possibility of a threshold[6] (Figure 2–4). Moreover, these data show that the dose/risk relationship may not be linear.

There is even some epidemiological evidence that exposure to ionizing radiation just a bit higher than background stimulates some biological defense mechanisms: A study of cancer rates and radon exposure showed a 15% lower

---

[4]Bond, Victor P., Wiepolski, L., and Shani, G., "Current Misinterpretations of the Linear Non-Threshold Hypothesis," *Health Physics* 70 (1996): 877.

[5]Bond, Victor P., "When Is a Dose Not a Dose?" Lauriston S. Taylor Lecture #15, National Council on Radiation Protection and Measurements, Bethesda, MD, January 1, 1992.

[6]Kossenko, M.M., "Cancer Mortality Among Techa River Residents and Their Offspring," *Health Physics* 71 (1996): 77–85.

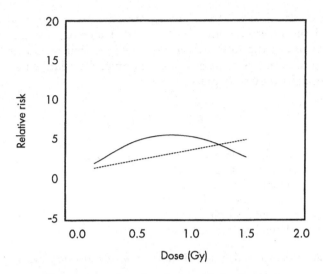

FIGURE 2–4. Sample relative risk for leukemia

incidence of radon-induced lung cancer than was predicted by the linear non-threshold theory,[7] though this is generally seen as scatter in the data. These studies include very large uncertainties, but they are the same uncertainties and have the same magnitude as studies that appear to confirm the linear non-threshold theory of radiation-induced carcinogenesis.

The linear nonthreshold theory is clearly controversial: There are strong arguments both for and against a threshold, and dose-response curves can be linear or linear-quadratic, or can show some other dependence. The strongest argument to retain the nonthreshold hypothesis is its conservatism. On the other hand, the strongest argument for recognizing a threshold is the existence of epidemiological data showing thresholds. As the populations that we have studied live out their lives, we get a retrospective demonstration of ionizing radiation dose vs. cancer incidence that will ultimately answer the remaining questions.

The *probability*, or frequency of occurrence, of adverse health effects in a population is written as

$$P = \frac{X}{N} \tag{2.1}$$

where   P = probability
         X = number of adverse health effects
         N = number of individuals in the population

---

[7]Cohen, Bernard L., "A Test of the Linear Non-Threshold Theory for Radon Induced Lung Cancer," *Health Physics* 66 *(Suppl)* (1994): 829.

If the adverse effect is death from cancer, and the cancer occurs after a long latency period, the adverse health effects are called latent cancer fatalities, or LCF.

Relative risk is the ratio of the probabilities that an adverse effect will occur in two different populations. For example, the relative risk of fatal lung cancer in smokers may be expressed as

$$\frac{P_s}{P_n} = \frac{(X_s/N_s)}{(X_n/N_n)} \tag{2.2}$$

where P_s = probability of fatal lung cancer in smokers
           P_n = probability of fatal lung cancer in nonsmokers
           X_s = fatal lung cancer in smokers
           X_n = fatal lung cancer in nonsmokers
           N_s = total number of smokers
           N_n = total number of nonsmokers

Relative risk of death is also called the *standard mortality ratio* (SMR), which is written as

$$SMR = \frac{D_s}{D_n} = \frac{P_s}{P_n} \tag{2.3}$$

where D_s = observed lung cancer deaths in a population of habitual smokers
           D_n = expected lung cancer deaths in a nonsmoking population of the same
                size

In this particular instance, the SMR is approximately 11/1 and is significantly greater than 1.[8]

Three important characteristics of epidemiological reasoning are illustrated by this example:

- Everyone who smokes heavily will not die of lung cancer.
- Some nonsmokers die of lung cancer.
- Therefore, one cannot unequivocally relate any given individual lung cancer death to cigarette smoking.[9]

Risk may be expressed in several ways:

- *Deaths per 100,000 persons.* In 1985 in the United States, 350,000 smokers died as a result of lung cancer and heart disease. In that year, the United

---

[8]In determining statistical significance, a test of significance (like the Fisher's test or the Student's t-test) appropriate to the population under consideration is applied. Such calculations are beyond the scope of this text.

[9]In spite of this principle of risk analysis, in 1990 the family of Rose Cipollino successfully sued cigarette manufacturers and advertisers, claiming that Ms. Cipollino had been enticed to smoke by advertising and that the cigarette manufacturers had concealed known adverse health effects. Ms. Cipollino died of lung cancer at the age of 59.

TABLE 2-1. Adult Deaths/100,000 Population

Cause of Death	Deaths/10^5 Population
Cardiovascular disease	408.0
Cancer	193.0
Chronic obstructive pulmonary disease	31.0
Motor vehicle accidents	18.6
Alcohol-related disease	11.3
Other causes	208.0
All causes	869.9

From National Center for Health Statistics, U.S. Department of Health and Human Services (1985).

States had a population of 226 million. The risk of death (from these two factors) associated with habitual smoking may thus be expressed as deaths per 100,000 population, or

$$\frac{(350,000)(100,000)}{226 \times 10^6} = 155 \qquad (2.4)$$

In other words, a habitual smoker in the United States has an annual risk of 155 in 100,000, or 1.55 in 1000, of dying of lung cancer or heart disease. The probability is 1.55 in 1000; the consequence is death from lung cancer or heart disease. Table 2-1 presents some typical statistics for the United States.

- *Deaths per 1000 deaths.* Using 1985 data again, there were 2,084,000 deaths in the United States that year. Of these, 350,000, or 168 deaths per 1000 deaths, were related to habitual smoking.
- *Loss of years of life or, for occupational risks, loss of work days or work years.* Loss of years of life depends on life expectancy, which differs considerably from one country to another. Average life expectancy in the United States is now 75 years; in Canada, 76.3 years; and in Ghana, 54 years.[10] Table 2-2 gives the loss of life expectancy from various causes of death in the United States.

These figures show that meaningful risk analyses can be conducted only with very large populations. Health risk that is considerably lower than the risks cited in Tables 2-1 and 2-2 may not be observed in small populations. Chapter 18 cites several examples of statistically valid risks from air pollutants.

---

[10]World Resources Institute, *World Resources 1987,* New York: Basic Books (1987).

TABLE 2–2. Loss of Life Expectancy in the United States from Various Causes of Death

Cause of Death	Loss of Life Expectancy (days)
Cardiovascular disease	2043 (5.6 yrs)
Cancer—all types	1247 (3.4 yrs)
Respiratory system cancer	343 (11.4 mos)
Chronic obstructive pulmonary disease	164 (5.4 mos)
Motor vehicle accidents	207 (6.9 mos)
Alcohol-related disease	365 (1 yr)
Abuse of controlled substances	125 (4.2 mos)
Air transportation accidents	3.7
Influenza	2.3

From Cohen, B.L., "Catalog of Risks Extended and Updated," *Health Physics* 61 (1991): 317.

Example 2.1

A butadiene plastics manufacturing plant is located in Beaverville, and the atmosphere is contaminated by butadiene, a suspected carcinogen. The cancer death rate in the community of 8000 residents is 36 people per year, and the total death rate is 106 people per year. Does Beaverville appear to be a healthy place to live, or is the cancer risk unusually high?

From Table 2–1, we see that the annual cancer death rate in the United States is 193 deaths/10^5 persons, and the death rate from all causes is 870 deaths/10^5 persons. The expected annual death rate in Beaverville from cancer is thus

$$\left(\frac{193 \text{ deaths}}{10^5 \text{ persons}} \right) (8000 \text{ persons}) = 15.4 \text{ deaths} \qquad (2.4a)$$

and the expected death rate from all causes is

$$\left(\frac{870 \text{ deaths}}{10^5 \text{ persons}} \right) (8000 \text{ persons}) = 69.6 \text{ deaths} \qquad (2.4b)$$

The annual SMR for cancer is thus

$$\text{SMR (cancer)} = \frac{36}{15.4} = 2.3 \qquad (2.4c)$$

For all causes of death, the annual SMR is

$$\text{SMR (total)} = \frac{106}{69.6} = 1.5 \qquad (2.4d)$$

Without performing a test of statistical significance, we may assume that the annual SMR for cancer is significantly greater than 1, and thus Beaverville displays excessive cancer deaths. Moreover, a Beaverville resident is about 1.5 times as likely to die in any given year from any cause as the average resident of the United States.

We may also calculate whether cancer deaths per 1000 deaths are higher in Beaverville than in the United States as a whole. In the United States, cancer deaths per 1000 deaths are

$$\left(\frac{193}{870}\right)(1000) = 221 \tag{2.4e}$$

while in Beaverville, cancer deaths per 1000 deaths are

$$\left(\frac{36}{106}\right)(1000) = 340 \tag{2.4f}$$

Thus, we may conclude further that a death in Beaverville in any given year is about 1.5 times more likely to be a cancer death than is the case in the United States as a whole.

Risk assessment usually compares risks because the absolute value of a particular risk is not very meaningful. EPA has adopted the concept of unit risk in discussion of potential risk. *Unit risk* is defined as the risk to an individual from exposure to a concentration of 1 μg/m³ of an airborne pollutant or 10^{-9} g/L of a waterborne pollutant. *Unit lifetime risk* is the risk to an individual from exposure to these concentrations for 70 years (a lifetime, as EPA defines it). *Unit occupational lifetime risk* implies exposure for 8 hours per day and 22 days per month every year, or 2000 hours per year for 47 years (a working lifetime).

EPA's concern with somatic risk from a number of hazardous substances is the carcinogenic potential of these substances, so the "consequence" part of the risk is given as LCFs. We can then write equations for the different expressions for unit risk and use these to calculate the estimated risk. In the example below, these calculations assume that risk increases linearly with time and concentration. EPA considers this a conservative assumption for low exposure to a carcinogen over a period of years. Nonlinear dose-response relationships imply more complex relationships between risk, concentration, and exposure time; examples of such more complex relationships will not be considered here.

For waterborne pollutants:

$$\text{Unit annual risk} = \frac{\text{LCF/year}}{10^{-9}\,\text{g/L}} \tag{2.5a}$$

$$\text{Unit lifetime risk} = \frac{\text{LCF}}{(10^{-9}\,\text{g/L})(70\,\text{yrs})} \tag{2.5b}$$

$$\text{Lifetime occupational risk} = \frac{\text{LCF}}{(10^{-9}\,\text{g/L})(47\,\text{yrs})(2000/8760)} \qquad (2.5c)$$

The factor 2000/8760 in Equation 2.5c is the fraction of hours per year spent in the workplace.

For airborne pollutants:

$$\text{Unit annual risk} = \frac{\text{LCF/year}}{10^{-6}\,\text{g/m}^3} \qquad (2.6a)$$

$$\text{Unit lifetime risk} = \frac{\text{LCF}}{(10^{-6}\,\text{g/m}^3)(70\,\text{yrs})} \qquad (2.6b)$$

$$\text{Unit lifetime occupational risk} = \frac{\text{LCF}}{(10^{-6}\,\text{g/m}^3)(47\,\text{yrs})(2000/8760)} \qquad (2.6c)$$

Example 2.2

EPA has calculated that unit lifetime risk from exposure to ethylene dibromide (EDB) in drinking water is 0.85 LCF per 10^5 persons. What risk is experienced by drinking water with an average EDB concentration of 5 pg/L for five years?

The risk may be estimated using either unit annual risk or unit lifetime risk. Since the unit lifetime risk is given, we may write

$$\text{Risk} = \frac{(5 \times 10^{-12}\,\text{g/L})(0.85\,\text{LCF})(5\,\text{yrs})}{(10^5)(10^{-9}\,\text{g/L})(70\,\text{yrs})} = 3.0 \times 10^{-9}\,\text{LCF} \qquad (2.7)$$

The answer is given to two significant figures because of the uncertainties in risk estimates. The estimated risk is that about three fatal cancers would be expected in a population of a billion people who drink water containing 5 pg/L EDB for five years. Although there is a popular tendency to translate this to an "individual risk" of "a chance of three in a billion having a fatal cancer," this statement of risk is less meaningful than the statement of population risk.

ECOSYSTEM RISK ASSESSMENT

Regulation of toxic or hazardous substances often requires an assessment of hazard or risk to some living species other than *homo sapiens*, or assessment of risk to an entire ecosystem. Methods for ecosystem risk assessment are now being developed.[11] Ecosystem risk assessment is done in the same general way

---

[11]See for example Suter, G.W., "Environmental Risk Assessment/Environmental Hazards Assessment: Similarities and Differences," in Landis, W.G., and van der Schalie, W.H., *Aquatic Toxicology and Risk Assessment* 13 (ASTM STP 1096) (1990): 5.

as human health risk assessment, except that identification of the species at risk and of the exposure pathway is a far more complex process than in human health risk assessment. Assessment *endpoints* are values of the ecosystem that are to be protected and are identified early in the analysis; these endpoints may include numbers of different species, life-cycle stages for a given species, reproductive patterns, or growth patterns. Identification of specific endpoints implies choices among potential target species. Ecosystem risk assessment is in its infancy, and details of its practice are beyond the scope of this textbook.

CONCLUSION

The best available control for nonthreshold pollutants will still entail a residual risk. Our industrial society needs accurate quantitative risk assessment to evaluate the protection afforded by various levels of pollution control. We must also remain aware that determination of safe levels of pollutants based on risk analysis is a temporary measure until the mechanism of the damage done by the pollutant is understood. At present, we can only identify apparent associations between most pollutants and a given health effect. We should note that analysis of epidemiological data and determination of significance of effects requires application of a test of statistical significance. There are a number of such tests in general use, but since their application is not central to the scope of this text they not considered here.

   Almost all of our knowledge of adverse health effects comes from occupational exposure, which is orders of magnitude higher than exposure of the general public. Doses to the public are usually so low that excess mortality, and even excess morbidity, are not identifiable. However, development of pollution control techniques continues to reduce risk. The philosophy, regulatory approaches, and design of environmental pollution control make up the remainder of this book.

PROBLEMS

   2.1   Using the data given in the chapter, calculate the expected deaths (from all diseases) for heavy smoking in the United States.

   2.2   Calculate the relative risks of smoking and alcoholism in the United States. Do you think a regulatory effort should be made to limit either smoking or the consumption of alcohol? Why and why not?

   2.3   EPA has determined the lifetime unit risk for cancer for low-energy ionizing radiation to be 3.9×10^{-4} per rem of radiation. The allowed level of airborne ionizing radiation (the EPA standard) above background is 10 mrem per source per year. Average nonanthropogenic background is about 100 mrem per year. How many fatal cancers attributable to ionizing radiation would result in the United States each year if the entire population were exposed at the level of the EPA standard? How many cancers may be attributed to background? If only 10% of the cancers were fatal each year, what percentage of the annual cancer deaths in the United States would be attributed to exposure to background radioactivity? (Note what "unit risk" means in this problem.)

2.4   The allowed occupational dose for ionizing radiation is 5 rem per year. By what factor does a worker exposed to this dose over a working lifetime increase her risk of cancer?

2.5   Workers in a chemical plant producing molded polyvinyl chloride plastics suffered from hemangioma, a form of liver cancer that is usually fatal. During the 20 years of the plant's operation, 20 employees out of 350—the total number of employees at the plant during those years—developed hemangioma. Does working in the plant present an excess cancer risk? Why? What assumptions need to be made?

2.6   Additional, previously unavailable data on hemangioma incidence indicates that among people who have never worked in the plastics industry, there are only 10 deaths per 100,000 persons per year from hemangioma. How does this change your answer to Problem 2.5?

2.7   The unit lifetime risk from airborne arsenic is 9.2×10^{-3} latent cancer fatalities (LCF). EPA regards an acceptable annual risk from any single source to be 10^{-6}. A copper smelter emits arsenic into the air, and the average concentration within a two-mile radius of the smelter is $5.5\ \mu g/m^3$. Is the risk from smelter arsenic emissions acceptable to EPA?

2.8   In the community of Problem 2.7, approximately 25,000 people live within a 2-mile radius of the smelter. Assuming that the residents live there throughout their lifetimes, how many excess LCFs can be expected per year in this population?

2.9   Using the data of Problems 2.7 and 2.8, estimate an acceptable workplace standard for ambient airborne arsenic.

2.10   What are the arguments for and against using the linear nonthreshold theory of carcinogenesis as a basis for regulating potential carcinogens? Consider both the notion of a threshold and the shape of the curve. What additional data would you collect to support the arguments?

LIST OF SYMBOLS

EDB	ethylene dibromide
EPA	U.S. Environmental Protection Agency
LCF	latent cancer fatalities
SMR	standard mortality ratio
TLV	threshold limit value

Chapter 3

Water Pollution

Although people now intuitively relate filth to disease, the transmission of disease by pathogenic organisms in polluted water was not recognized until the middle of the nineteenth century. The Broad Street pump handle incident demonstrated dramatically that water can carry diseases.

A British public health physician named John Snow, assigned to control the spread of cholera, noticed a curious concentration of cholera cases in one part of London. Almost all of the people affected drew their drinking water from a community pump in the middle of Broad Street. However, people who worked in an adjacent brewery were not affected. Snow recognized that the brewery workers' apparent immunity to cholera occurred because the brewery drew its water from a private well and not from the Broad Street pump (although the immunity might have been thought due to the health benefits of beer). Snow's evidence convinced the city council to ban the polluted water supply, which was done by removing the pump handle so that the pump was effectively unusable. The source of infection was cut off, the cholera epidemic subsided, and the public began to recognize the public health importance of drinking water supplies.

Until recently, polluted drinking water was seen primarily as a threat to public health because of the transmission of bacterial waterborne disease. In less developed countries, and in almost any country in time of war, it still is. In the United States and other developed countries, however, water treatment and distribution methods have almost eradicated bacterial contamination. Most surface water pollution is harmful to aquatic organisms and causes possible public health problems (primarily from contact with the water). Groundwater can be contaminated by various hazardous chemical compounds that can pose serious health risks. In this chapter we discuss the sources of water pollution and the effect of this pollution on streams, lakes, and oceans.

SOURCES OF WATER POLLUTION

Water pollutants are categorized as *point source* or *nonpoint source*, the former being identified as all dry weather pollutants that enter watercourses through pipes or channels. Storm drainage, even though the water may enter watercourses by way of pipes or channels, is considered nonpoint source pollution.

Other nonpoint source pollution comes from farm runoff, construction sites, and other land disturbances, discussed further in Chapter 10.

Point source pollution comes mainly from industrial facilities and municipal wastewater treatment plants. The range of pollutants is vast, depending only on what gets "thrown down the drain."

1   *Oxygen-demanding substances,* such as might be discharged from milk processing plants, breweries, or paper mills, as well as municipal wastewater treatment plants, make up one of the most important types of pollutant because these materials decompose in the watercourse and can deplete the water's oxygen and create anaerobic conditions. *Suspended solids* also contribute to oxygen depletion; in addition, they create unsightly conditions and can cause unpleasant odors. *Nutrients,* mainly nitrogen and phosphorus, can promote accelerated eutrophication, and some *bioconcentrated metals* can adversely affect aquatic ecosystems as well as make the water unusable for human contact or consumption.

Heat is also an industrial waste that is discharged into water; heated discharges may drastically alter the ecology of a stream or lake. Although local heating can have beneficial effects such as freeing harbors from ice, the primary effect is deleterious: lowering the solubility of oxygen in the water, because gas solubility in water is inversely proportional to temperature, and thereby reducing the amount of dissolved oxygen (DO) available to gill-breathing species. As the level of DO decreases, metabolic activity of aerobic aquatic species increases, thus increasing oxygen demand.

2   *Municipal wastewater* is as important a source of water pollution as industrial waste. A century ago, most discharges from municipalities received no treatment whatsoever. Since that time, the population and the pollution contributed by municipal discharge have both increased, but treatment has increased also. We define a *population equivalent* of municipal discharge as equivalent to the amount of untreated discharge contributed by a given number of people. For example, if a community of 20,000 people has 50% effective sewage treatment, the population equivalent is

$$(0.5)(20,000) = 10,000$$

Similarly, if each individual contributes 0.2 lb solids/day into wastewater, and an industry discharges 1000 lb/day, the industry has a population equivalent of 1000/0.2, or 5000 persons.

The sewerage systems in older U.S. cities have aggravated the wastewater discharge situation. When these cities were first built, engineers realized that sewers were necessary to carry off both stormwater and sanitary wastes, and they usually designed a single system to carry both discharges to the nearest appropriate body of water. Such systems are known as *combined sewers.* As years passed, city populations increased, and the need for sewage treatment became apparent, *separate sewer* systems were built: one system to carry sanitary sewage to the treatment facility and the other to carry off stormwater runoff.

Almost all of the cities with combined sewers have built treatment plants that can treat *dry weather flow*—the sanitary wastes when there is no stormwater runoff. As long as it does not rain, the plants can handle the flow and provide sufficient treatment; however, rain increases the flow to many times the dry weather flow, and most of it must be bypassed directly into a river, lake, or bay. The overflow will contain sewage as well as stormwater, and can be a significant pollutant to the receiving water. Attempts to capture and store the excess flow for subsequent treatment are expensive, but the cost of separating combined sewer systems is prohibitive.

Agricultural wastes, should they flow directly into surface waters, have a collective population equivalent of about 2 billion. Feedlots where large numbers of animals are penned in relatively small spaces provide an efficient way to raise animals for food. They are usually located near slaughterhouses and thus near cities. Feedlot drainage (and drainage from intensive poultry cultivation) creates an extremely high potential for water pollution. Aquaculture has a similar problem because wastes are concentrated in a relatively small space.

Sediment from land erosion may also be classified as a pollutant. Sediment consists of mostly inorganic material washed into a stream as a result of land cultivation, construction, demolition, and mining operations. Sediment interferes with fish spawning because it can cover gravel beds and block light penetration, making food harder to find. Sediment can also damage gill structures directly.

Pollution from petroleum compounds ("oil pollution") first came to public attention with the *Torrey Canyon* disaster in 1967. The huge tanker, loaded with crude oil, plowed into a reef in the English Channel, even though maps showed the submerged reefs. Despite British and French attempts to burn it, almost all of the oil leaked out and fouled French and English beaches. Eventually, straw to soak up the oil and detergents to disperse it helped remove the oil from the beaches, but the detergents were found to be the cleanup method more harmful to the coastal ecology.

By far the most notorious recent incident has been the *Exxon Valdez* spill in Prince William Sound in Alaska. Oil in Alaska is produced in the Prudhoe Bay region in northern Alaska and piped down to the tanker terminal in Valdez on the southern coast. On 24 March 1989, the *Exxon Valdez*, a huge oil tanker loaded with crude oil, veered off course and hit a submerged reef, spilling about 11 million gallons of oil into Prince William Sound, devastating the fragile ecology. About 40,000 birds died, including about 150 bald eagles. The final toll on wildlife will never be known, but the effect of the spill on the local fishing economy can be calculated, and it exceeds $100 million. The cleanup by Exxon cost about $2 billion.

While oil spills as large as the *Exxon Valdez* spill get a lot of publicity, it is estimated that there are about 10,000 serious oil spills in the United States every year, and many more minor spills from routine operations that do not make headlines. The effect of some of these spills may never be known.

The acute effect of oil on birds, fish, and microorganisms is well catalogued. The subtle effects of oil on other aquatic life is not so well understood and is potentially more harmful. For example, anadromous fish such as salmon, which find their home stream by the smell or taste of the water, can become so confused by the presence of strange hydrocarbons that they will refuse to enter their spawning stream.

Acid mine drainage has polluted surface waters since the beginning of ore mining. Sulfur-laden water leaches from mines, including old and abandoned mines as well as active ones, and contains sulfur compounds that oxidize to sulfuric acid on contact with air. The resulting acidity of the stream or lake into which this water drains is often high enough to kill the aquatic ecosystem.

The effects of water pollution can be best understood in the context of an aquatic ecosystem, by studying one or more specific interactions of pollutants with that ecosystem.

ELEMENTS OF AQUATIC ECOLOGY

Plants and animals in their physical environment make up an *ecosystem*. The study of ecosystems is *ecology*. Although we often draw lines around a specific ecosystem to be able to study it more fully (e.g., a farm pond) and thereby assume that the system is completely self-contained, this is obviously not true. One of the tenets of ecology is that "everything is connected with everything else."

Three categories of organism make up an ecosystem. The *producers* take energy from the sun and nutrients like nitrogen and phosphorus from the soil and produce high-energy chemical compounds by the process of photosynthesis. The energy from the sun is stored in the molecular structure of these compounds. Producers are often referred to as being in the first *trophic* (growth) level and are called *autotrophs* by the *heterotrophs*.

The second category of organism in an ecosystem is the *consumers,* who use the energy stored by photosynthesis by ingesting the high-energy compounds. Consumers in the second trophic level use the energy of the producers directly. There may be several more trophic levels of consumers, each using the level below it as an energy source. A simplified ecosystem showing various trophic levels is illustrated in Figure 3–1, which also shows the progressive use of energy through the trophic levels.

The third category of organism, the *decomposers* or decay organisms, use the energy in animal wastes and dead animals and plants, thereby converting the organic compounds to stable inorganic compounds. The residual inorganics (e.g., nitrates) are then nutrients for the producers, with the sun as the source of energy.

Ecosystems exhibit a flow of both energy and nutrients. Energy flow is in only one direction: from the sun and through each trophic level. Nutrient flow, on the other hand, is cyclic: Nutrients are used by plants to make high-energy molecules that are eventually decomposed to the original inorganic nutrients, ready to be used again.

The entire food web, or ecosystem, stays in dynamic balance, with adjustments being made as required. This balance is called *homeostasis*. For example, a drought may produce little grass, starving field mice and exposing them to predators like owls.

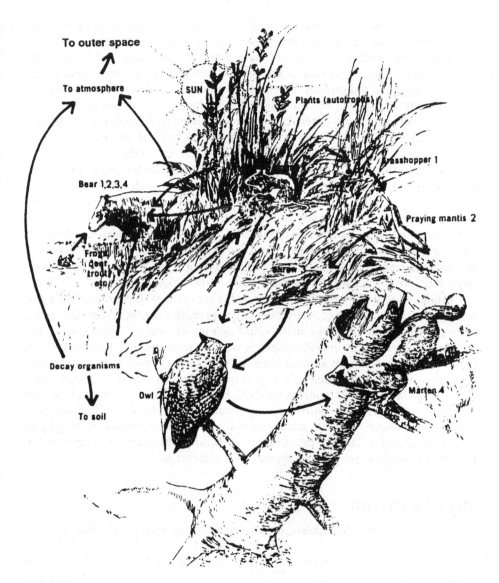

To outer space

To atmosphere

SUN

Plants (autotrophs)

Grasshopper 1

Bear 1,2,3,4

Praying mantis 2

Frogs, deer, toads, etc.

Shrew

Decay organisms

Owl 3

Marten 4

To soil

FIGURE 3–1. A typical terrestrial ecosystem. The numbers refer to trophic level above the autotrophic, and the arrows show progressive loss of energy. [From Turk, A., et al., *Environmental Science*, Philadelphia: W.B. Saunders (1974). Used with permission.]

The field mice spend more time inside their burrows, eat less, and thus allow the grass to reseed for the following year. External perturbations may upset and even destroy an ecosystem. In the previous example, use of a herbicide to kill the grass (instead of merely thinning it) might also destroy the field mouse population, since the mice would be more exposed to predatory attack.

Once the field mice, the source of food for the predatory owls, are gone, the owls eventually starve also and the ecosystem collapses.

Most ecosystems can absorb a certain amount of damage, but sufficiently large perturbations may cause irreversible damage. The ongoing attempt to limit the logging of old growth forests in the Pacific Northwest is an attempt to limit the damage to the forest ecosystem to what it can accommodate. The amount of perturbation a system can absorb is related to the concept of the *ecological niche*. The combination of function and habitat of an organism in an ecosystem is its niche. A niche is an organism's best accommodation to its environment. In the example given previously, if there are two types of grass that the mice could eat, and the herbicide destroys only one, the mice would still have food and shelter, and the ecosystem could survive. This simple example illustrates another important ecological principle: The stability of an ecosystem is proportional to the number of organisms capable of filling various niches. A jungle is a more stable ecosystem than the Alaskan tundra, which is very fragile. Another fragile system is that of the deep oceans, a fact that must be considered before the oceans are used as waste repositories.

Inland waterways tend to be fairly stable ecosystems, but are certainly not totally resistant to destruction by outside perturbations. Other than the direct effect of toxic materials like metals and refractory organic compounds, the most serious effect on inland waters is depletion of dissolved (free) oxygen (DO). All higher forms of aquatic life exist only in the presence of oxygen, and most desirable microbiologic life also requires oxygen. Natural streams and lakes are usually *aerobic* (containing DO). If a watercourse becomes *anaerobic* (absence of oxygen), the entire ecology changes and the water becomes unpleasant and unsafe. The DO concentration in waterways and the effect of pollutants are closely related to the concept of decomposition and biodegradation, part of the total energy transfer system that sustains life.

λ BIODEGRADATION

Plant growth, or photosynthesis, may be represented by the equation

$$CO_2 + H_2O \; \frac{\text{sunlight}}{\text{nutrients}} \Rightarrow HCOH + O_2 \tag{3.1}$$

In this representation formaldehyde (HCOH) and oxygen are produced from carbon dioxide and water, with sunlight as the source of energy and chlorophyll as a catalyst.[1] If the formaldehyde-oxygen mixture is ignited, it explodes, and the energy released in the explosion is the energy that was stored in the hydrogen-oxygen bonds of formaldehyde.

---

[1]Of course, formaldehyde is not the usual end product of photosynthesis, but is given here as an example of both how organic molecules are formed in photosynthesis and how the energy stored in it may be recovered.

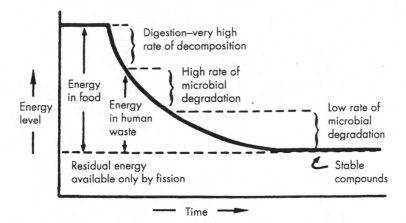

FIGURE 3–2. Energy loss in biodegradation. [Adapted from McGauhey, P.H., *Engineering Management of Water Quality*, New York: McGraw-Hill (1968).]

As discussed above, plants (producers) use inorganic chemicals as nutrients and, with sunlight as an energy source, build high-energy compounds. Consumers eat and metabolize (digest) these compounds, releasing some of the energy for the consumer to use. The end products of metabolism (excrement) become food for decomposers and are degraded further, but at a much slower rate than metabolism. After several such steps, very low energy compounds remain that can no longer be used by microorganism decomposers as food. Plants then use these compounds to build more high-energy compounds by photosynthesis, and the process starts over. The process is shown symbolically in Figure 3–2.

Many organic materials responsible for water pollution enter watercourses at a high energy level. The biodegradation, or gradual use of energy, of the compounds by a chain of organisms causes many water pollution problems.

AEROBIC AND ANAEROBIC DECOMPOSITION

Decomposition or biodegradation may take place in one of two distinctly different ways: aerobic (using free oxygen) and anaerobic (in the absence of free oxygen). If formaldehyde could decompose aerobically, the equation for decomposition would be the reverse of Equation 3.2, or

$$HCOH + O_2 \rightarrow CO_2 + H_2O + energy \tag{3.2}$$

Generally, the basic equation for *aerobic decomposition* of complex organic compounds of the form $C_xH_yN_z$ is

$$\left(\frac{3x + y + z}{2}\right)O_2 + C_xH_yN_z \rightarrow xCO_2 + \frac{y}{2}H_2O + zNO + products \tag{3.3}$$

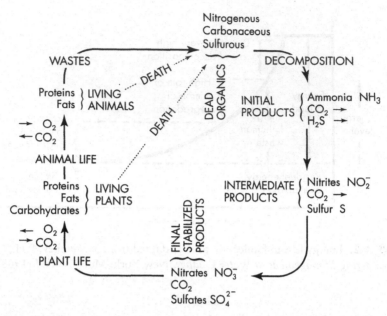

FIGURE 3–3. Aerobic nitrogen, carbon, and sulfur cycles. [Adapted from McGauhey, P.H., *Engineering Management of Water Quality*, New York: McGraw-Hill (1968).]

Carbon dioxide and water are always two of the end products of aerobic decomposition. Both are stable, low in energy, and used by plants in photosynthesis (plant photosynthesis is a major CO_2 sink for the earth). Sulfur compounds (like the mercaptans in mammal excrement) are oxidized to SO_4^{2-}, the sulfate ion, and phosphorus is oxidized to PO_4^{3-}, orthophosphate. Nitrogen is oxidized through a series of compounds ending in nitrate, in the progression

$$\text{Organic nitrogen} \rightarrow NH_3 \text{ (ammonia)} \rightarrow NO_2^- \text{ (nitrite)} \rightarrow NO_3^- \text{ (nitrate)}$$

Because of this distinctive progression, nitrogen has been, and still is, used as an indicator of water pollution. A schematic representation of the aerobic cycle for carbon, sulfur, and nitrogen is shown in Figure 3–3. This figure shows only the basic phenomena and greatly simplifies the actual steps and mechanisms.

Anaerobic decomposition is performed by a completely different set of microorganisms, to which oxygen is toxic. The basic equation for anaerobic biodegradation is

$$C_xH_yN_z \rightarrow CO_2 + CH_4 + NH_3 + \text{partly stable compounds} \tag{3.4}$$

Many of the end products of the reaction are biologically unstable. Methane (CH_4), for example, a high-energy gas commonly called marsh gas,[2] is physi-

---

[2]When methane is burned as a fossil fuel it is called "natural gas."

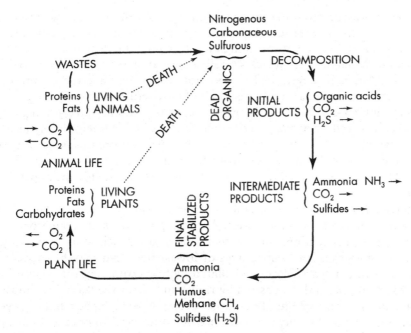

FIGURE 3–4. Anaerobic nitrogen, carbon, and sulfur cycles. [Adapted from McGauhey, P.H., *Engineering Management of Water Quality*, New York: McGraw-Hill (1968).]

cally stable but can be decomposed biologically. Ammonia (NH_3) can be oxidized, and sulfur is anaerobically biodegraded to evil-smelling sulfhydryl compounds like hydrogen sulfide (H_2S). Figure 3–4 is a schematic representation of anaerobic decomposition. Note that the left half of the cycle, photosynthesis by plants, is identical to the aerobic cycle.

Biologists often speak of certain compounds as hydrogen acceptors. When energy is released from high-energy compounds a C-H or N-H bond is broken and the freed hydrogen must be attached somewhere. In aerobic decomposition, oxygen serves the purpose of a hydrogen scavenger or hydrogen acceptor and forms water. In anaerobic decomposition, oxygen is not available. The next preferred hydrogen acceptor is NH_3, since in the absence of oxygen ammonia cannot be oxidized to nitrite or nitrate. If no appropriate nitrogen compound is available, sulfur accepts hydrogen to form H_2S, the compound responsible for the notorious rotten egg smell.

EFFECT OF POLLUTION ON STREAMS

When a high-energy organic material such as raw sewage is discharged to a stream, a number of changes occur downstream from the point of discharge. As the organic components of the sewage are oxidized, oxygen is used at a greater rate than upstream from the sewage discharge, and the DO in the stream decreases markedly. The rate of reaeration, or solution of oxygen from the air, also increases, but is often not great enough to prevent a total depletion of oxygen in the stream. When the stream DO is totally depleted, the stream is said to become anaerobic. Often, however, the DO does not drop to zero and the stream recovers without a period of anaerobiosis. Both of these situations are depicted graphically in Figure 3–5. The dip in DO is referred to as a *dissolved oxygen sag curve*.

The dissolved oxygen sag curve can be described mathematically as a dynamic balance between the use of oxygen by the microorganisms (deoxygenation) and the supply of oxygen from the atmosphere (reoxygenation). The mathematical derivation of the oxygen sag curve is included in the appendix to this chapter.

Stream flow is of course variable, and the critical DO levels can be expected to occur when the flow is the lowest. Accordingly, most state regulatory agencies base their calculations on a statistical low flow, such as a 7-day, 10-year low flow: the seven consecutive days of lowest flow that may be expected to occur once in ten years. This is calculated by first estimating the lowest 7-day flow for each year and then assigning ranks: m = 1 for the least flow (most severe) to m = n for the greatest flow (least severe), where n is the number of years considered. The probability of occurrence of a flow equal to or more than a particular low flow is

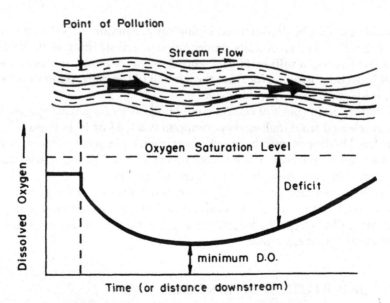

FIGURE 3–5. Dissolved oxygen downstream from a source of organic pollution. The curve shows a DO sag without anaerobic conditions.

$$P = \frac{m}{n + 1} \tag{3.5}$$

and is graphed against the flow. The 10-year low flow is then read from the graph at $m/(n + 1) = 0.1$.

Example 3.1
Calculate the 10-year, 7-day low flow given the data below.

Year	Lowest Flow 7 Consecutive Days (m^3/s)	Ranking (m)	$m/(n + 1)$	Lowest Flow in Order of Severity (m^3/s)
1965	1.2	1	1/14 = 0.071	0.4
1966	1.3	2	2/14 = 0.143	0.6
1967	0.8	3	3/14 = 0.214	0.6
1968	1.4	4	4/14 = 0.285	0.8
1969	0.6	5	5/14 = 0.357	0.8
1970	0.4	6	6/14 = 0.428	0.8
1971	0.8	7	7/14 = 0.500	0.9
1972	1.4	8	8/14 = 0.571	1.0
1973	1.2	9	9/14 = 0.642	1.2
1974	1.0	10	10/14 = 0.714	1.2
1975	0.6	11	11/14 = 0.786	1.3
1976	0.8	12	12/14 = 0.857	1.4
1977	0.9	13	13/14 = 0.928	1.4

These data are plotted in Figure 3–6, and the minimum 7-day, 10-year low flow is read at $m/(n + 1) = 0.1$ as 0.5 m^3/s.

When the rate of oxygen use overwhelms the rate of oxygen resupply, the stream may become anaerobic. An anaerobic stream is easily identifiable by the presence of floating sludge and bubbling gas. The gas is formed because oxygen is no longer available to act as the hydrogen acceptor, and ammonia, hydrogen sulfide, and other gases are formed. Some of the gases formed dissolve in water, but others can attach themselves as bubbles to sludge (solid black or dark benthic deposits) and buoy the sludge to the surface. In addition, the odor of H_2S will advertise the anaerobic condition for some distance, the water is usually black or dark, and fungus grows in long slimy filaments that cling to rocks and gracefully wave streamers downstream.

The outward evidence of an anaerobic stream is accompanied by adverse effects on aquatic life. Types and numbers of species change drastically downstream from the pollution discharge point. Increased turbidity, settled solid

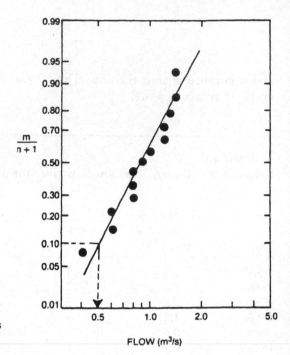

FIGURE 3–6.
Plot of 10-year, 7-day low flows
for Example 3.1

matter, and low DO all contribute to a decrease in fish life. Fewer and fewer species of fish are able to survive, but those that do find food plentiful and often multiply in large numbers. Carp and catfish can survive in waters that are quite foul and can even gulp air from the surface. Trout, on the other hand, need very pure, cold, oxygen-saturated water and are notoriously intolerant of pollution.

Numbers of other aquatic species are also reduced, and the remaining species like sludge worms, bloodworms, and rat-tailed maggots abound, often in staggering numbers—as many as 50,000 sludge worms per square foot.

Figure 3–7 illustrates the distribution of both species and numbers of organisms downstream from a source of pollution. The diversity of species may be quantified by an index, such as

$$d = \sum_{i=1}^{s} \left(\frac{n_i}{n}\right) \log_{10}\left(\frac{n_i}{n}\right) \tag{3.6}$$

where d = diversity index
           n_i = number of individuals in the ith species
           n = total number of individuals in all S species

Table 3–1 shows the results of a study in which the diversity index was calculated above and below a sewage outfall.

These reactions of a stream to pollution occur when a rapidly decomposable organic material is the waste. The stream will react much differently to in-

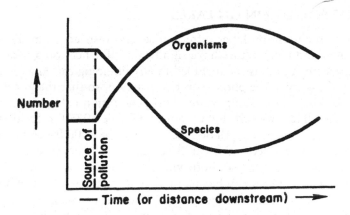

FIGURE 3–7. The number of species and the total number of organisms downstream from a point of organic pollution

TABLE 3–1. Diversity of Aquatic Organisms

Location	Diversity Index (d)
Above the outfall	2.75
Immediately below the outfall	0.94
Downstream	2.43
Further downstream	3.80

organic waste, as from a metal-plating plant. If the waste is toxic to aquatic life, both the kind and total number of organisms will decrease below the outfall. The DO will not fall and might even rise. There are many types of pollution, and a stream will react differently to each (Figure 3–8). When two or more wastes are involved the situation is even more complicated.

FIGURE 3–8
Typical variations in nitrogen compounds downstream from a point of organic pollution

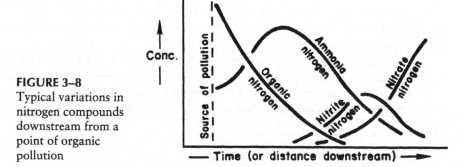

EFFECT OF POLLUTION ON LAKES

The effect of pollution on lakes differs in several respects from the effect on streams. Light and temperature have significant influences on a lake, more so than on a stream, and must be included in any limnological[3] analysis. Light is the source of energy in the photosynthetic reaction, so that the penetration of light into the lake water is important. This penetration is logarithmic; for example, at a depth of 1 foot the light intensity may be 10,000 ft-candles (a measure of light intensity); at a depth of 2 feet, it might be 1000 ft-candles; at 3 feet, 100 ft-candles, and at 4 feet, 10 ft-candles. Light usually penetrates only the top two feet of a lake; hence, most photosynthetic reactions occur in that zone.

Temperature and heat can have a profound effect on a lake. Water is at a maximum density at 4°C; water both colder and warmer than this is less dense, and therefore ice floats. Water is also a poor conductor of heat and retains it quite well.

Lake water temperature usually varies seasonally. Figure 3–9 illustrates these temperature-depth relationships. During the winter, assuming that the lake does not freeze, the temperature is often constant with depth. As the weather warms in spring, the top layers begin to warm. Since warmer water is less dense, and water is a poor conductor of heat, a distinct temperature gradient known as *thermal stratification* is formed. These strata are often very stable and last through the summer months. The top layer is called the *epilimnion*; the middle, the *metalimnion*; and the bottom, the *hypolimnion*. The inflection point in the curve is called the *thermocline*. Circulation of water occurs only within a zone, and thus there is only limited transfer of biological or chemical material (including DO) across the boundaries. As the colder weather approaches, the top layers begin to cool, become more dense, and sink. This creates circulation within the lake, known as *fall turnover*. A spring turnover may also occur.

The biochemical reactions in a natural lake may be represented schematically as in Figure 3–10. A river feeding the lake would contribute carbon, phosphorus, and nitrogen, either as high-energy organics or as low-energy compounds. The *phytoplankton* or *algae* (microbial free-floating plants) take C, P, and N and, using sunlight as a source of energy, make high-energy compounds. Algae are eaten by *zooplankton* (tiny aquatic animals), which are in turn eaten by larger aquatic life such as fish. All of these life forms defecate, contributing a pool of *dissolved organic carbon*. This pool is further fed by the death of aquatic life. Bacteria use dissolved organic carbon and produce CO_2, in turn used by algae. CO_2 in addition to that dissolved directly is provided from the respiration of the fish and zooplankton, as well as the CO_2 dissolved directly from the air.

The supply of C, P, and N coming into an unpolluted lake is small enough to limit algae production, and productivity of the entire ecological system is limited. When large amounts of C, P, and N are introduced into the lake, however, they promote uncontrolled growth of algae in the epilimnion, since the algae can

---

[3]Limnology is the study of lakes.

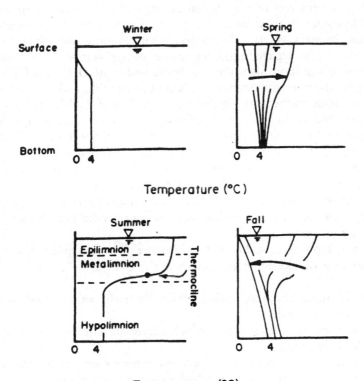

FIGURE 3–9. Typical temperature-depth relationships

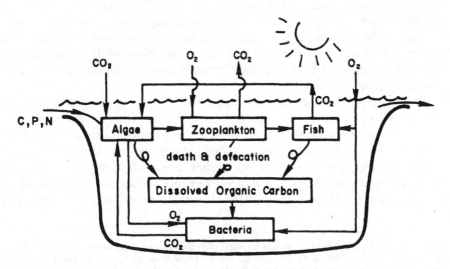

FIGURE 3-10. Schematic representation of lake ecology [Courtesy of Donald Francisco.]

assimilate nutrients very rapidly. When the algae die, they drop to the lake bottom (the hypolimnion) and become a source of carbon for decomposing bacteria. Aerobic bacteria will use all available DO in decomposing this material, and DO may thereby be depleted enough to cause the hypolimnion to become anaerobic. As more and more algae die, and more and more DO is used in their decomposition, the metalimnion may also become anaerobic, and aerobic biological activity would be concentrated in the upper few feet of the lake, the epilimnion.

The aerobic biological activity produces turbidity, decreasing light penetration and in turn limiting photosynthetic algal activity in the surface layers. The amount of DO contributed by the algae is therefore decreased. Eventually, the epilimnion also becomes anaerobic, all aerobic aquatic life disappears, and the algae concentrate on the lake surface because there is only enough light available for photosynthesis. The algal concentration forms large green mats, called *algal blooms*. When the algae in these blooms die and ultimately fill up the lake, a peat bog is formed.

The entire process is called *eutrophication*. It is the continually occurring natural process of lake aging and occurs in three stages:

- The *oligotrophic* stage, during which both the variety and number of species grow rapidly.
- The *mesotrophic* stage, during which a dynamic equilibrium exists among species in the lake.
- The *eutrophic* stage, during which less complex organisms take over and the lake appears to become gradually choked with weeds.

Natural eutrophication may take thousands of years. If enough nutrients are introduced into a lake system, as may happen as a result of human activity, the eutrophication process may be shortened to as little as a decade. The addition of phosphorus, in particular, can speed eutrophication, since phosphorus is often the *limiting nutrient* for algae: the particular nutrient that limits algal growth. The limiting nutrient for a system is that element the system requires the smallest amount of; consequently, growth depends directly on the amount of that nutrient.[4]

Where do these nutrients originate? One source is excrement, since all human and animal wastes contain C, N, and P. Synthetic detergents and fertilizers are a much greater source. About half of the phosphorus in U.S. lakes is estimated to come from agricultural runoff; about one-fourth, from detergents; and the remaining one-fourth, from all other sources. It seems unfortunate that the presence of phosphates in detergents has received so much unfavorable attention when runoff from fertilized land is a much more important source of P.

---

[4]The addition of a limiting nutrient acts for algal growth much as stepping on the gas pedal limits the speed of your car. All of the components are available to make the car go faster, but it can't speed up until you "give it more gas." The gas pedal is a constraint, or limit, against higher speed. Dumping excess phosphorus into a lake is like floorboarding the gas pedal.

Conversion to nonphosphate detergents is of limited value when other phosphate sources are not controlled.

Phosphate concentrations between 0.01 mg/L and 0.1 mg/L appear to be enough to accelerate eutrophication. Sewage treatment plant effluents may contain between 5 mg/L and 10 mg/L of phosphorus as phosphate, and a river draining farm country may carry from 1 mg/L to 4 mg/L. High phosphorus concentration is not a problem in a moving stream, in which algae are continually flushed out and do not accumulate. Eutrophication occurs mainly in lakes, ponds, estuaries, and sometimes in very sluggish rivers. Phosphorus is not always the culprit in accelerated eutrophication. Generally, a P:N:C ratio of 1:16:100 is required for algal growth; that is, algae need 16 parts N and 100 parts C for every part P. P is the limiting nutrient if N and C are in excess of this ratio. However, in some lakes, P can be present in excess and N can act as the limiting nutrient. Evidence suggests that nitrogen limits growth in brackish waters like bays and estuaries. Interaction among the many chemical pollutants present, rather than any single chemical, is often to blame for accelerated eutrophication.

Actual profiles in a lake for a number of parameters are shown in Figure 3–11. The foregoing discussion clarifies why a lake is warmer on top than lower down, how DO can drop to zero, and why N and P are highly concentrated in the lake depths while algae bloom on the surface.

HEAVY METALS AND TOXIC SUBSTANCES

In 1970, Barry Commoner (Commoner 1970) and other scientists alerted the nation to the growing problem of mercury contamination of lakes, streams, and marine waters. The manufacture of chlorine and lye from brine, called the chlor-alkali process, was identified as a major source of mercury contamination. Elemental mercury is methylated by aquatic organisms, and methylated mercury finds its way into fish and shellfish and thus into the human food chain. Methyl mercury is a powerful neurological poison. Methyl mercury poisoning was first identified in Japan in the 1950s as "Minamata disease." Mercury-containing effluent from the Minamata Chemical Company was found to be the source of mercury in food fish.

Arsenic, copper, lead, and cadmium are often deposited in lakes and streams from the air near emitting facilities. These substances may also enter waterways from runoff from slag piles, mine drainage, and industrial effluent. Effluent from electroplating contains a number of heavy metal constituents. Heavy metals, copper in particular, may be toxic to fish as well as harmful to human health.

In the past quarter century, a considerable incidence of surface water contamination by hazardous and carcinogenic organic compounds was reported in the United States. The sources of contamination include effluent from petrochemical industries and agricultural runoff, which contains both pesticide and fertilizer residues. Trace quantities of chlorinated hydrocarbon compounds in

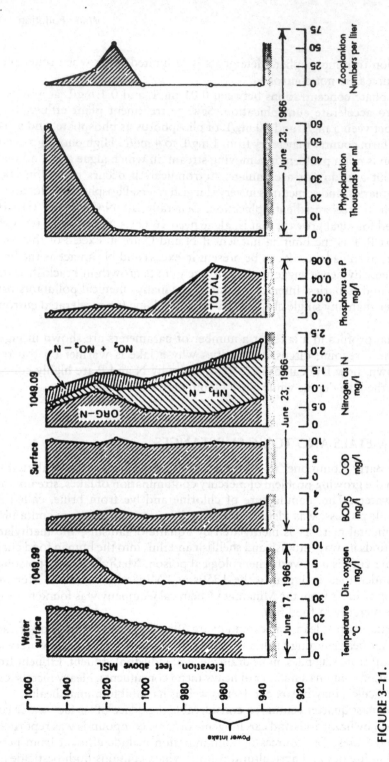

FIGURE 3–11.
Water quality profiles for a water supply reservoir [From Berthouex, P. and Rudd, D. *Strategy of Pollution Control*, New York: John Wiley (1977).]

drinking water may also be attributed to the chlorination of organic residues by chlorine added as a disinfectant.

EFFECT OF POLLUTION ON OCEANS

Not many years ago, the oceans were considered infinite sinks; the immensity of the seas and oceans seemed impervious to assault. However, we now recognize seas and oceans as fragile environments and are able to measure detrimental effects. Ocean water is a complicated chemical solution and appears to have changed very little over millions of years. Because of this constancy, however, marine organisms have become specialized and intolerant to environmental change. Oceans are thus fragile ecosystems, quite susceptible to pollution.

A relief map of the ocean bottom reveals two major areas: the continental shelf and the deep oceans. The continental shelf, especially near major estuaries, is the most productive in terms of food supply. Because of its proximity to human activity, it receives the greatest pollution load. Many estuaries have become so badly polluted that they are closed to commercial fishing. The Baltic and Mediterranean seas are in danger of becoming permanently damaged.

Ocean disposal of wastewater is severely restricted in the United States, but many major cities all over the world still discharge all untreated sewage into the ocean.[5] Although the sewage is carried a considerable distance from shore by pipeline and discharged through diffusers to achieve maximum dilution, the practice remains controversial, and the long-term consequences are much in doubt.

CONCLUSION

Water pollution stems from many sources and causes, only a few of which are discussed here. Rivers and streams demonstrate some capacity to recover from the effects of certain pollutants, but lakes, bays, ponds, and sluggish rivers may not recover. The oceans are far more sensitive to pollutants than was thought. The effects of pollutants on oceans and groundwater, and the effects of inorganic poisons like heavy metals, also do not receive detailed discussion here.

PROBLEMS

3.1 Figure 3–3 shows the aerobic cycle for nitrogen, carbon, and sulfur. Phosphorus should have been included in the cycle since it exists as organically bound phosphorus in living and dead tissue and decomposes to polyphosphates such as $(P_2O_7)^{4-}$ and $(P_3O_{10})^{5-}$, and finally to orthophosphates such as PO_4^{3-}. Draw a phosphorus cycle similar to Figure 3–3.

---

[5]The citizens of Victoria, British Columbia, voted in 1992 to continue to dispose of untreated municipal sewage in the Strait of Juan de Fuca in Puget Sound (*Seattle Times*, December 10, 1992).

3.2   Some researchers have suggested that the empirical analysis of some algae gives it the chemical composition $C_{106}H_{181}O_{45}N_{16}P$. Suppose that analysis of lake water yields the following: C = 62 mg/L, N = 1.0 mg/L, P = 0.01 mg/L. Which elements are the limiting nutrients for the growth of algae in this lake?

3.3   A stream feeding a lake has an average flow of 1 ft³/sec and a phosphate concentration of 10 mg/L. The water leaving the lake has a phosphate concentration of 5 mg/L.

- How much phosphate in metric tons is deposited in the lake each year (1000 kg = 1 metric ton)?
- Where does this phosphorus go, since the outflowing concentration is less than the inflowing concentration?
- Would the average phosphate concentration be higher near the surface of the lake or near the bottom?
- Would you expect eutrophication of the lake to be accelerated? Why?

3.4   Show how the compound thiodiazine—$C_{21}H_{26}N_2S_2$—decomposes anaerobically and how these end products in turn decompose aerobically to stabilized sulfur and nitrogen compounds.

3.5   If an industrial plant discharges an effluent with a solids concentration at a rate of 5000 lb/day, and if each person contributes 0.2 lb/day, what is the population equivalent of the waste?

3.6   The temperature soundings for a lake are as follows:

Depth (ft)	Water Temperature (°F)
0 (surface)	80
4	80
8	60
16	40
24	40
30	40

Plot depth vs. temperature and label the hypolimnion, epilimnion, and thermocline.

3.7   Sketch the DO sag curves you would expect in a stream from the following wastes. Assume the stream flow equals the flow of wastewater. (Do not calculate.)

Waste Source	BOD (mg/L)	Suspended Solids (SS) (mg/L)	Phosphorus (mg/L)
Dairy	2000	100	40
Brick manufacturing	5	100	10
Fertilizer manufacturing	25	5	200
Electroplating plant	0	100	10

3.8   Starting with nitrogenous dead organic matter, follow N around the aerobic and anaerobic cycles by writing down all the various forms of nitrogen.

3.9   Suppose a stream with a velocity of 1 ft/sec, a flow of 10 million gallons/day (mgd), and an ultimate carbonaceous BOD of 5 mg/L is hit with treated sewage at 5 mgd with an ultimate carbonaceous BOD (L_0) of 60 mg/L. The temperature of the stream water is 20°C, at the point of sewage discharge the stream is 90% saturated with oxygen, and the wastewater is at 30°C and has no oxygen (see Table 3–1). Measurements show the deoxygenation constant $k_1' = 0.5$ and the reoxygenation $k_2' = 0.6$, both as days^{-1}. Calculate: (a) the oxygen deficit one mile downstream, (b) the minimum DO (the lowest part of the sag curve), and (c) the minimum DO (or maximum deficit) if the ultimate carbonaceous BOD of the treatment plant's effluent is 10 mg/L. You may write a computer program or use a spreadsheet program to solve this problem.

LIST OF SYMBOLS

D	deficit in DO, in mg/L
D_0	initial DO deficit, in mg/L
DO	dissolved (free) oxygen
d	diversity index
H	depth of stream flow, in m
k_1	deoxygenation constant, ($\log_{10}$) sec^{-1}
k_1'	deoxygenation constant, ($\log_e$) sec^{-1}
k_2	reoxygenation constant, ($\log_{10}$) sec^{-1}
k_2'	reoxygenation constant, ($\log_e$) sec^{-1}
L_0	ultimate biochemical oxygen demand, in mg/L
m	rank assigned to low flows
n	number of years in low flow records
n_i	number of individuals in species i
ppm	parts per million
Q_s	stream flow, in mgd or m^3/sec
Q_p	pollutant flow, in mgd or m^3/sec
T	temperature, in °C
t	time, in sec
t_c	critical time, time when minimum DO occurs, in sec
v	velocity, in m/sec
y	oxygen used, in mg/L
z	oxygen required for decomposition, in mg/L

Appendix

Mathematical Description of the Dissolved Oxygen Sag Curve

The effect of a certain waste on a stream's oxygen level may be estimated mathematically.[6] The basic assumption is that there is an oxygen balance at any point in the stream that depends on (1) how much oxygen is being used by the microorganisms, and (2) how much oxygen is supplied to the water through reaeration. The rate of oxygen use, or oxygen depletion, may be expressed as

$$\text{Rate of deoxygenation} = -k_1'z \tag{3.7}$$

where z = amount of oxygen still required at any time t, or the biochemical demand for oxygen remaining in the water, in mg/L (or ppm)

           k_1' = deoxygenation constant, a function of the type of waste material decomposing, temperature, stream velocity, etc., in days^{-1}

In this equation, the concentration of DO is expressed in mg/L, a common way of expressing the concentration of chemicals in water. In water, mg/L becomes equivalent to parts per million (ppm) by assuming that the dissolved substances have the same density as water. This assumption is generally valid for low concentrations.

The value of k_1', the rate constant, is measured in the laboratory, as discussed in the next chapter. Integrating Equation 3.7 yields

$$z = L_0 e^{-k_1't} \tag{3.8}$$

where L_0 is the *ultimate carbonaceous oxygen demand*. Since the long-term need for oxygen is L_0 and the amount of oxygen still *needed* at any time t is z, the amount of oxygen *used* at any time t is

$$y = L_0 - z \tag{3.9}$$

This relationship is shown in Figure 3–A1. The term y is defined later in this text as the *biochemical oxygen demand* (BOD) and expressed as

$$y = L_0(1 - e^{-k_1't}) \tag{3.10}$$

---

[6]Streeter, H.W., and Phelps, E.B., "A Study of the Pollution and Natural Purification of the Ohio River," *Public Health Bulletin* 146, Washington, DC: USPHS (1925).

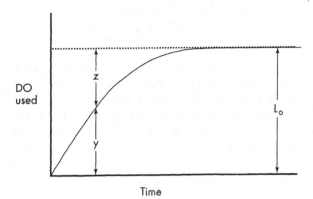

FIGURE 3–A1
Dissolved oxygen used at
any time t(y) plus the DO
still needed at time t(z) is
equal to the ultimate
oxygen demand (L_0).

Although y is commonly termed oxygen *demand*, it is more correctly described as the DO *used*. In this text, we use the terms DO *demand* and DO *used* synonymously.

The reoxygenation of a stream may be expressed as

$$\text{Rate of reoxygenation} = k_2'D \qquad (3.11)$$

where D = deficit in DO, or the difference between saturation (the maximum DO the water can hold) and the actual DO, in mg/L (or ppm)
  k_2' = reoxygenation constant, in days^{-1}

The value of k_2' is obtained by studying the stream using a tracer and may be estimated from a table like Table 3–A1. If this cannot be done, a generalized expression (O'Connor 1958) may be used:

$$k_2' = \frac{3.9v^{1/2}\sqrt{(1.037)^{(T-20)}}}{H^{3/2}} \qquad (3.11a)$$

TABLE 3–A1. Reaeration Constants

Type of Watercourse	k_2' at 20°C,[a] (days^{-1})
Small ponds or backwaters	0.10–0.23
Sluggish streams	0.23–0.35
Large streams, low velocity	0.35–0.46
Large streams, normal velocity	0.46–0.69
Swift streams	0.69–1.15
Rapids	>1.15

[a]For temperatures other than 20°C, $k_2'(T) = k_2'(20°C)(1.024)^{T-20}$.

From O'Connor, D.J., and Dobbins, W.E., "Mechanisms of Reaeration of Natural Streams," *ASCE Transactions* 153 (1958), p. 641.

where T = temperature of the water, in °C
            H = average depth of flow, in m
            v = mean stream velocity, in m/sec

For a stream loaded with organic material, the simultaneous deoxygenation and reoxygenation form the DO sag curve, first developed by Streeter and Phelps in 1925. The shape of the oxygen sag curve, as shown in Figure 3–5, is the sum of the rate of oxygen use and the rate of oxygen supply. Immediately downstream from a pollution discharge into a stream, the rate of use will exceed the reaeration rate and the DO concentration will fall sharply. As the discharged sewage is oxidized, and fewer high-energy organic compounds are left, the rate of use will decrease, the supply will begin to catch up with the use, and the DO will once again reach saturation. This may be expressed mathematically as

$$\frac{dD}{dt} = k_1'z - k_2'D \tag{3.12}$$

where all terms are defined as above. The rate of change in the oxygen deficit D depends on the concentration of decomposable organic matter, or the need by the microorganisms for oxygen (z) and the oxygen deficit at any time t. The need for oxygen at time t is given by Equation 3.7. Integrating Equation 3.12 yields

$$D = \frac{k_1'L_0}{k_2' - k_1'} (e^{-k_1't} - e^{-k_2't}) + D_0 e^{-k_2't} \tag{3.13}$$

where D_0 = initial oxygen deficit, at the point of discharge, after the stream flow has mixed with the discharged material, in mg/L
            D = oxygen deficit at any time t, in mg/L

The initial oxygen deficit is given by

$$D_0 = \frac{D_sQ_s + D_pQ_p}{Q_s + Q_p} \tag{3.14}$$

where D_s = oxygen deficit in the stream directly upstream from the point of discharge, in mg/L
            Q_s = stream flow above the discharge, in m³/s
            D_p = oxygen deficit in the pollutant stream, in mg/L
            Q_p = flow rate of pollutant, in m³/sec

The ultimate carbonaceous BOD at the start of the DO sag curve must be the ultimate carbonaceous BOD immediately below the outfall, calculated in proportion to the flow as

$$L_0 = \frac{L_sQ_s + L_pQ_p}{Q_s + Q_p} \tag{3.15}$$

where L_s = ultimate BOD in the stream immediately upstream from the point of
discharge, in mg/L
Q_s = stream flow above the discharge, in m³/sec
L_p = ultimate BOD of the waste, in mg/L
Q_p = flow rate of the pollutant, in m³/s

The deficit equation is often expressed in common logarithms:

$$D = \frac{k_1'L_0}{k_2' - k_1'} (10^{-k_1't} - 10^{-k_2't}) + D_0 10^{-k_2't} \qquad (3.16)$$

since

$$e^{-k't} = 10^{-k't} \text{ when } k = 0.43k' \qquad (3.17)$$

The most serious water quality concern is the location in the sag curve
where the oxygen deficit is the greatest, or where the DO concentration is the
least. By setting $dD/dt = 0$, we can solve for the time when this minimum DO
occurs, the *critical time*, as

$$t_c = \frac{1}{k_2' - k_1'} \ln \left[\frac{k_2'}{k_1'} \left(1 - \frac{D_0(k_2' - k_1')}{k_1'L_0}\right) \right] \qquad (3.18)$$

where t_c is the time downstream when the DO concentration is the lowest.

Example 3.A1
Assume that a large stream has a reoxygenation constant, k_2', of 0.4/day and a
flow velocity of 5 mi/hr, and that at the point of a pollution discharge the stream
is saturated with oxygen at 10 mg/L. The wastewater flow rate is very small
compared with the stream flow, so the mixture is assumed to be saturated with
DO and to have an oxygen demand of 20 mg/L. The deoxygenation constant, k_1',
is 0.2/day. What is the DO level 30 miles downstream?
    Stream velocity = 5 mi/hr; hence, it takes 30/5 = 6 hours to travel 30 miles.
Thus

$$t = 6 \text{ hr/24 hr/day} = 0.25 \text{ day}$$

and
$$D_0 = 0$$
since the stream is saturated.

$$D = \frac{(0.2)(20)}{0.4 - 0.2} \left(e^{-(0.2)(0.25)} - e^{-(0.4)(0.25)}\right) = 1.0 \text{ mg/L} \qquad (3.19)$$

The DO is thus the saturation level minus the deficit, or 10 − 1.0 = 9.0 mg/L.

Chapter 4

Measurement of Water Quality

Quantitative measurements of pollutants are obviously necessary before water pollution can be controlled. However, measurement of these pollutants is fraught with difficulties. Sometimes specific materials responsible for the pollution are not known. Moreover, these pollutants are generally present at low concentrations, and very accurate methods of detection are required.

Only a representative sample of the analytical tests available to measure water pollution is discussed in this chapter. A complete volume of analytical techniques used in water and wastewater engineering is compiled as *Standard Methods for the Examination of Water and Wastewater*.[1] This volume, now in its 20th edition, is the result of a need for standardizing test techniques. It is considered definitive in its field and has the weight of legal authority.

Many water pollutants are measured in terms of milligrams of the substance per liter of water (mg/L). In older publications pollutant concentrations are expressed as parts per million (ppm), a weight/weight parameter.[2] If the liquid involved is water, ppm is identical with mg/L, since one liter (L) of water weighs 1000 grams (g). For pollutants present in very low concentrations (<10 mg/L), ppm is approximately equal to mg/L. However, because of the possibility that some wastes have specific gravity different from water, mg/L is preferred to ppm. A third commonly used parameter is percent, a weight/weight relationship. Note that 10,000 ppm = 1% and is equal to 10,000 mg/L only when 1 mL = 1 g.

SAMPLING

Some tests require the measurement to be conducted in the stream since the process of obtaining a sample may change the measurement. For example, if it is necessary to measure the dissolved oxygen (DO) in a stream, the measure-

---

[1] American Public Health Association, *Standard Methods for the Examination of Water and Wastewater* 20th ed., Water Pollution Control Federation, American Water Works Association (1996).
[2] For air pollutants, however, ppm is a volume/volume measurement (L of pollutant per 10^6 L of air) and is not the same as mg/L.

ment should be conducted right in the stream, or the sample must be extracted with great care to ensure that no transfer of oxygen from the air and water (in or out) occurs.

Most tests may be performed on a water sample taken from the stream. The process by which the sample is obtained, however, may greatly influence the result. The three basic types of samples are grab, composite, and flow weighted composite.

The grab sample, as the name implies, measures water quality at only one sampling point. Its value is that it accurately represents the water quality at the moment of sampling, but it says nothing about the quality before or after the sampling. The composite sample is obtained by taking a series of grab samples and mixing them together. The flow weighted composite is obtained by taking each sample so that the volume of the sample is proportional to the flow at that time. The last method is especially useful when daily loadings to wastewater treatment plants are calculated. Whatever the technique or method, however, the analysis can only be as accurate as the sample, and often the sampling methods are far more sloppy than the analytical determination.

DISSOLVED OXYGEN

Probably the most important measure of water quality is the dissolved oxygen (DO). Oxygen, although poorly soluble in water, is fundamental to aquatic life. Without free DO, streams and lakes become uninhabitable to gill-breathing aquatic organisms. Dissolved oxygen is inversely proportional to temperature, and the maximum oxygen that can be dissolved in water at most ambient temperatures is about 10 mg/L. The saturation value decreases rapidly with increasing water temperature, as shown in Table 4–1. The balance between saturation and depletion is therefore tenuous.

The amount of oxygen dissolved in water is usually measured either with an oxygen probe or by iodometric titration. The latter method is the Winkler test for DO, developed about 100 years ago and the standard against which all other measurements are compared.

The simplest (and historically the first) probe is shown in Figure 4–1. The principle of operation is that of a galvanic cell. If lead and silver electrodes are put into an electrolyte solution with a microammeter between, the reaction at the lead electrode will be

$$Pb + 2OH^- \rightarrow PbO + H_2O + 2e^- \qquad (4.1)$$

At the lead electrode (the cathode), electrons are liberated that travel through the microammeter to the silver electrode (the anode), where the following reaction takes place:

$$2e^- + 1/2O_2 + H_2O \rightarrow 2OH^- \qquad (4.1a)$$

TABLE 4–1. Solubility of Oxygen in Water

Water Temperature (°C)	Saturation Concentration of Oxygen in Water (mg/L)
0	14.6
2	13.8
4	13.1
6	12.5
8	11.9
10	11.3
12	10.8
14	10.4
16	10.0
18	9.5
20	9.2
22	8.8
24	8.5
26	8.2
28	8.0
30	7.6

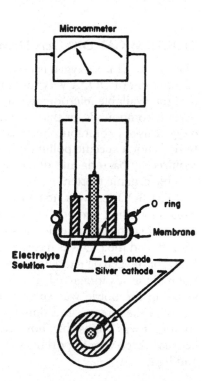

FIGURE 4–1
Schematic diagram of a galvanic cell oxygen probe

The overall reaction, balancing the electrons, is then

$$Pb + 2OH^- \rightarrow PbO + H_2O + 2e^- \tag{4.2}$$

$$2e^- + \frac{1}{2}O_2 + H_2O \rightarrow 2OH \tag{4.3}$$

$$Pb + \frac{1}{2}O_2 \rightarrow PbO \tag{4.4}$$

Unless free DO is available the reaction does not occur, and the microammeter does not register any current. The meter must be constructed and calibrated so that the electricity recorded is proportional to the concentration of oxygen in the electrolyte solution.

In the commercial models, the electrodes are insulated from each other with nonconducting plastic and are covered with a permeable membrane with a few drops of electrolyte between the membrane and electrodes. The amount of oxygen that travels through the membrane is proportional to the DO concentration. A high DO concentration in the water creates a strong push to get through the membrane, whereas a low DO concentration forces only limited O_2 through to participate in the reaction and thereby create electrical current. Thus the current is proportional to the oxygen level in solution.

BIOCHEMICAL OXYGEN DEMAND

The *rate* at which oxygen is used is perhaps even more important than the determination of DO. A very low rate of use would indicate either clean water, that the available microorganisms are uninterested in consuming the available organic compounds, or that the microorganisms are dead or dying. The rate of oxygen use is commonly referred to as biochemical oxygen demand (BOD). BOD is not a specific pollutant but rather a measure of the amount of oxygen required by bacteria and other microorganisms engaged in stabilizing decomposable organic matter.

The BOD test was first used for measuring the oxygen consumption in a stream by filling two bottles with stream water, measuring the DO in one bottle, and placing the other in the stream. In a few days, the second bottle was retrieved, and the DO was measured. The difference in the oxygen levels was the BOD, or oxygen demand, in milligrams of oxygen used per liter of sample. This test had the advantage of being very specific for the stream in question since the water in the bottle was subjected to the same environmental factors as the water in the stream, and thus was an accurate measure of DO usage in that stream. It was impossible, however, to compare the results in different streams because three very important variables were not constant: temperature, time, and light.

Temperature has a pronounced effect on oxygen uptake, because metabolic activity increases significantly at higher temperatures. The time allotted for the test is also important, since the amount of oxygen used increases with time. Light is another important variable since most natural waters contain algae and oxygen that can be replenished in the bottle if light is available. Different amounts of light affect the final oxygen concentration.

The BOD test has been standardized by requiring the test to be run in the dark at 20°C for five days. The 5-day BOD, or BOD_5, is the oxygen used by microorganisms in the water sample during the first five days after sampling. The choice of five days has led to some frivolous speculation—for example, samples prepared on Monday are finished on Friday, leaving the weekend free. Scientifically, five days is a compromise between running the test long enough to get reproducible results and running it so long that anaerobic material and mold in the BOD bottles interferes.

The oxidation of BOD is an exponential decay curve, as is shown for example in Figure 3–2, and the decay constant is usually such that most of the BOD is oxidized in the first five days. Two-day, 10-day, or any other day BOD can be determined. One measure used in some cases is ultimate BOD, or the O_2 demand after a very long time.

The BOD test is usually carried out in standard BOD bottles (about 300-mL volume) as shown in Figure 4–2. The test is begun by measuring the initial DO or by saturating the samples with DO at the temperature at which they will be stored, usually 20°C. DO is then measured in several samples every day for five days, and the results produce curves like those shown in Figure 4–3. In this example, Sample A has an initial DO of 8 mg/L, and in five days this drops to 2 mg/L. The BOD is therefore 8 − 2 = 6 mg/L.

Sample B also has an initial DO of 8 mg/L, but the oxygen is used so fast that it drops to zero by the second day. Since there is no measurable DO left after five days, the BOD of sample B must be more than 8 − 0 = 8 mg/L, but we don't know how much more since the organisms in the sample might have used more DO if it had been available. Samples like this require repeating the 5-day BOD test after diluting the sample.

FIGURE 4–2
A biochemical oxygen demand (BOD) bottle

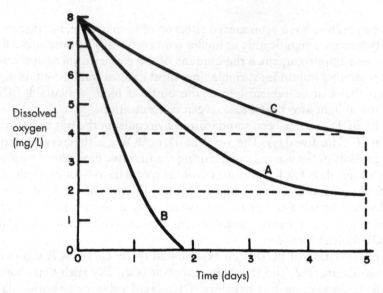

FIGURE 4–3. Typical oxygen uptake curves in a BOD test

Suppose sample C shown on the graph of Figure 4–3 is sample B diluted by 1:10. The BOD of sample B is therefore

$$\frac{8 - 4}{0.1} = 40 \text{ mg/L} \tag{4.5}$$

It is possible to measure the BOD of any organic material (e.g., sugar) and thus estimate its influence on a stream, even though the material in its original state might not contain the necessary organisms. *Seeding* is a process in which the microorganisms that oxidize the BOD are added to the BOD bottle. Measurement of very low BOD concentrations is also facilitated by seeding.

Suppose we use the water previously described in curve A as seed water, since it obviously contains microorganisms (it has a 5-day BOD of 6 mg/L). We now put 100 mL of an unknown solution into a bottle and add 200 mL of seed water, thus filling the 300-mL bottle. Assuming that the initial DO of this mixture is 8 mg/L and the final DO is 1 mg/L, the total oxygen used is 7 mg/L. However, some of this is due to the seed water, since it also has a BOD, and only a portion is due to the decomposition of the unknown material. The oxygen use due to the seed water is

$$6 \times \left(\frac{2}{3}\right) = 4 \text{ mg/L} \tag{4.6}$$

since only two-thirds of the bottle is seed water, which has a BOD of 6 mg/L. The remaining oxygen use, 7 − 4 = 3 mg/L, must be due to the unknown material.

BOD may be calculated using Equation 4.7, which combines the seeding and dilution methods:

$$\text{BOD (mg/L)} = \frac{(I - F) - (I' - F')(X/Y)}{D} \qquad (4.7)$$

where
I = initial DO of the bottle with sample and seeded dilution water
F = final DO of the bottle with sample and seeded dilution water
I′ = initial DO of seeded dilution water
F′ = final DO of seeded dilution water
X = mL of seeded dilution water in the sample bottle
Y = total mL in the bottle
D = dilution of the sample

Example 4.1
Calculate the BOD$_5$ of a water sample, given the following data:

Temperature of sample is 20°C
Initial DO is saturation
Dilution is 1:30, with seeded dilution water
Final DO of seeded dilution water is 8 mg/L
Final DO bottle with sample and seeded dilution water is 2 mg/L
Volume of BOD bottle is 300 mL

From Table 4–1, DO saturation at 20°C is 9.2 mg/L; hence, this is the initial DO. Since the BOD bottle contains 300 mL, a 1:30 dilution with seeded water contains 10 mL of sample and 290 mL of seeded dilution water, and, by Equation 4.7,

$$\text{BOD}_5 = \frac{(9.2 - 2) - (9.2 - 8)(290/300)}{0.033} = 18.3 \text{ mg/L} \qquad (4.8)$$

BOD is a measure of oxygen use, or potential oxygen use. An effluent with a high BOD may be harmful to a stream if the oxygen consumption is great enough to cause anaerobic conditions. Obviously, a small trickle going into a great river will have negligible effect, regardless of the BOD concentration involved. Similarly, a large flow into a small stream may seriously affect the stream even though the BOD concentration might be low. The BOD of most domestic sewage is about 250 mg/L, and many industrial wastes run as high as 30,000 mg/L. The potential detrimental effect of an untreated dairy waste that might have a BOD of 20,000 mg/L is quite obvious.

As discussed in Chapter 3, the BOD curve can be modeled using Equation 3.10 (reproduced here as Equation 4.9):

$$y = L_0(1 - e^{-k_1't})$$ (4.9)

where $y = BOD_t$ = the amount of oxygen demanded by the organisms at any time t, in mg/L

L_0 = ultimate demand for oxygen, in mg/L

k_1' = rate constant (deoxygenation constant of Chapter 3), in days^{-1}

t = time in days

If, instead of stopping a BOD test after five days, we allow the reactions to proceed and measure the DO each day, we might get a curve like that shown in Figure 4–4. Note that some time after five days the curve turns sharply upward. This discontinuity is due to the demand for oxygen by the microorganisms that decompose nitrogenous organic compounds to stable nitrate, NO_3^-. The curve is thus divided into nitrogenous and carbonaceous BOD areas. Ultimate BOD, as shown in this figure, includes both nitrogenous and carbonaceous BOD.

For streams and rivers with travel times greater than about five days, the ultimate demand for oxygen must include the nitrogenous demand. Although the use of BOD_{ult} (carbonaceous plus nitrogenous) in DO sag calculations is not strictly accurate, it is often assumed that the ultimate BOD may be calculated as

$$BOD_{ult} = a(BOD_5) + b(KN)$$ (4.10)

where   KN = Kjeldahl nitrogen (organic plus ammonia), in mg/L

a, b = constants

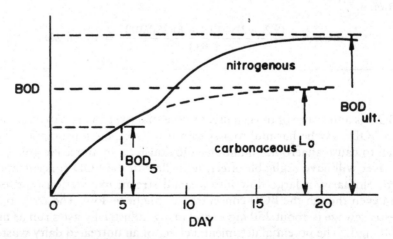

FIGURE 4–4. Long-term BOD. Note that BOD_{ult} here includes both ultimate carbonaceous BOD (L_0) and ultimate nitrogenous BOD.

The state of North Carolina, for example, uses a = 1.2 and b = 4.0 for calculating the ultimate BOD, which is then substituted for L_0 in the DO sag equation.

CHEMICAL OXYGEN DEMAND

Among many drawbacks of the BOD test, the most important is that it takes five days to run. If the organic compounds are oxidized chemically instead of biologically, the test can be shortened considerably. Such oxidation is accomplished with the chemical oxygen demand (COD) test. Because nearly all organic compounds are oxidized in the COD test and only some are decomposed during the BOD test, COD values are always higher than BOD values. One example of this is wood pulping waste, in which compounds such as cellulose are easily oxidized chemically (high COD) but are very slow to decompose biologically (low BOD).

TURBIDITY

Water that is not clear but "dirty," in the sense that light transmission is inhibited, is considered *turbid*. Turbidity can be caused by many materials, some of which were discussed in Chapter 3. In the treatment of water for drinking purposes, turbidity is of great importance, first because of aesthetic considerations and second because pathogenic organisms can hide on (or in) the tiny colloidal particles.

The standard for calibrating turbidimeters is defined as

$$1 \text{ mg/L of } SiO_2 = 1 \text{ normalized turbidity unit} \qquad (4.11)$$

Turbidimeters are photometers that measure the intensity of scattered light. Opaque particles scatter light, so the scattered light measured at right angles to a beam of incident light is proportional to the turbidity.

COLOR AND ODOR

Color and odor are both important measurements in water treatment. Along with turbidity they are called physical parameters of drinking water quality. Color and odor are important from the standpoint of aesthetics. If water looks colored or smells bad, people instinctively avoid using it, even though it might be perfectly safe from the public health aspect. Both color and odor may be and often are caused by organic substances such as algae or humic compounds.

Color is measured by comparison with standards. Colored water made with potassium chloroplatinate when tinted with cobalt chloride closely resembles the color of many natural waters. When multicolored industrial wastes are involved, such color measurement is meaningless.

Odor is measured by successive dilutions of the sample with odor-free water until the odor is no longer detectable. This test is obviously subjective and depends entirely on the olfactory senses of the tester.

pH

The pH of a solution is a measure of hydrogen ion concentration, which in turn is a measure of its acidity. Pure water dissociates slightly into equal concentrations of hydrogen and hydroxyl (OH^-) ions.

$$H_2O \Leftrightarrow H^+ + OH^- \tag{4.12}$$

An excess of hydrogen ions makes a solution acidic, whereas a dearth of H^+ ions, or an excess of hydroxyl ions, makes it basic. The equilibrium constant for this reaction, K_w, is the product of H^+ and OH^- concentrations and is equal to 10^{-14}. This relationship may be expressed as

$$[H^+][OH^-] = K_w = 10^{-14} \tag{4.13}$$

where $[H^+]$ and $[OH^-]$ are the concentrations of hydrogen and hydroxyl ions, respectively, in moles per liter. Considering Equation 4.12 and solving Equation 4.13, in pure water,

$$[H^+] = [OH^-] = 10^{-7} \text{ moles/L} \tag{4.14}$$

The hydrogen ion concentration is so important in aqueous solutions that an easier method of expressing it has been devised. Instead of as moles as per liter, we define a quantity pH as the negative logarithm of $[H^+]$, so that

$$pH = -\log_{10}[H^+] = \log_{10} \frac{1}{[H^+]} \tag{4.15}$$

or

$$[H^+] = 10^{-pH} \tag{4.16}$$

For a neutral solution, $[H^+]$ is 10^{-7}, or pH = 7. For larger hydrogen ion concentrations, then, the pH of the solution is < 7. For example, if the hydrogen ion concentration is 10^{-4}, the pH = 4 and the solution is acidic. In this solution, we see that the hydroxyl ion concentration is $10^{-14}/10^{-4} = 10^{-10}$. Since $10^{-4} \gg 10^{-10}$, the solution contains a large excess of H^+ ions, confirming that it is indeed acidic. A solution containing a dearth of H^+ ions would have $[H^+] < 10^{-7}$, or pH > 7, and would be basic. The pH range of dilute solutions is from 0 (very acidic; 1 mole of H^+ ions per liter) to 14 (very alkaline). Solutions containing more than 1 mole of H+ ions per liter have negative pH.

The measurement of pH is now almost universally by electronic means. Electrodes that are sensitive to hydrogen ion concentration (strictly speaking, the hydrogen ion activity) convert the signal to electric current. pH is important in almost all phases of water and wastewater treatment. Aquatic organisms are sensitive to pH changes, and biological treatment requires either pH control or monitoring. In water treatment as well as in disinfection and corrosion control, pH is important in ensuring proper chemical treatment. Mine drainage often involves the formation of sulfuric acid (high H^+ concentration), which is extremely detrimental to aquatic life. Continuous acid deposition from the atmosphere (see Chapter 18) may substantially lower the pH of a lake.

ALKALINITY

A parameter related to pH is alkalinity, or the buffering capacity of the water against acids. Water that has a high alkalinity can accept large doses of an acid without lowering the pH significantly. Waters with low alkalinity, such as rainwater, can experience a drop in the pH with only a minor addition of hydrogen ion.

In natural waters much of the alkalinity is provided by the carbonate/bicarbonate buffering system. Carbon dioxide (CO_2) dissolves in water and is in equilibrium with the bicarbonate and carbonate ions.

$$CO_2 \text{ (gas)} \Leftrightarrow CO_2 \text{ (dissolved)}$$

$$CO_2 \text{ (dissolved)} + H_2O \Leftrightarrow H_2CO_3$$

$$H_2CO_3 \Leftrightarrow H^+ + HCO_3^-$$

$$HCO_3^- \Leftrightarrow H^+ + CO_3^{2-}$$

(4.16a)

where
H_2CO_3 = carbonic acid
HCO_3^- = bicarbonate ion
CO_3^{2-} = carbonate ion

Any change that occurs in the components of this equation influences the solubility of CO_2. If acid is added to the water, the hydrogen ion concentration is increased, and this combines with both the carbonate and bicarbonate ions, driving the carbonate and bicarbonate equilibria to the left, releasing carbon dioxide into the atmosphere. The added hydrogen ion is absorbed by readjustment of all the equilibria, and the pH does not change markedly. Only when all of the carbonate and bicarbonate ions are depleted will the additional acid added to the water cause a drop in pH. The effect of alkalinity on the pH of a water sample is shown in Figure 4–5.

SOLIDS

Wastewater treatment is complicated by the dissolved and suspended inorganic material the wastewater contains. In discussion of water treatment, both dissolved and suspended materials are called solids. The separation of these solids from the water is one of the primary objectives of treatment.

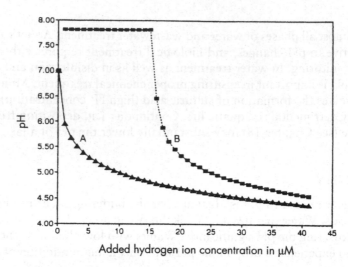

FIGURE 4–5. Effect of alkalinity in buffering against a pH with the addition of an acid: (A) acid added to deionized water; (B) acid added to monobasic phosphate buffer solution

Strictly speaking, in wastewater anything other than water is classified as solid. The usual definition of solids, however, is the residue after evaporation at 103°C (slightly higher than the boiling point of water). The solids thus measured are known as total solids. Total solids may be divided into two fractions: the total dissolved solids (TDS) and the total suspended solids (TSS). The difference is illustrated in the following example:

A teaspoonful of table salt dissolves in a glass of water, forming a water-clear solution. However, the salt remains behind if the water evaporates. Sand, however, does not dissolve and remains as sand grains in the water and forms a turbid mixture. The sand also remains behind if the water evaporates. The salt is an example of a dissolved solid, whereas the sand is a suspended solid.

Suspended solids are separated from dissolved solids by filtering the water through a filter paper as seen in Figure 4–6. The suspended material is retained on the filter paper, while the dissolved fraction passes through. If the initial dry weight of the filter paper is known, the subtraction of this from the total weight of the filter and the dried solids caught in the filter paper yields the weight of suspended solids, expressed in milligrams per liter.

Solids may be classified in another way: those that are volatilized at a high temperature and those that are not. The former are known as volatile solids, the latter as fixed solids. Volatile solids are usually organic compounds. Obviously, at 600°C, the temperature at which the combustion takes place, some of the inorganics are decomposed and volatilized, but this is not considered a serious drawback. The relationship between total solids and total volatile solids is illustrated by Example 4.2.

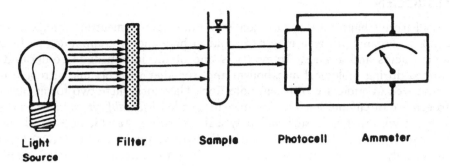

FIGURE 4–6. Elements of a filter photometer

Example 4.2
Given the following data:

- Weight of a dish = 48.6212 g.
- 100 mL of sample is placed in a dish and evaporated. Weight of the dish and dry solids = 48.6432 g.
- The dish is then placed in a 600°C furnace, then cooled. Weight = 48.6300 g.

Find the total, fixed, and volatile solids.

$$\text{Total Solids} = \frac{(\text{dish} + \text{dry solids}) - (\text{dish})}{\text{sample volume}} \qquad (4.16b)$$

$$= \frac{48.6432 - 48.6212}{100}$$

$$= (220)10^{-6} \text{ g/mL}$$

$$= (220)10^{-3} \text{ mg/mL}$$

$$= 220 \text{ mg/L}$$

$$\text{Fixed Solids} = \frac{(\text{dish} + \text{unburned solids}) - (\text{dish})}{\text{sample volume}} \qquad (4.16c)$$

$$= \frac{48.6300 - 48.6212}{100}$$

$$= 88 \text{ mg/L}$$

$$\text{Volatile Solids} = \text{Total Solids} - \text{Total Fixed Solids} \qquad (4.16d)$$

$$= 220 - 88 = 132 \text{ mg/L}$$

Measurement of the volatile fraction of suspended material, the *volatile suspended solids*, is made by burning the suspended solids and weighing them again. The loss in weight is interpreted as the volatile suspended solids.

NITROGEN

Recall from Chapter 3 that nitrogen is an important element in biological re-actions. Organic nitrogen may be bound in high-energy compounds such as amino acids and amines. Ammonia is one of the intermediate compounds formed during biological metabolism and, together with organic nitrogen, is considered an indicator of recent pollution. Therefore, these two forms of ni-trogen are often combined in one measure, called *Kjeldahl nitrogen*, after the scientist who first suggested the analytical procedure. Aerobic decomposition (oxidation) eventually produces nitrite (NO_2^-) and finally nitrate (NO_3^-) from organically bound nitrogen and ammonia. A high-nitrate and low-ammonia ni-trogen therefore suggests that pollution occurred, but some time before.

These forms of nitrogen can all be measured analytically by colorimetric techniques. In colorimetry, the ion in question combines with a reagent to form a colored compound; the color intensity is proportional to the original con-centration of the ion. For example, ammonia can be measured by adding an ex-cess amount of a compound called *Nessler reagent* (potassium mercuric iodine, K_2HgI_4) to the unknown sample. Nessler reagent combines with ammonia in solution to form a yellow-brown colloid. The amount of colloid formed is pro-portional to the concentration of ammonium ions in the sample.

The color is measured photometrically. A photometer, illustrated in Figure 4–6, consists of a light source, a filter, the sample, and a photocell. The filter allows only those wavelengths of light to pass through that the compounds being measured will absorb. Light from the light source passes through the sam-ple to the photocell, which converts light energy into electric current. An in-tensely colored sample will absorb a considerable amount of light and allow only a limited amount of light to pass through and thus create little current. On the other hand, a sample containing very little of the chemical in question will be lighter in color and allow almost all of the light to pass through, and will set up a substantial current.

The intensity of light transmitted by the colored solution obeys the *Beer-Lambert Law*:

$$\frac{I}{I_0} = e^{-acx} \tag{4.17}$$

or

$$\ln \frac{I_0}{I} = acx \tag{4.18}$$

where   I = intensity of light after it has passed through the sample
   I_0 = intensity of light incident on the sample
   a = "absorption coefficient" of the colored compound
   x = path length of light through the sample
   c = concentration

A photometer, as shown in Figure 4–6, measures the difference between the in-tensity of light passing through the sample (I in Equation 4.17) and the intensity of light passing through clear distilled water (I_0). This difference may be read out as I_0/I on a logarithmic scale; $\ln(I_0/I)$ is called the *absorbance*. The absorbance of samples containing known ammonia concentrations is plotted against the known concentrations, and the absorbance of an unknown sample is then compared with that of these standards, as in Example 4.3.

Example 4.3
Several known samples and an unknown sample were treated with Nessler reagent, and the color was measured with a photometer. Find the ammonia concentration of the unknown sample.

Sample	Absorbance
Standards: 0 mg/L of ammonia	
(Distilled water)	0
1 mg/L of ammonia	0.06
2 mg/L of ammonia	0.12
3 mg/L of ammonia	0.18
4 mg/L of ammonia	0.24
Unknown sample	0.15

A plot (Figure 4–7) of ammonia concentration of the standards vs. absorbance results in a straight line (Beer-Lambert Law is adhered to). We may then enter this chart at 15% absorbance (the unknown) and read the concentration of ammonia in our unknown as 2.5 mg/L.

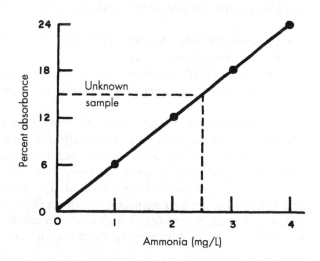

FIGURE 4–7
Calculation using colorimetric standards

PHOSPHATES

The importance of phosphorus compounds in the aquatic environment was discussed in Chapter 3. Phosphorus in wastewater may be either inorganic or organic. Although the greatest single source of inorganic phosphorus is synthetic detergents, organic phosphorus is found in food and human waste as well. All phosphates in nature will, by biological action, eventually revert to inorganic forms to be used again by the plants in making high-energy material. Total phosphates may be measured by first boiling the sample in acid solution, which converts all the phosphates to their inorganic forms. From that point the test is colorimetric, like the tests for nitrogen, using a chemical that when combined with phosphates produces a color directly proportional to the phosphate concentration.

BACTERIOLOGICAL MEASUREMENTS

From the public health standpoint, the bacteriological quality of water is as important as the chemical quality. A large number of infectious diseases may be transmitted by water, among them typhoid and cholera. However, it is one thing to declare that water must not be contaminated by *pathogens* (disease-causing organisms) and another to determine the existence of these organisms. First, there are many pathogens. Each has a specific detection procedure and must be screened individually. Second, the concentration of these organisms, although large enough to spread disease, may be so small as to make their detection impossible, like the proverbial needle in a haystack.

How then can we measure for bacteriological quality? The answer lies in the concept of *indicator organisms* that, while not particularly harmful, indicate the possible presence of bacteria that are pathogenic. The indicator most often used is a group of microbes of the family *Escherichia coli (E. coli)*, often called coliform bacteria, which are organisms normal to the digestive tracts of warm-blooded animals. In addition, *E. coli* are

- Plentiful and hence not difficult to find
- Easily detected with a simple test
- Generally harmless except in unusual circumstances
- Hardy, surviving longer than most known pathogens

Coliforms have thus become universal indicator organisms. The presence of coliforms does not prove the presence of pathogens. If a large number of coliforms are present, there is a good chance of recent pollution by wastes from warm-blooded animals, and therefore the water may contain pathogenic organisms.

This last point should be emphasized. The presence of coliforms does not prove that there are pathogenic organisms in the water, but indicates that such organisms might be present. A high coliform count is thus suspicious, and the water should not be consumed, even though it may be safe.

Coliforms are measured by first filtering the sample through a sterile micropore filter by suction, thereby capturing any coliforms on the filter. The filter is then placed in a petri dish containing a sterile agar that soaks into the filter and promotes the growth of the coliforms while inhibiting other organisms. After 24 or 48 hours of incubation the number of shiny black dots, indicating coliform colonies, is counted. If we know how many milliliters of sample poured through the filter, the concentration of coliforms may be expressed as coliforms/mL.

VIRUSES

Because of their minute size and extremely low concentration and the need to culture them on living tissues, pathogenic (or animal) viruses are fiendishly difficult to measure. Moreover, there are as yet no standards for viral quality of water supplies, as there are for pathogenic bacteria.

One possible method of overcoming this difficulty is to use an indicator organism, much like the coliform group is used as an indicator for bacterial contamination. This can be done by using a *bacteriophage*—a virus that attacks only a certain type of bacterium. For example, *coliphages* attack coliform organisms and, because of their association with wastes from warm-blooded animals, seem to be an ideal indicator. The test for coliphages is performed by inoculating a petri dish containing an ample supply of a specific coliform with the wastewater sample. Coliphages will attack the coliforms, leaving visible spots, or plaques, that can be counted, and an estimation can be made of the number of coliphages per unit volume.

HEAVY METALS

Chapter 3 introduces the issue of heavy metals in industrial effluent. Heavy metals such as arsenic and mercury can harm fish even at low concentrations. Consequently, the method of measuring these ions in water must be very sensitive. The method of choice is *atomic absorption spectrophotometry*, in which a solution of lanthanum chloride is added to the sample, and the treated sample is sprayed into a flame using an atomizer. Each metallic element in the sample imparts a characteristic color to the flame, whose intensity is then measured spectrophotometrically.

TRACE TOXIC ORGANIC COMPOUNDS

Very low concentrations of chlorinated hydrocarbons and other agrichemical residues in water can be assayed by *gas chromatography*. Oil residues in water are generally measured by extracting the water sample with Freon and then evaporating the Freon and weighing the residue from this evaporation.

CONCLUSION

Only a few of the most important tests used in water pollution control are discussed in this chapter. More than 500 analytical procedures have been documented, many of which can be performed only with special equipment and by skilled technicians. Understanding this, and realizing the complexity, variations, and objectives of some of the measurements of water pollution, how would you answer someone who brings a jug of water to your office, sets it on your desk, and asks, "Can you tell me how polluted this water is?"

PROBLEMS

4.1   Given the following BOD test results:

Initial DO = 8 mg/L

Final DO   = 0 mg/L

Dilution   = 1/10

what can you say about

a. BOD_5?
b. BOD ultimate?
c. COD?

4.2   If you had two bottles full of lake water and kept one in the dark and the other in daylight, which would have a higher DO after a few days? Why?

4.3   Name three substances you need to seed if you want to measure their BOD.

4.4   The following data were obtained for a sample:

Total solids = 4000 mg/L

Suspended solids = 5000 mg/L

Volatile suspended solids = 2000 mg/L

Fixed suspended solids = 1000 mg/L

Which of these numbers is questionable and why?

4.5   A water has a BOD_5 of 10 mg/L. The initial DO in the BOD bottle was 8 mg/L, and the dilution was 1/10. What is the final DO in the BOD bottle?

4.6   Draw a typical BOD curve. Label the (a) ultimate carbonaceous BOD (L_0), (b) ultimate nitrogenous BOD, (c) ultimate BOD (BOD_{ult}), and (d) 5-day BOD. On the same graph, plot the BOD curve if the test had been run at 30°C instead of at the usual temperature. Also plot the BOD curve if a substantial amount of toxic materials were added to the sample.

4.7   Consider the following data from an undiluted, unseeded BOD test:

Day	DO (mg/L)	Day	DO (mg/L)
0	9	10	6
2	9	12	6
4	9	14	4
6	8	16	3
8	7	18	3

a. Based on only this result, what is the (a) BOD_5, (b) ultimate carbonaceous BOD (L_0), and (c) ultimate BOD (BOD_{ult})?
b. Why was there so little oxygen used during the first few days?
c. If the sample had been seeded, would the final DO (at 18 days) have been higher or lower and why?

4.8   The BOD test for a waste diluted 1/10 produced the following curve:

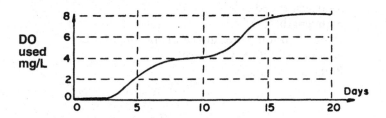

a. What is the BOD_5 of the waste?
b. What is the ultimate carbonaceous BOD?
c. What is the ultimate BOD from nitrification alone?

4.9   A wastewater is estimated to have a BOD of 200 mg/L.

a. What dilution is necessary for this BOD to be measured by the usual technique?
b. If the initial and final DO of the test thus conducted is 9.0 and 4.0 mg/L, and the dilution water has a BOD of 1.0 mg/L, what is the dilution?

LIST OF SYMBOLS

a	absorption coefficient
BOD	biochemical oxygen demand, in mg/L
BOD_5	5-day BOD
BOD_{ult}	ultimate BOD: carbonaceous plus nitrogenous
c	concentration
D	dilution (volume of sample/total volume)
DO	dissolved oxygen
F	final BOD of sample, in mg/L
F'	final BOD of seeded dilution water, in mg/L
I	initial BOD of sample, in mg/L
I'	initial BOD of seeded dilution water, in mg/L
I	intensity of light
I_0	intensity of light incident on sample
L_0	ultimate carbonaceous BOD
t	time in days
TDS	total dissolved solids
TSS	total suspended solids
x	path length of light
X	seeded dilution water in sample bottle, in mL
Y	volume of BOD bottle, in mL
y	BOD at any time, t

Chapter 5

Water Supply

A supply of water is critical to the survival of life as we know it. People need water to drink, animals need water to drink, and plants need water to drink. The basic functions of society require water: cleaning for public health, consumption for industrial processes, and cooling for electrical generation. In this chapter, we discuss water supply in terms of:

- The hydrologic cycle and water availability
- Groundwater supplies
- Surface water supplies
- Water transmission

The direction of our discussion is that sufficient water supplies exist, but many areas are water poor while others are water rich. Adequate water supply requires engineering the supply and its transmission from one area to another, keeping in mind the environmental effects of water transmission systems. In many cases, moving the population to the water may be less environmentally damaging than moving the water. This chapter concentrates on measurement of water supply, and the following chapter discusses treatment methods available to clean up the water once it reaches areas of demand.

THE HYDROLOGIC CYCLE AND WATER AVAILABILITY

The hydrologic cycle is a useful starting point for the study of water supply. This cycle, illustrated in Figure 5–1, includes precipitation of water from clouds, infiltration into the ground or runoff into surface water, followed by evaporation and transpiration of the water back into the atmosphere.

The rates of precipitation and evaporation/transpiration help define the baseline quantity of water available for human consumption. *Precipitation* is the term applied to all forms of moisture falling to the ground, and a range of instruments and techniques have been developed for measuring the amount and intensity of rain, snow, sleet, and hail. The average depth of precipitation over a given region, on a storm, seasonal, or annual basis, is required in many water availability studies. Any open receptacle with vertical sides is a common rain gauge, but varying

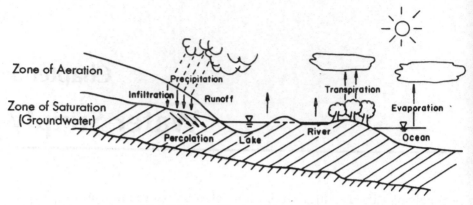

FIGURE 5–1. The hydrologic cycle

wind and splash effects must be considered if amounts collected by different gauges are to be compared.

√Evaporation and transpiration are the movement of water back to the atmosphere from open water surfaces and from plant respiration. The same meteorological factors that influence evaporation are at work in the transpiration process: solar radiation, ambient air temperature, humidity, and wind speed. The amount of soil moisture available to plants also affects the transpiration rate. Evaporation is measured by measuring water loss from a pan. Transpiration can be measured with a *phytometer*, a large vessel filled with soil and potted with selected plants. The soil surface is hermetically sealed to prevent evaporation; thus moisture can escape only through transpiration. The rate of moisture escape is determined by weighing the entire system at intervals up to the life of the plant. Phytometers cannot simulate natural conditions, so results have limited value. However, they can be used as an index of water demand by a crop under field conditions and thus relate to calculations that help an engineer determine water supply requirements for that crop. Because it is often not necessary to distinguish between evaporation and transpiration, the two processes are often linked as *evapotranspiration*, or the total water loss to the atmosphere.

GROUNDWATER SUPPLIES

Groundwater is an important direct source of supply that is tapped by wells, as well as a significant indirect source since surface streams are often supplied by subterranean water.

Near the surface of the earth, in the *zone of aeration*, soil pore spaces contain both air and water. This zone, which may have zero thickness in swamplands and be several hundred feet thick in mountainous regions, contains three types of moisture. After a storm, *gravity water* is in transit through the larger soil pore spaces. *Capillary water* is drawn through small pore spaces by capillary action and is available for plant uptake. *Hygroscopic moisture* is held in place by molecular forces during all except the driest climatic conditions. Moisture from the zone of aeration cannot be tapped as a water supply source.

In the *zone of saturation*, located below the zone of aeration, the soil pores are filled with water, and this is what we call *groundwater.* A stratum that contains a substantial amount of groundwater is called an *aquifer*. At the surface between the two zones, called the *water table* or *phreatic surface*, the hydrostatic pressure in the groundwater is equal to the atmospheric pressure. An aquifer may extend to great depths, but because the weight of overburden material generally closes pore spaces, little water is found at depths greater than 600 m (2000 ft). The amount of water that will drain freely from an aquifer is known as *specific yield*.

The flow of water out of a soil can be illustrated using Figure 5–2. The flow rate must be proportional to the area through which flow occurs times the velocity, or

$$Q = Av \tag{5.1}$$

where Q = flow rate, in m³/sec
           A = area of porous material through which flow occurs, in m²
           v = superficial velocity, in m/sec

The superficial velocity is of course not the actual velocity of the water in the soil, since the volume occupied by the soil solid particles greatly reduces the available area for flow. If a is the area available for flow, then

$$Q = Av = av' \tag{5.2}$$

where v' = actual velocity of water flowing through the soil
           a = area available for flow

Solving for v',

$$v' = \frac{Av}{a} \tag{5.3}$$

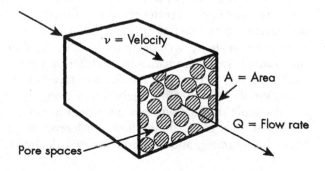

FIGURE 5–2
Flow through soil

If a sample of soil is of some length L, then

$$v' = \frac{Av}{a} = \frac{AvL}{aL} = \frac{v}{porosity} \tag{5.4}$$

since the total volume of the soil sample is AL and the volume occupied by the water is aL.

Water flowing through the soil at a velocity v' loses energy, just as water flowing through a pipeline or an open channel does. This energy loss per distance traveled is defined as

$$\text{energy loss} = \frac{\Delta h}{\Delta L} \tag{5.5}$$

where   h = energy, measured as elevation of the water table in an unconfined
              aquifer or as pressure in a confined aquifer, in m
         L = horizontal distance in direction of flow, in m

The symbol (delta) simply means "a change in," as in "a change in length, L." Thus this equation means that there is a change (loss) of energy, h, as water flows through the soil some distance, L.

In an unconfined aquifer, the drop in the elevation of the water table with distance is the slope of the water table in the direction of flow. The elevation of the water surface is the potential energy of the water, and water flows from a higher elevation to a lower elevation, losing energy along the way. Flow through a porous medium such as soil is related to the energy loss using the Darcy equation,

$$Q = KA \frac{\Delta h}{\Delta L} \tag{5.6}$$

where   K = coefficient of permeability, in m/day
         A = cross-sectional area, in m^2

The Darcy equation makes intuitive sense, in that the flow rate (Q) increases with increasing area (A) through which the flow occurs and with the drop in pressure, $\Delta h/\Delta L$. The greater the driving force (the difference in upstream and downstream pressures), the greater the flow. The fudge factor, K, is the *coefficient of permeability*, an indirect measure of the ability of a soil sample to transmit water; it varies dramatically for different soils, ranging from about 0.0005 m/day for clay to over 5000 m/day for gravel. The coefficient of permeability is measured commonly in the laboratory using *permeameters*, which consist of a soil sample through which a fluid such as water is forced. The flow rate is measured for a given driving force (difference in pressures) through a known area of soil sample, and the permeability calculated.

Example 5.1

Problem. A soil sample is installed in a permeameter as shown in Figure 5–3. The length of the sample is 0.1 m, and it has a cross-sectional area of 0.05 m². The water pressure on the upflow side is 2.5 m and on the downstream side it is 0.5 m. A flow rate of 2.0 m³/day is observed. What is the coefficient of permeability?

Solution. The pressure drop is the difference between the upstream and downstream pressures, or Δh = 2.5 – 0.5 = 2.0 m. Using the Darcy equation, and solving for K,

$$K = \frac{Q}{A \dfrac{\Delta h}{\Delta L}} = \frac{2.0}{0.05 \times \dfrac{2}{0.1}} = 2 \text{ m/day} \qquad (5.7)$$

If a well is sunk into an unconfined aquifer, shown in Figure 5–4, and water is pumped out, the water in the aquifer will begin to flow toward the well. As the water approaches the well, the area through which it flows gets progressively smaller, and therefore a higher superficial (and actual) velocity is required. The higher velocity of course results in an increasing loss of energy, and the energy gradient must increase, forming a *cone of depression.* The reduction in the water table is known in groundwater terms as a *drawdown.* If the rate of water flowing toward the well is equal to the rate of water being pumped out of the well, the condition is at equilibrium, and the drawdown remains constant.

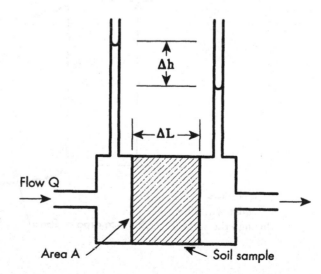

FIGURE 5–3
Permeameter used for measuring coefficient of permeability using the Darcy equation

Flow Q

Δh

←ΔL→

Area A

Soil sample

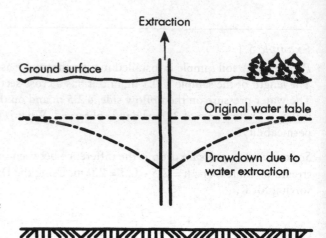

FIGURE 5–4
Drawdown in water table
due to pumping from
a well

If, however, the rate of water pumping is increased, the radial flow toward the well has to compensate, and this results in a deeper cone or drawdown.

Consider a cylinder, shown in Figure 5–5, through which water flows toward the center. Using Darcy's equation,

$$Q = KA \frac{\Delta h}{\Delta L} = K(2\pi rh) \frac{\Delta h}{\Delta r} \qquad (5.8)$$

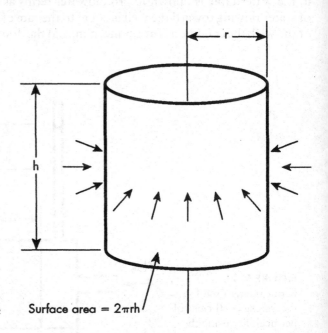

FIGURE 5–5
Cylinder with flow
through the surface

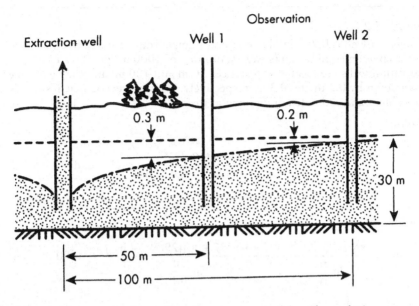

FIGURE 5–6. Two monitoring wells define the extent of drawdown during extraction.

where r is the radius of the cylinder, and $2\pi rh$ is the cross-sectional surface area of the cylinder. If water is pumped out of the center of the cylinder at the same rate as water is moving in through the cylinder surface area, the above equation can be integrated to yield

$$Q = \frac{\pi K(h_1^2 - h_2^2)}{\ln \frac{r_1}{r_2}} \tag{5.9}$$

where h_1 and h_2 are the height of the water table at radial distances r_1 and r_2 from the well.

This equation can be used to estimate the pumping rate for a given drawdown any distance away from a well, using the water level measurements in two observation wells in an unconfined aquifer, as shown in Figure 5–6. Also, knowing the diameter of a well, it is possible to estimate the drawdown at the well, the critical point in the cone of depression. If the drawdown is depressed all the way to the bottom of the aquifer, the well "goes dry"—it cannot pump water at the desired rate. Although the derivations of the above equations are for an unconfined aquifer, the same situation would occur for a confined aquifer, where the pressure would be measured by observation wells.

Example 5.2

Problem. A well is 0.2 m in diameter and pumps from an unconfined aquifer 30 m deep at an equilibrium (steady state) rate of 1000 m³ per day. Two observation wells are located at distances 50 m and 100 m, and they have been drawn down by 0.2 m and 0.3 m, respectively. What is the coefficient of permeability and estimated drawdown at the well?

Solution

$$K = \frac{Q \ln \frac{r_1}{r_2}}{\pi(h_1^2 - h_2^2)} = \frac{1000 \ln(100/50)}{3.14[(24.8)^2 - (29.7^2)]} = 37.1 \text{ m/day} \tag{5.10}$$

If the radius of the well is 0.2/2 = 0.1 m, this can be plugged into the same equation, as

$$Q = \frac{\pi K(h_1^2 - h_2^2)}{\ln \frac{r_1}{r_2}} = \frac{3.14 \times 37.1 \times [(29.7)^2 - h_2^2]}{\ln \frac{50}{0.1}} = 1000 \tag{5.11}$$

and solving for h_2,

$$h_2 = 28.8 \text{ m} \tag{5.11a}$$

Since the aquifer is 30 m deep, the drawdown at the well is 30 − 28.8 = 1.2 m.

Multiple wells in an aquifer can interfere with each other and cause excessive drawdown. Consider the situation in Figure 5–7, where a single well creates a

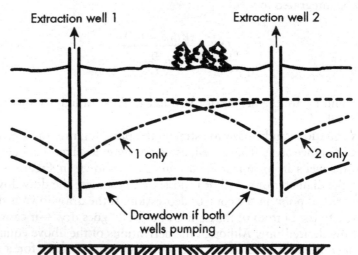

FIGURE 5–7. Multiple wells and the effect of extraction on the groundwater table

cone of depression. If a second production well is installed, the cones will overlap, causing greater drawdown at each well. If many wells are sunk into an aquifer, the combined effect of the wells can deplete the groundwater resources and all wells will "go dry."

The reverse is also true, of course. Suppose one of the wells is used as an injection well, then the injected water flows from this well into the others, building up the groundwater table and reducing the drawdown. The judicious use of extraction and injection wells is one way that the flow of contaminants from hazardous waste or refuse dumps can be controlled, as discussed further in Chapter 15.

Finally, many assumptions are made in the above discussion. First, we assume that the aquifer is homogeneous and infinite—that is, it sits on a level aquaclude and the permeability of the soil is the same at all places for an infinite distance in all directions. The well is assumed to penetrate the entire aquifer and is open for the entire depth of the aquifer. Finally, the pumping rate is assumed to be constant. Clearly, any of these conditions may cause the analysis to be faulty, and this model of aquifer behavior is only the beginning of the story. Modeling the behavior of groundwater is a complex and sophisticated science.

SURFACE WATER SUPPLIES

Surface water supplies are not as reliable as groundwater sources since quantities often fluctuate widely during the course of a year or even a week, and water quality is affected by pollution sources. If a river has an average flow of 10 cubic feet per second (cfs), this does not mean that a community using the water supply can depend on having 10 cfs available at all times.

The variation in flow may be so great that even a small demand cannot be met during dry periods and so storage facilities must be built to save water during wetter periods. Reservoirs should be large enough to provide dependable supplies. However, reservoirs are expensive and, if they are unnecessarily large, represent a waste of community resources.

One method of estimating the proper reservoir size is to use a *mass curve* to calculate historical storage requirements and then to calculate risk and cost using statistics. Historical storage requirements are determined by summing the total flow in a stream at the location of the proposed reservoir and plotting the change of total flow with time. The change of water demand with time is then plotted on the same curve. The difference between the total water flowing in and the water demanded is the quantity that the reservoir must hold if the demand is to be met. The method is illustrated by Example 5.3.

Example 5.3

A reservoir is needed to provide a constant flow of 15 cfs. The monthly stream flow records, in total cubic feet, are

Month

J	F	M	A	M	J	J	A	S	O	N	D

Million ft³ of water

50	60	70	40	32	20	50	80	10	50	60	80

The storage requirement is calculated by plotting the cumulative stream flow as in Figure 5–8. Note that the graph shows 50 million ft³ for January, 60 + 50 = 110 million ft³ for February, 70 + 110 million ft³ for March, and so on.

The demand for water is constant at 15 cfs, or

$$15 \times 10^6 \ \frac{ft^3}{sec} \times 60 \ \frac{sec}{min} \times 60 \ \frac{min}{hr} \times 24 \ \frac{hr}{day} \times 30 \ \frac{days}{month}$$

$$= 38.8 \times 10^6 \ \frac{ft^3}{month} \tag{5.15}$$

This constant demand is represented in Figure 5–8 as a straight line with a slope of 38.8×10^6 ft³/month, and is plotted on the curved supply line. Note that the stream flow in May was lower than the demand, and this was the start of a drought lasting until June. In July the supply increased until the reservoir could be filled up again, late in August. During this period the reservoir had to make up the difference between demand and supply, and the capacity needed for this time was 60×10^6 ft³. A second drought, from September to November, required 35×10^6 ft³ of capacity. The municipality therefore needs a reservoir with a capacity of 60×10^6 ft³ to draw water from throughout the year.

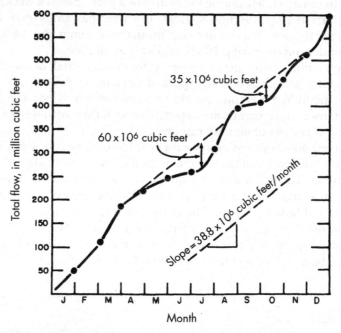

FIGURE 5–8. Mass curve for determining required reservoir capacity

A mass curve like Figure 5–8 is not very useful if only limited stream flow data are available. Data for one year yield very little information about long-term variations. The data in Example 5.3 do not indicate whether the 60 million ft^3 deficit was the worst drought in 20 years, an average annual drought, or one that occurred during an unusually wet year.

Long-term variations may be estimated statistically when actual data are not available. Water supplies are often designed to meet demands of 20-year cycles, and about once in 20 years the reservoir capacity will not be adequate to offset the drought. The community may choose to build a larger reservoir that will prove inadequate only every 50 years, for example. A calculation comparing the additional capital investment to the added benefit of increased water supply will assist in making such a decision. One calculation method requires first assembling required reservoir capacity data for a number of years, ranking these data according to the drought severity, and calculating the drought probability for each year. If the data are assembled for n years and the rank is designated by m, with m = 1 for the largest reservoir requirement during the most severe drought, the probability that the supply will be adequate for any year is given by m/(n + 1). For example, if storage capacity will be inadequate, on the average, one year out of every 20 years,

$$m/(n + 1) \approx 1/20 = 0.05 \qquad (5.13)$$

If storage capacity will be inadequate, on the average, one year out of every 100 years,

$$m/(n + 1) \approx 1/100 = 0.01 \qquad (5.14)$$

The calculation of storage is illustrated in Example 5.4.

Example 5.4

A reservoir is needed to supply water demand for 9 out of 10 years. The required reservoir capacities were determined by the method illustrated in the example below.

Year	Required Reservoir Capacity (ft$^3 \times 10^6$)	Year	Required Reservoir Capacity (ft$^3 \times 10^6$)
1961	60	1971	53
1962	40	1972	62
1963	85	1973	73
1964	30	1974	80
1965	67	1975	50
1966	46	1976	38
1967	60	1977	34
1968	42	1978	28
1969	90	1979	40
1970	51	1980	45

These data must now be ranked, with the highest required capacity, or worst drought, ranked 1; the next highest, 2; and so on. Data were collected for 21 years, so that n = 21 and n + 1 = 22. Next, m/(n + 1) is calculated for each drought:

Rank	Capacity $(m^3 \times 10^6)$	m/(n + 1)	Rank	Capacity $(m^3 \times 10^6)$	m/(n + 1)
1	90	0.05	11	50	0.52
2	85	0.1	12	46	0.57
3	80	0.14	13	45	0.61
4	73	0.19	14	42	0.66
5	67	0.23	15	40	0.71
6	62	0.28	16	40	0.76
7	60	0.33	17	38	0.81
8	60	0.38	18	34	0.85
9	53	0.43	19	30	0.90
10	51	0.48	20	28	0.95

These data are plotted in Figure 5–9. A semi-log plot often yields an acceptable straight line. If the reservoir capacity is required to be adequate 9 years out of 10, it may be inadequate 1 year out of 10. Entering Figure 5–9 at m/(n + 1) ≈ 1/10 = 0.01, we find that

$$m/n + 1 \approx 2/21 = 0.1 \tag{5.15}$$

The 10% probability of adequate capacity requires a reservoir capacity of 82 million cubic meters. Had the community only required adequate capacity one year out of five, m/(n + 1) = 0.2 and, from Figure 5–9, a reservoir capacity of 71 million cubic feet would have sufficed.

This procedure is a *frequency analysis* of a recurring natural event. The frequencies chosen for investigation were once in 10 years and once in 5 years, or a "10-year drought" and a "5-year drought," but droughts occurring 3 years in a row and then not again for 30 years still constitute "10-year droughts." Planning for a 30-year drought will result in the construction of a large expensive reservoir; planning for a 10-year drought will result in the construction of a smaller less-expensive reservoir.

WATER TRANSMISSION

Water can be transported from a ground or surface supply either directly to the water users in a community or initially to a water treatment facility. Water is transported by different types of conduits, including

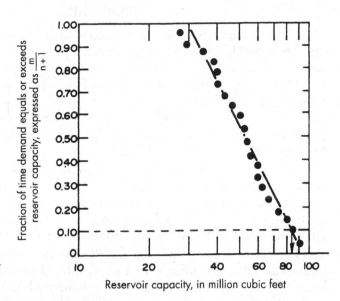

FIGURE 5–9
Frequency analysis of
reservoir capacity

- Pressure conduits: tunnels and pipelines
- Gravity-flow conduits: channels and canals

The location of the river or well field as well as the location of the water treatment facility defines the length of these conduits. Long, gentle slopes allow canals and aqueducts to be used, but in most instances, pressurized systems are constructed for water transmission from the water supply watershed. The water then enters a water treatment facility where it is cleaned into potable water and subsequently distributed to the community of residential, commercial, and industrial users through a system of pressurized pipes. Because the demand for water is variable, we use more water during the daylight hours and for random fire control; for example, this distribution system must include storage facilities to even out the fluctuations.

CONCLUSION

As the hydrologic cycle indicates, water is a renewable resource because of the driving force of energy from the sun. The earth is not running out of water, though enough water or enough clean water may not be available in some areas because of climate and water use. Both groundwater and surface water supplies are available to varying degrees over the entire earth's surface and can be protected by sound engineering and environmental judgment. The next chapter addresses methods of preparing and treating water for distribution and consumption once the supply has been provided.

PROBLEMS

5.1   A storage reservoir is needed to ensure a constant flow of 15 cfs to a city. The monthly stream flow records are

| Month | | | | | | | | | | | |
| J | F | M | A | M | J | J | A | S | O | N | D |

Million ft³ of water

| | | | | | | | | | | | |
| 60 | 70 | 85 | 50 | 40 | 25 | 55 | 85 | 20 | 55 | 70 | 90 |

Calculate the storage requirement for this year. How large must the reservoir be to provide the 15 cfs all year long?

5.2   If a faucet drips at a rate of 2 drops per second, and it takes 25,000 drops to make one gallon of water, how much water is lost each day? Each year? If water costs $5.00 per 1000 gallons, how long will the water leak until its cost equals the 75 cents in parts needed to fix the leak (if you fix it yourself)? How long if you call a plumber for $40.00 minimum rate plus 75 cents in parts?

5.3   Two adjacent landowners drill wells into what appears to be a continuous unconfined aquifer, with a water table elevation of 300 ft. An impervious layer underlies the aquifer at an elevation of 250 ft. Assume that both wells reach down to an elevation of 270 ft. The ground level varies, but is between 360 and 420 ft elevation. Draw a picture of the drawdown if

   a. Only one of the wells is drilled and starts pumping
   b. Both wells are drilled and start pumping
   c. One well pumps so much that it "goes dry"

5.4   A well is 0.1 m in diameter and pumps from an unconfined aquifer 50 m deep. The well point (the low end of the pipe) is in the middle of the aquifer (25 m below the water table). The well pumps at an equilibrium (steady rate) rate of 1500 m³ per day. Two observation wells are located at distances 60 m and 100 m, and they have been drawn down by 0.2 m and 0.3 m, respectively. What is the coefficient of permeability and estimated drawdown at the well?

5.5   In problem 5.4, how much can be pumped (how high can the rate of water flow be) before the well "goes dry"?

LIST OF SYMBOLS

A	area, in m²
a	area available for flow
cfs	cubic feet per second
Δ	a change in
h	energy, in m
L	horizontal distance, in m
k	coefficient of permeability
Q	flow rate, in m³/sec
r	radius of cylinder
v	velocity, in m/sec
v′	actual velocity

Chapter 6

Water Treatment

Many aquifers and isolated surface waters are high in water quality and may be pumped from the supply and transmission network directly to any number of end uses, including human consumption, irrigation, industrial processes, and fire control. However, clean water sources are the exception in many parts of the world, particularly regions where the population is dense or where there is heavy agricultural use. In these places, the water supply must receive varying degrees of treatment before distribution.

Impurities enter water as it moves through the atmosphere, across the earth's surface, and between soil particles in the ground. These background levels of impurities are often supplemented by human activities. Chemicals from industrial discharges and pathogenic organisms of human origin, if allowed to enter the water distribution system, may cause health problems. Excessive silt and other solids may make water aesthetically unpleasant and unsightly. Heavy metal pollution, including lead, zinc, and copper, may be caused by corrosion of the very pipes that carry water from its source to the consumer.

The method and degree of water treatment are site specific. Although water from public water systems is used for other uses, such as industrial consumption and firefighting, the cleanest water that is needed is for human consumption and therefore this requirement defines the degree of treatment. Thus, we focus on treatment techniques that produce *potable water*, or water that is both safe and pleasing.

A typical water treatment plant is diagrammed in Figure 6–1. It is designed to remove odors, color, and turbidity as well as bacteria and other contaminants. Raw water entering a treatment plant usually has significant turbidity caused by colloidal clay and silt particles. These particles carry an electrostatic charge that keeps them in continual motion and prevents them from colliding and sticking together. Chemicals like alum (aluminum sulfate) are added to the water both to neutralize the particles electrically and to aid in making them "sticky" so that they can coalesce and form large particles called flocs. This process is called coagulation and flocculation and is represented in stages 1 and 2 in Figure 6–1.

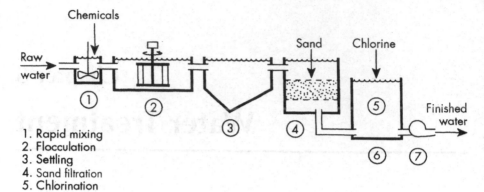

1. Rapid mixing
2. Flocculation
3. Settling
4. Sand filtration
5. Chlorination
6. Clear well storage
7. Pumping to distribution system

FIGURE 6–1. Movement of water through a water treatment facility

COAGULATION AND FLOCCULATION

Naturally occurring silt particles suspended in water are difficult to remove because they are very small, often colloidal in size, and possess negative charges; thus they are prevented from coming together to form large particles that can more readily be settled out. However, the charged layers surrounding the particles form an energy barrier between the particles. The removal of these particles by settling requires reduction of this energy barrier by neutralizing the electric charges and by encouraging the particles to collide with each other. The charge neutralization is called coagulation, and the building of larger flocs from smaller particles is called flocculation.

One means of accomplishing this end is to add trivalent cations to the water. These ions would snuggle up to the negatively charged particle and, because they possess a stronger charge, displace the monovalent cations. The effect of this would be to reduce the net negative charge and thus lower the repulsive force seen in Figure 6–2. In this condition, the particles will not repel each other and, upon colliding, will stick together. A stable colloidal suspension has thus been made into an unstable colloidal suspension.

The usual source of trivalent cations in water treatment is alum (aluminum sulfate). Alum has an additional advantage in that some fraction of the aluminum may form aluminum oxides/hydroxides, represented simply as

$$Al^{+++} + 3OH^- \rightarrow AlOH_3\downarrow$$

These complexes are sticky and heavy and will greatly assist in the clarification of the water in the settling tank if the unstable colloidal particles can be made to come into contact with the floc. This process is enhanced through the operation known as flocculation.

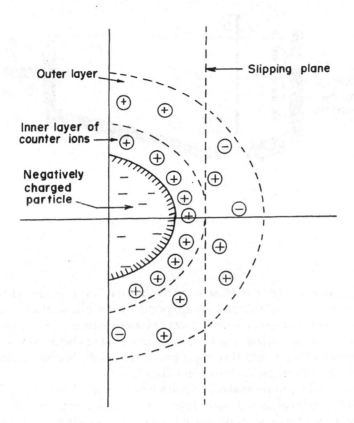

FIGURE 6–2. A colloidal particle is negatively charged and attracts positive counter ions to its surface.

SETTLING

When the flocs have been formed they must be separated from the water. This is invariably done in gravity-settling tanks that allow the heavier-than-water particles to settle to the bottom. Settling tanks are designed to minimize turbulence and allow the particles to fall to the bottom. The two critical elements of a settling tank are the entrance and exit configurations because this is where turbulence is created and where settling can be disturbed. Figure 6–3 shows one type of entrance and exit configuration used for distributing the flow entering and leaving the water treatment settling tank.

The particles settling to the bottom become what is known as *alum sludge.* Alum sludge is not very biodegradable and will not decompose at the bottom of the tank. After some time, usually several weeks, the accumulation of alum sludge at the bottom of the tank is such that it has to be removed. Typically,

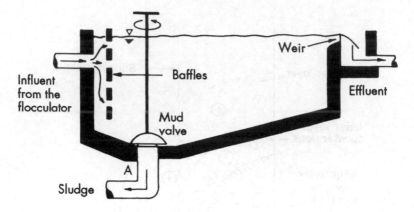

FIGURE 6–3. Settling tank used in water treatment

the sludge exits through a mud valve at the bottom and is wasted either into a sewer or to a sludge holding and drying pond. In contrast to alum sludge from water treatment, sludges collected in wastewater treatment plants can remain in the bottom of the settling tanks only a matter of hours before starting to produce odoriferous gases and floating some of the solids. Settling tanks used in wastewater treatment are discussed in Chapter 8.

Settling tanks can be analyzed by assuming an ideal settling tank. In this tank an imaginary column of water enters at one end, moves through the tank, and exits at the other end. Solid particles within this column settle to the bottom, and all those that reach the bottom before the column reaches the far end of the tank are assumed to be removed (settled out). If a solid particle enters the tank at the top of the column and settles at a velocity of v_o, it should have settled to the bottom as the imaginary column of water exits the tank, having moved through the tank at velocity v.

Consider now a particle entering the settling tank at the water surface. This particle has a settling velocity of v_o and a horizontal velocity v. In other words, the particle is just barely removed (it hits the bottom at the last instant). Note that if the same particle enters the settling tank at any other height, such as height h, its trajectory always carries it to the bottom. Particles having this velocity are termed *critical particles* in that particles with lower settling velocities are not all removed. For example, the particle having velocity v_s, entering the settling tank at the surface, will not hit the bottom and escape the tank. However, if this same particle enters at some height h, it should just barely hit the bottom and be removed. Any of these particles having a velocity v_s that happen to enter the settling tank at height h or lower are thus removed, and those entering above h are not. Since the particles entering the settling tank are assumed to be equally distributed, the proportion of those particles with a velocity of v_s removed is equal to h/H, where H is the height of the settling tank.

The retention time of a settling tank is the amount of time necessary to fill the tank at some given flow rate, or the amount of time an average water particle spends in the tank. Mathematically, the residence time is defined as

$$\bar{t} = \frac{V}{Q} \tag{6.1}$$

or the volume divided by the flow rate.

Although the water leaving a settling tank is essentially clear, it still may contain some turbidity and may carry pathogenic organisms. Thus, additional polishing is performed with a rapid sand filter.

FILTRATION

The movement of water into the ground and through soil particles, and the cleansing action the particles have on contaminants in the water, were discussed in Chapter 5. Picture the extremely clear water that bubbles up from "underground streams" as spring water. Soil particles help filter the groundwater, and this principle is applied to water treatment. In almost all cases, filtration is performed by a rapid sand filter.

As the sand filter removes the impurities, the sand grains get dirty and must be cleaned. The process of rapid sand filtration therefore involves two operations: filtration and backwashing. Figure 6–4 shows a cutaway of a slightly

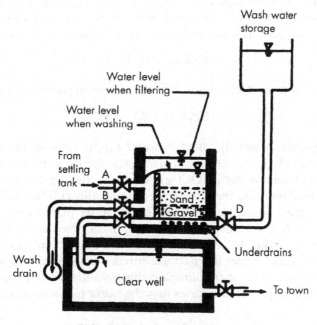

FIGURE 6–4. Rapid sand filter

simplified version of the rapid sand filter. Water from the settling basins enters the filter and seeps through the sand and gravel bed, through a false floor, and out into a clear well that stores the finished water. Valves A and C are open during filtration.

The cleaning process is done by reversing the flow of water through the filter. The operator first shuts off ine flow of water to the filter, closing valves A and C, then opens valves D and B, which allow wash water (clean water stored in an elevated tank or pumped from the clear well) to enter below the filter bed. This rush of water forces the sand and gravel bed to expand and jolts individual sand particles into motion, rubbing against their neighbors. The light colloidal material trapped within the filter is released and escapes with the wash water. After 10 to 30 minutes of washing, the wash water is shut off and filtration is resumed.

Filter beds might contain filtration media other than sand. Crushed coal, for example, is often used in combination with sand to produce a dual media filter which can achieve greater removal efficiencies.

DISINFECTION

After filtration, the finished water is disinfected, often with chlorine (step 5 in Figure 6–1). Disinfection kills the remaining microorganisms in the water, some of which may be pathogenic. Chlorine gas from bottles or drums is fed in correct proportions to the water to obtain a desired level of chlorine in the finished water. When chlorine comes in contact with organic matter, including microorganisms, it oxidizes this material and is in turn itself reduced. Chlorine gas is rapidly hydrolyzed in water to form hypochlorous acid, by the reaction

$$Cl_2 + H_2O \Leftrightarrow HOCl + H^+ + Cl^- \tag{6.5}$$

The hypochlorous acid itself ionizes further to the hypochlorous ion:

$$HOCl \Leftrightarrow OCl^- + H^+ \tag{6.6}$$

At the temperatures usually found in water supply systems, the hydrolysis of chlorine is usually complete in a matter of seconds, while the ionization of HOCl is instantaneous. Both HOCl and OCl⁻ are effective disinfectants and are called *free available chlorine* in water. Free available chlorine kills pathogenic bacteria and thus disinfects the water. Many water plant operators prefer to maintain a *residual* of chlorine in the water; that is, have some available chlorine left over once the chlorine has reacted with the currently available organics. Then, if organic matter like bacteria enters the distribution system, there is sufficient chlorine present to eliminate this potential health hazard. Tasting chlorine in drinking water indicates that the water has maintained its chlorine residual.

Chlorine may have adverse secondary effects. It is thought to combine with trace amounts of organic compounds in the water to produce chlorinated organic compounds that may be carcinogenic or have other adverse health effects. Some studies have shown an association between bladder and rectal cancer and consumption of chlorinated drinking water, indicating that there may be some risk of carcinogenesis. Disinfection by ozonation, the bubbling of ozone through the water, avoids the risk of side effects from chlorination, but ozone disinfection does not leave a residual in the water.

A number of municipalities also add fluorine to drinking water because it has been shown to prevent tooth decay in children and young adults. The amount of fluorine added is so small that it does not participate in the disinfection process.

From the clear well (step 6 in Figure 6–1) the water is pumped to the distribution system, a closed network of pipes, all under pressure. Users tap into these pipes to obtain potable water. Similarly, commercial and industrial facilities use the clean water for a variety of applications.

CONCLUSION

Water treatment is often necessary if surface water supplies, and sometimes groundwater supplies, are to be available for human use. Because the vast majority of cities use one water distribution system for households, industries, and fire control, large quantities of water often must be made available to satisfy the highest use, which is usually drinking water.

But does it make sense to produce drinkable water and then use it for other purposes, such as lawn irrigation? Growing demands for water have prompted serious consideration of dual water supplies: one high-quality supply for drinking and other personal use and one of lower quality, perhaps reclaimed from wastewater, for urban irrigation, firefighting, and similar applications. The growing use of bottled water for drinking is an example of a dual supply. In many parts of the world the population concentrations have stretched the supply of potable water to the limit, and either dual systems will be necessary or people will not be allowed to move to areas that have limited water supply. The availability of potable water often dictates land use and the migration of populations.

PROBLEMS

6.1 Propose some use and disposal options for water treatment sludges collected in the settling tanks following flocculation basins. Remember that these sludges consist mostly of aluminum oxides and clay.

6.2 For the house or dormitory where you live, suggest which water uses require potable water and which require a lower water quality. What *minimum* requirements must be met for the lower water-quality supply?

6.3   A settling tank has dimensions of 3 m deep, 10 m long, and 4 m wide. The flow entering the tank is 10 cubic meters per minute.

a. What is the residence time in this tank?
b. What is the velocity of a *critical particle* in this settling tank?

6.4   A settling tank is to settle out a slurry that has particles with a settling velocity of 0.01 m/min. An engineer decides that a tank of 5 m wide, 15 m long, and 2 m deep is adequate. What is the maximum allowable water flow rate into the tank if 100% removal of particles is to be achieved?

6.5   A settling tank has a residence time of two hours. It is 4 m wide, 4 m deep, and 10 m long. What is the critical settling velocity of the particles to be settled out?

6.6   The order of the processes in Figure 6–1 is important because this arrangement generally provides for the highest degree of treatment. Speculate what would happen to the water quality if

a. disinfection occurred as the first step? (Remember how chlorine reacts!)
b. filtration preceded coagulation/flocculation and settling?
c. settling preceded coagulation/flocculation?

6.7   Draw a sketch of a sand filter and show how the valves are manipulated to backwash the system.

Chapter 7
Collection of Wastewater

The "Shambles" is a street or area in many medieval English cities, like London and York. During the eighteenth and nineteenth centuries, Shambles were commercialized areas, with meat packing as a major industry. The butchers of the Shambles would throw all of their waste into the street, where it was washed away by rainwater into drainage ditches. The condition of the street was so bad that it contributed its name to the English language originally as a synonym for butchery or a bloody battlefield.

In old cities, drainage ditches like those at the Shambles were constructed for the sole purpose of moving stormwater out of the cities. In fact, discarding human excrement into these ditches was illegal in London. Eventually, the ditches were covered over and became what we now know as *storm sewers*. As water supplies developed and the use of the indoor water closet increased, the need for transporting domestic wastewater, called *sanitary waste*, became obvious. In the United States, sanitary wastes were first discharged into the storm sewers, which then carried both sanitary waste and stormwater and were known as *combined sewers*. Eventually a new system of underground pipes, known as *sanitary sewers,* was constructed for removing the sanitary wastes. Cities and parts of cities built in the twentieth century almost all built separate sewers for sanitary waste and stormwater.

ESTIMATING WASTEWATER QUANTITIES

Domestic wastewater (sewage) comes from various sources within the home, including the washing machine, dishwasher, shower, sinks, and of course the toilet. The toilet, or water closet (WC), as it is still known in Europe, has become a standard fixture of modern urban society. As important as this invention is, however, there is some dispute as to its inventor. Some authors[1] credit John Bramah with its invention in 1778; others[2] recognize it as the brainchild of Sir John Harrington in 1596. The latter argument is strengthened by Sir John's original description of the device, although there is no record of his donating

---

[1]Kirby, R.S., et. al., *Engineering in History*, New York: McGraw-Hill (1956).
[2]Reyburn, W., *Flushed with Pride,* London: McDonald (1969).

his name to the invention. The first recorded use of that euphemism is found in a 1735 regulation at Harvard University that decreed, "No Freshman shall go to the Fellows' John."

The term *sewage* is used here to mean only domestic wastewater. Domestic wastewater flows vary with the season, the day of the week, and the hour of the day. Figure 7–1 shows typical daily flow for a residential area. Note the wide variation in flow and strength. Typically, average sewage flows are in the range of 100 gallons per day per person, but especially in smaller communities that average can range widely.

Sewers also commonly carry industrial wastewater. The quantity of industrial wastes may usually be established by water use records, or the flows may be measured in manholes that serve only a specific industry, using a small flow meter. Industrial flows also often vary considerably throughout the day, the day of the week, and the season.

In addition to sewage and industrial wastewater, sewers carry groundwater and surface water that seeps into the pipes. Since sewer pipes can and often do have holes in them (due to faulty construction, cracking by roots, or other causes), groundwater can seep into the sewer pipe if the pipe is lower than the top of the groundwater table. This flow into sewers is called *infiltration*. Infiltration is least for new, well-constructed sewers, but can be as high as 500

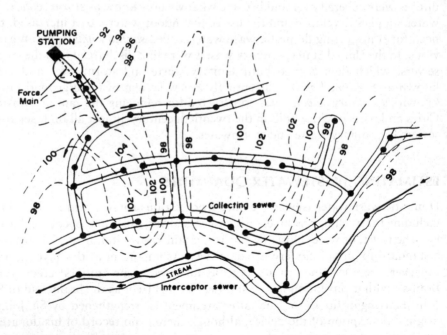

FIGURE 7–1. Typical wastewater collection system layout. [Adapted from Clark, J., Viessman, W., and Hammer, M., *Water Supply and Sewerage,* New York: IEP (1977).]

m^3/km-day (200,000 gal/mi-day). For older systems, 700 m^3/km-day (300,000 gal/mi-day) is the commonly estimated infiltration. Infiltration flow is detrimental since the extra volume of water must go through the sewers and the wastewater treatment plant. It should be reduced as much as possible by maintaining and repairing sewers and keeping sewerage easements clear of large trees whose roots can severely damage the sewers.

Inflow is stormwater collected unintentionally by the sanitary sewers. A common source of inflow is a perforated manhole cover placed in a depression, so that stormwater flows into the manhole. Sewers laid next to creeks and drainageways that rise up higher than the manhole elevation, or where the manhole is broken, are also a major source. Illegal connections to sanitary sewers, such as roof drains, can substantially increase the wet weather flow over the dry weather flow. The ratio of dry weather flow to wet weather flow is usually between 1:1.2 and 1:4.

For these reasons, the sizing of sewers is often difficult, since not all of the expected flows can be estimated and their variability is unknown. The more important the sewer and the more difficult is to replace it, the more important it is to make sure that it is sufficiently large to be able to handle all the expected flows for the foreseeable future.

SYSTEM LAYOUT

Sewers collect wastewater from residences and industrial establishments. A system of sewers installed for the purpose of collecting wastewater is known as a *sewerage system* (*not* a sewage system). Sewers almost always operate as open channels or gravity flow conduits. Pressure sewers are used in a few places, but these are expensive to maintain and are useful only when there are severe restrictions on water use or when the terrain is such that gravity flow conduits cannot be efficiently maintained.

A typical system for a residential area is shown in Figure 7–1. Building connections are usually made with clay or plastic pipe, 6 inches in diameter, to the collecting sewers that run under the street. *Collecting sewers* are sized to carry the maximum anticipated peak flows without surcharging (filling up) and are ordinarily made of plastic, clay, cement, concrete, or cast iron pipe. They discharge into *intercepting sewers*, or *interceptors*, that collect from large areas and discharge finally into the wastewater treatment plant.

Collecting and intercepting sewers must be constructed with adequate slope for adequate flow velocity during periods of low flow, but not so steep a slope as to promote excessively high velocities when flows are at their maximum. In addition, sewers must have manholes, usually every 120 to 180 m (400 to 600 ft) to facilitate cleaning and repair. Manholes are necessary whenever the sewer changes slope, size, or direction. Typical manholes are shown in Figure 7–2.

Gravity flow may be impossible, or uneconomical, in some locations so that the wastewater must be pumped. This requires the installation of pumping

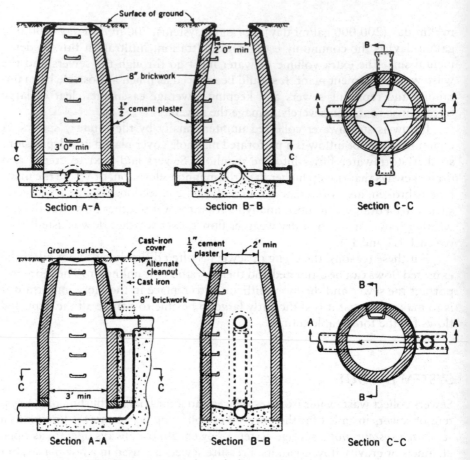

FIGURE 7–2. Typical manholes used in sewerage systems

stations at various locations throughout the system. The pumping station collects wastewater from a collecting sewer and pumps it to a higher elevation by means of a force main. The end of a force main is always into a manhole.

A power outage would render the pumps inoperable, and eventually the sewage would back up into homes. As you can imagine, this would be highly undesirable; therefore, a good system layout minimizes pumping stations and/or provides auxiliary power.

CONCLUSION

Sewers have been a part of civilized settlements for thousands of years, and in the modern United States we have become accustomed to and even complacent about the sewers that serve our communities. They never seem to fail, and there never seems to be a problem with them. Most important, we can dump whatever we want to down the drain, and it just disappears.

Of course, it doesn't just disappear. It flows through the sewer and ends up in a wastewater treatment plant, the subject of the next chapter. The stuff we often thoughtlessly dump down the drain can in fact cause serious problems in wastewater treatment and may even cause health problems in future drinking water supplies. Therefore, we must be cognizant of what we flush down the drain and recognize that it does not just disappear.

PROBLEMS

7.1   Illegal connections are sometimes made to sanitary sewers. Suppose a family of four, living in a home with a roof area of 70 ft × 40 ft, connects the roof drain to the sewer. If rain falls at the rate of 1 in/hr, what percent increase will there be in the flow from their house over the dry weather flow? The dry weather flow is 50 gal/person/day.

7.2   A transoceanic flight on a Boeing 747 with 430 persons aboard takes seven hours. Estimate the weight of water necessary to flush the toilets if each flush uses two gallons. Make any assumptions necessary and state them. What fraction of the total payload (people) does the flush water represent? How can this weight be reduced, since discharging, for obvious reasons, is illegal?

7.3   Suggest a design for a water treatment system for the proposed space station.

7.4   Estimate the amount of water storage a town (population 10,000) must have to satisfy its firefighting requirements.

7.5   Develop five suggestions for water conservation in a 10-story apartment building.

Chapter 8

Wastewater Treatment

As civilization developed and cities grew, domestic sewage and industrial waste were eventually discharged into drainage ditches and sewers and the entire contents emptied into the nearest watercourse. For major cities, this discharge was often enough to destroy even a large body of water. As Samuel Taylor Coleridge described the city of Cologne (Köln), Germany:

> In Köln, a town of monks and bones
> And pavements fanged with murderous stones
> And rags, and bags, and hideous wenches;
> I counted two and seventy stenches,
> All well defined, and several stinks!
>
> Ye Nymphs that reign o'er sewers and sinks,
> The river Rhine, it is well known,
> Doth wash your city of Cologne;
> But tell me Nymphs! What power divine
> Shall henceforth wash the river Rhine?

During the nineteenth century, the River Thames was so grossly polluted that the House of Commons stuffed lye-soaked rags into cracks in the windows of Parliament to reduce the stench.

Sanitary engineering technology for treating wastewater to reduce its impact on watercourses, pioneered in the United States and England, eventually became economically, socially, and politically feasible. This chapter reviews these systems from the earliest simple treatment systems to the advanced systems used today. The discussion begins by reviewing those wastewater characteristics that make disposal difficult, showing why wastewater cannot always be disposed of onsite, and demonstrating the necessity of sewers and centralized treatment plants.

WASTEWATER CHARACTERISTICS

Discharges into a sanitary sewerage system consist of domestic wastewater (sewage), industrial discharge, inflow, and infiltration. The last two add to the total wastewater volume, but are generally not of concern in waste-

TABLE 8–1. Characteristics of Typical Domestic Wastewater

Parameter	Typical Value for Domestic Sewage
BOD	250 mg/L
SS	220 mg/L
Phosphorus	8 mg/L
Organic and ammonia nitrogen	40 mg/L
pH	6.8
Chemical oxygen demand	500 mg/L
Total solids	270 mg/L

water disposal. Industrial discharges vary widely with the size and type of industry and the amount of treatment applied before discharge into sewers. In the United States, the trend has been to mandate increasing pretreatment of wastewater (see Chapter 11) in response to both regulations limiting discharges and the imposition of local sewer surcharges.

Biochemical oxygen demand (BOD) is reduced and suspended solids (SS) are removed by wastewater treatment, but heavy metals, motor oil, refractory organic compounds, radioactive materials, and similar exotic pollutants are not readily handled this way. Communities usually severely restrict the discharge of such substances by requiring pretreatment of industrial wastewater.

Domestic sewage varies substantially over time and from one community to the next, and no two municipal wastewaters are the same. For illustrative purposes, however, there is some advantage in talking about "average" wastewaters. Table 8–1 shows typical values for the most important parameters of domestic wastewater.

ONSITE WASTEWATER DISPOSAL

In many smaller communities, sewers are both impractical and unnecessary. In such situations, wastewater is treated and disposed of onsite, that is, at the same location as the house.

The original onsite system is of course the pit privy, glorified in song and fable.[1] The privy, still used in camps and temporary residences and in many less industrialized countries, consists of a pit about 2 m (6 feet) deep into which

---

[1] A literary work on this theme is "The Passing of the Backhouse" by James Whitcomb Riley:

But when the crust was on the snow and sullen skies were gray,
In sooth the building was no place where one would wish to stay.
We did our duties promptly there, one purpose swayed the mind.
We tarried not nor lingered long on what we left behind.
The torture of that icy seat would make a Spartan sob . . .

human excrement is deposited. When a pit fills up, it is covered and a new one is dug. The composting toilet that accepts both human excrement and food waste, and produces a useful compost, is a logical extension of the pit privy. In a dwelling with a composting toilet, wastewater from other sources like washing is discharged separately into a tank and seepage field.

By far the greatest number of households with onsite disposal systems use a form of the septic tank and tile field. As shown in Figure 8–1, a septic tank consists of a concrete box that removes the solids in the waste and promotes partial decomposition. The solid particles settle out and eventually fill the tank, thus necessitating periodic cleaning. The water overflows into a tile drain field that promotes the seepage of discharged water.

A tile field consists of pipe laid in about a 1-m (3-ft) deep trench, end on end but with short gaps between each section of pipe. The effluent from the septic tank flows into the tile field pipes and seeps into the ground through these gaps. Alternatively, seepage pits consisting of gravel and sand may be used for promoting adsorption of effluent into the ground. The most important consideration in designing a septic tank and tile field system is the ability of the ground to absorb the effluent. Percolation tests, used to measure the suitability of the ground for the fields, are conducted in the following way:

1. A hole is dug about 6 to 12 in² and as deep as the proposed tile field trench.
2. The soil is scratched to remove smeared surfaces and to provide a more natural soil interface, and some gravel is put in the bottom of the pit.
3. The pit is filled with water and allowed to stand overnight.
4. The next day, the pit is filled with water to 6 inches above the gravel, and the drop in the water level after 30 minutes is measured.
5. The percolation rate is calculated in inches per minute.

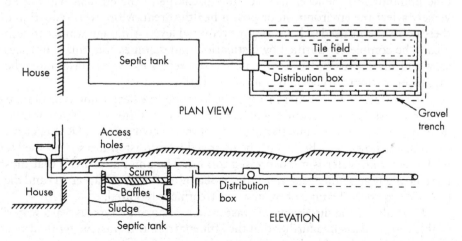

FIGURE 8–1. Septic tank and tile field used for onsite wastewater disposal

TABLE 8–2. Adsorption Area Requirements for Private Residences

Percolation Rate (in/min)	Required Adsorption Field Area per Bedroom (ft²)
Greater than 1	70
Between 1 and 0.5	85
Between 0.5 and 0.2	125
Between 0.2 and 0.07	190
Between 0.07 and 0.03	250
Less than 0.03	Unsuitable ground

The U.S. Public Health Service and all county and local departments of health have established guidelines for sizing the tile fields or seepage pits. Typical standards are shown in Table 8–2. Many areas in the United States have soils that percolate poorly, and septic tank/tile fields are not appropriate. In fact, onsite disposal in many locations has been discouraged and, in some regions, prohibited. Community growth, planning for future growth, and the legislated mandate to assess environmental impact are making onsite disposal in many locations obsolete. A better way to move human waste out of a congested community is to use water as a carrier and channel the wastewater to a central treatment facility.

CENTRAL WASTEWATER TREATMENT

The objective of wastewater treatment is to reduce the concentrations of specific pollutants to the level at which the discharge of the effluent will not adversely affect the environment or pose a health threat. Moreover, reduction of these constituents need only be to some required level. Although water can technically be completely purified by distillation and deionization, this is unnecessary and may actually be detrimental to the receiving water. Fish and other organisms cannot survive in deionized or distilled water.

For any given wastewater in a specific location, the degree and type of treatment are variables that require engineering decisions. Often the degree of treatment depends on the assimilative capacity of the receiving water. DO sag curves can indicate how much BOD must be removed from wastewater so that the DO of the receiving water is not depressed too far. The amount of BOD that must be removed is an *effluent standard* (discussed more fully in Chapter 11) and dictates in large part the type of wastewater treatment required.

To facilitate the discussion of wastewater, assume a "typical wastewater" (Table 8–1) and assume further that the effluent from this wastewater treatment must meet the following effluent standards:

BOD ≤ 15 mg/L

SS ≤ 15 mg/L

P ≤ 1 mg/L

Additional effluent standards could have been established, but for illustrative purposes we consider only these three. The treatment system selected to achieve these effluent standards includes

- Primary treatment: physical processes that remove nonhomogenizable solids and homogenize the remaining effluent.
- Secondary treatment: biological processes that remove most of the biochemical demand for oxygen.
- Tertiary treatment: physical, biological, and chemical processes to remove nutrients like phosphorus and inorganic pollutants, to deodorize and decolorize effluent water, and to carry out further oxidation.

PRIMARY TREATMENT

The most objectionable aspect of discharging raw sewage into watercourses is the floating material. Thus screens were the first form of wastewater treatment used by communities, and they are used today as the first step in treatment plants. Typical screens, shown in Figure 8–2, consist of a series of steel bars that might be about 2.5 cm apart. A screen in a modern treatment plant removes materials that might damage equipment or hinder further treatment. In some older treatment plants screens are cleaned by hand, but mechanical cleaning equipment is used in almost all new plants. The cleaning rakes are activated when screens get sufficiently clogged to raise the water level in front of the bars.

In many plants, the second treatment step is a *comminutor,* a circular grinder designed to grind the solids coming through the screen into pieces about 0.3 cm or less in diameter. A typical comminutor design is shown in Figure 8–3.

The third treatment step is the removal of grit or sand from the wastewater. Grit and sand can damage equipment like pumps and flow meters and must be removed. The most common *grit chamber* is a wide place in the channel where the flow is slowed enough to allow the dense grit to settle out. Sand is about 2.5 times denser than most organic solids and thus settles much faster. The objective of a grit chamber is to remove sand and grit without removing organic material. Organic material must be treated further in the plant, but the separated sand may be used as fill without additional treatment.

Most wastewater treatment plants have a *settling tank* (Figures 8–4 and 8–5) after the grit chamber, to settle out as much solid material as possible. Accordingly, the retention time is long and turbulence is kept to a minimum.

The solids settle to the bottom of the tank and are removed through a pipe, while the clarified liquid escapes over a *V-notch weir* that distributes the liquid discharge equally all the way around a tank. Settling tanks are also called

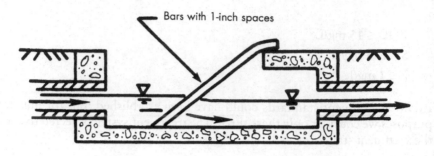

FIGURE 8–2. Bar screen used in wastewater treatment. The top picture shows a manually cleaned screen; the bottom picture represents a mechanically cleaned screen [Photo courtesy Envirex.]

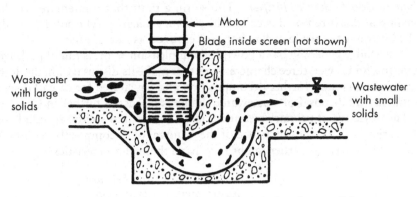

FIGURE 8–3. A comminutor used to grind up large solids

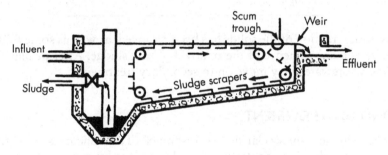

FIGURE 8–4. Rectangular settling tank

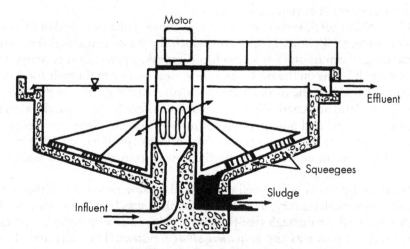

FIGURE 8–5. Circular settling tank

sedimentation tanks or *clarifiers*. The settling tank that immediately follows screening and grit removal is called the *primary clarifier*. The solids that drop to the bottom of a primary clarifier are removed as *raw sludge*.

Raw sludge generally has a powerfully unpleasant odor, is full of pathogenic organisms, and is wet, three characteristics that make its disposal difficult. It must be stabilized to retard further decomposition and dewatered for ease of disposal. Treatment and disposal of wastewater sludge is discussed further in Chapter 9.

The objective of primary treatment is the removal of solids, although some BOD is removed as a consequence of the removal of decomposable solids. The wastewater described earlier might now have these characteristics:

	Raw Wastewater	After Primary Treatment
BOD_5, mg/L	250	175
SS, mg/L	220	60
P, mg/L	8	7

A substantial fraction of the solids has been removed, as well as some BOD and a little P, as a consequence of the removal of raw sludge. After primary treatment the wastewater may move on to secondary treatment.

SECONDARY TREATMENT

Water leaving the primary clarifier has lost much of the solid organic matter but still contains high-energy molecules that decompose by microbial action, creating BOD. The demand for oxygen must be reduced (energy wasted) or else the discharge may create unacceptable conditions in the receiving waters. The objective of secondary treatment is to remove BOD, whereas the objective of primary treatment is to remove solids.

The *trickling filter*, shown in Figure 8–6, consists of a filter bed of fist-sized rocks or corrugated plastic blocks over which the waste is trickled. The name is something of a misnomer since no filtration takes place. A very active biological growth forms on the rocks, and these organisms obtain their food from the waste stream dripping through the rock bed. Air either is forced through the rocks or circulates automatically because of the difference between the air temperature in the bed and ambient temperatures. Trickling filters use a rotating arm that moves under its own power, like a lawn sprinkler, distributing the waste evenly over the entire bed. Often the flow is recirculated and a higher degree of treatment attained.

Trickling filtration was a well-established treatment system at the beginning of the twentieth century. In 1914, a pilot plant was built for a different system that bubbled air through free-floating aerobic microorganisms, a process which became known as the *activated sludge system*. The activated sludge

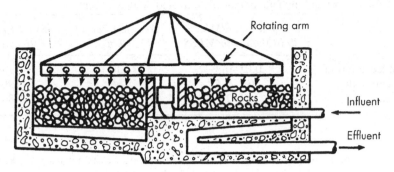

FIGURE 8–6. Trickling filter

process differs from trickling filtration in that the microorganisms are suspended in the liquid.

An activated sludge system, as shown in the block diagram in Figure 8–7, includes a tank full of waste liquid from the primary clarifier and a mass of microorganisms. Air bubbled into this aeration tank provides the necessary oxygen for survival of the aerobic organisms. The microorganisms come in contact with dissolved organic matter in the wastewater, adsorb this material, and ultimately decompose the organic material to CO_2, H_2O, some stable compounds, and more microorganisms.

When most of the organic material, that is, food for the microorganisms, has been used up, the microorganisms are separated from the liquid in a settling tank, sometimes called a *secondary* or *final clarifier*. The microorganisms remaining in the settling tank have no food available, become hungry, and are thus activated—hence the term *activated sludge*. The clarified liquid escapes over a weir and may be discharged into the receiving water. The settled microorganisms, now called

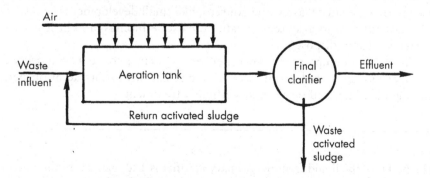

FIGURE 8–7. Block diagram of an activated sludge system

TABLE 8–3. Loadings and Efficiencies of Activated Sludge Systems

Process	Loading: F/M = lb BOD/day/lb MLSS	Aeration Period (hr)	BOD Removal Efficiency (%)
Extended aeration	0.05–0.2	30	95
Conventional	0.2–0.5	6	90
High rate	1–2	4	85

return activated sludge, are pumped back to the head of the aeration tank, where they find more food in the organic compounds in the liquid entering the aeration tank from the primary clarifier, and the process starts over again. Activated sludge treatment is a continuous process, with continuous sludge pumping and clean-water discharge.

Activated sludge treatment produces more microorganisms than necessary and if the microorganisms are not removed, their concentration will soon increase and clog the system with solids. Some of the microorganisms must therefore be wasted and the disposal of such waste activated sludge is one of the most difficult aspects of wastewater treatment.

Activated sludge systems are designed on the basis of loading, or the amount of organic matter, or food, added relative to the microorganisms available. The food-to-microorganism (F/M) ratio is a major design parameter. Both F and M are difficult to measure accurately, but may be approximated by influent BOD and SS in the aeration tank, respectively. The combination of liquid and microorganisms undergoing aeration is known as *mixed liquor,* and the SS in the aeration tank are *mixed liquid suspended solids* (MLSS). The ratio of influent BOD to MLSS, the F/M ratio, is the loading on the system, calculated as pounds (or kg) of BOD per day per pound or kg of MLSS.

Relatively small F/M, or little food for many microorganisms, and a long aeration period (long retention time in the tank) result in a high degree of treatment because the microorganisms can make maximum use of available food. Systems with these features are called *extended aeration systems* and are widely used to treat isolated wastewater sources, like small developments or resort hotels. Extended aeration systems create little excess biomass and little excess activated sludge to dispose of.

Table 8–3 compares extended aeration systems, conventional secondary treatment systems, and "high-rate" systems that have short aeration periods and high loading, and result in less efficient treatment.

Example 8.4

The BOD_5 of the liquid from the primary clarifier is 120 mg/L at a flow rate of 0.05 mgd. The dimensions of the aeration tank are 20 ft × 10 ft × 20 ft, and the MLSS = 2000 mg/L. Calculate the F/M ratio.

$$\text{lb BOD/day} = \left(120\ \frac{mg}{L}\right)(0.05\ \text{mgd})\left(3.8\ \frac{L}{gal}\right)\left(\frac{1\ lb}{454\ g}\right) = 50\ \text{lb/day} \qquad (8.1)$$

$$\text{lb MLSS} = (20 \times 10 \times 20)\ ft^3\left(2000\ \frac{mg}{L}\right)\left(3.8\ \frac{L}{gal}\right)\left(7.48\ \frac{gal}{ft^3}\right)\left(\frac{1\ lb}{454\ g}\right)$$
$$= 229\ \text{lb} \qquad (8.2)$$

$$\frac{F}{M} = \frac{50}{229} = 0.22\ \frac{\text{lb BOD/day}}{\text{lb MLSS}} \qquad (8.3)$$

Another process modification is contact stabilization, or biosorption, in which the sorption and bacterial growth phases are separated by a settling tank. Contact stabilization provides for growth at high solids concentrations, thus saving tank space. An activated sludge plant can often be converted to a contact stabilization plant when tank volume limits treatment efficiency. The two principal means of introducing sufficient oxygen into the aeration tank are by bubbling compressed air through porous diffusers or by beating air in mechanically.

The success of the activated sludge system also depends on the separation of the microorganisms in the final clarifier. When the microorganisms do not settle out as anticipated, the sludge is said to be a *bulking sludge*. Bulking is often characterized by a biomass composed almost totally of filamentous organisms that form a kind of lattice structure within the sludge flocs which prevents settling.[2] A trend toward poor settling may be the forerunner of a badly upset and ineffective system. The settleability of activated sludge is most often described by the sludge volume index (SVI), which is reasoned by allowing the sludge to settle for 30 minutes in a 1-L cylinder. The volume occupied by the settled sludge is divided by the original suspended solids concentration.

$$\text{SVI} = \frac{(1000)\ (\text{volume of sludge after 30 min, mL})}{\text{mg/L of SS}} \qquad (8.4)$$

Example 8.5
A sample of sludge has an SS concentration of 4000 mg/L. After settling for 30 minutes in a 1-L cylinder, the sludge occupies 400 mL. Calculate the SVI.

$$\text{SVI} = \frac{(1000)\ (400\ mL)}{4000\ mg/L} = 100 \qquad (8.5)$$

---

[2]You may picture this as a glass filled with cotton balls. When water is poured into the glass, the cotton filaments are not dense enough to settle to the bottom of the glass.

If the SVI is 100 or lower, the sludge solids settle rapidly and the sludge returned to the final clarifier can be expected to be at a high solids concentration. SVIs above 200, however, indicate bulking sludges and can lead to poor treatment.

Some causes of poor settling are improper or varying F/M ratios, fluctuations in temperature, high concentrations of heavy metals, or deficiencies in nutrients. Cures include chlorination, changes in air supply, or dosing with hydrogen peroxide to kill the filamentous microorganisms. When sludge does not settle, the return activated sludge is thin because SS concentration is low, and the microorganism concentration in the aeration tank drops. Since there are fewer microorganisms to handle the same food input, the BOD removal efficiency is reduced.

The performance of an activated sludge system depends on the performance of the final clarifier. If this settling tank cannot achieve the required return sludge solids, the solids in the aeration tank (MLSS) will drop and the treatment efficiency will be reduced. Final clarifiers act as settling tanks for flocculent settling and as thickeners. Their design requires consideration of both solids loading and overflow rate.

Secondary treatment of wastewater usually includes a biological step, like activated sludge, that removes a substantial part of the BOD and the remaining solids. The typical wastewater that we began with now has the following approximate water quality:

	Raw Wastewater	After Primary Treatment	After Secondary Treatment
BOD, mg/L	250	175	15
SS, mg/L	220	60	15
P, mg/L	8	7	6

The effluent from secondary treatment meets the previously established effluent standards for BOD and SS. Only phosphorus content remains high. The removal of inorganic compounds, including inorganic phosphorus and nitrogen compounds, requires advanced or tertiary wastewater treatment.

TERTIARY TREATMENT

Primary and secondary (biological) treatments are a part of conventional wastewater treatment plants. However, secondary treatment plant effluents are still significantly polluted. Some BOD and suspended solids remain, and neither primary nor secondary treatment is effective in removing phosphorus and other nutrients or toxic substances. A popular advanced treatment for BOD removal is the polishing pond, or oxidation pond, commonly a large lagoon into which the secondary effluent flows. Such ponds have a long retention time, often measured in weeks. An oxidation pond and the reactions that take place in it are shown in Figure 8–8.

Oxidation ponds, as their name implies, are designed to be aerobic; hence, light penetration for algal growth is important and a large pond surface area is needed. Ponds can provide complete treatment and a sufficiently large oxidation pond may be the only treatment step for a small waste flow. When the rate of oxidation in a pond is too great and oxygen availability becomes limiting, the pond may be forcibly aerated by either diffusive or mechanical aerators. Such ponds are called aerated lagoons and are widely used in treating industrial effluent.

BOD may also be removed by activated carbon adsorption, which has the added advantage of removing some inorganic as well as organic compounds. An activated carbon column is a completely enclosed tube, which dirty water is pumped into at the bottom and clear water exits at the top. Microscopic crevices in the carbon catch and hold colloidal and smaller particles. As the carbon column becomes saturated, the pollutants must be removed from the carbon and the carbon reactivated, usually by heating it in the absence of oxygen. Reactivated or regenerated carbon is somewhat less efficient than using virgin carbon, some of which must always be added to ensure effective performance.

Nitrogen compounds may be removed from wastewater in two ways. Even after secondary treatment, most of the nitrogen exists as ammonia. Increasing the pH produces the reaction

$$NH_4^+ + OH^- \rightarrow NH_3\uparrow + H_2O \tag{8.6}$$

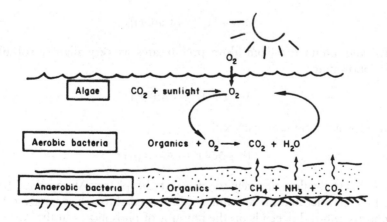

FIGURE 8–8. An oxidation pond

Much of the dissolved ammonia gas may then be expelled from the water into the atmosphere. The ammonium ion in the wastewater may also be oxidized completely to nitrate by bacteria like *Nitrobacter* and *Nitrosomonas*, in a process called *nitrification*.

$$2NH_4^+ + 3O_2 \xrightarrow{\text{Nitrosomonas}} 2NO_2^- + 2H_2O + 4H^-.$$

$$2NO_2^- + O_2 \xrightarrow{\text{Nitrobacter}} 2NO_3^-$$

(8.7)

These reactions are slow and require long retention times in the aeration tank as well as sufficient DO. If the flow rate is too high, the slow-growing microorganisms are washed out of the aeration tank.

Once the ammonia has been oxidized to nitrate, it may be reduced by a broad range of facultative and anaerobic bacteria like *Pseudomonas*. This denitrification requires a source of carbon, and methanol (CH_3OH) is often used for that purpose.

$$6NO_3^- + 2CH_3OH \rightarrow 6NO_2^- + 2CO_2\uparrow + 4H_2O$$

$$6NO_2^- + 3CH_3OH \rightarrow 3N_2\uparrow + 3CO_2\uparrow + 3H_2O + 6OH^-$$

(8.8)

Phosphate may be removed chemically or biologically. The most popular chemical methods use lime, $Ca(OH)_2$, and alum, $Al_2(SO_4)_3$. Under alkaline conditions, the calcium ion will combine with phosphate to form calcium hydroxyapatite, a white insoluble precipitate that is settled out and removed from the wastewater. Insoluble calcium carbonate ($CaCO_3$) is also formed and removed. The calcium may be reclaimed by burning in a furnace.

$$CaCO_3 \rightarrow CO_2 + CaO$$

(8.9)

Quicklime (CaO) is then slaked by adding water and forming lime, which may be reused for phosphorus removal.

$$CaO + H_2O \rightarrow Ca(OH)_2$$

(8.10)

The aluminum ion from alum precipitates as very slightly soluble aluminum phosphate,

$$Al^{3+} + PO_4^{3-} \rightarrow AlPO_4\downarrow$$

(8.11)

and also forms aluminum hydroxide,

$$Al^{3+} + 3OH^- \rightarrow Al(OH)_3\downarrow$$

(8.12)

which forms sticky flocs that help to settle out phosphates. Alum is usually added in the final clarifier and the amount of alum needed to achieve a given level of phosphorus removal depends on the amount of phosphorus in the wastewater.

Biological phosphorus removal is becoming increasingly popular since it does not require the addition of chemicals. In this process, the aeration tank in the activated sludge system is subdivided into zones, some of which are not aerated. In these zones the aerobic microorganisms become sorely stressed because of the lack of oxygen. If these microorganisms are then transferred to an aerated zone, they try to make up for lost time and assimilate organic matter (as well as phosphorus) at a rate much higher than they ordinarily would. Once the microorganisms have adsorbed the phosphorus, they are removed as waste activated sludge, thus carrying with them high concentrations of phosphorus. Using such sequencing of nonaerated and aerated zones, it is possible to remove as much as 90% of the phosphorus.

DISINFECTION

EPA and state effluent rules require that municipal wastewater treatment plant effluents be disinfected before they are discharged to receiving bodies of water. Chlorine is commonly used for this purpose and a *chlorine contact chamber* is constructed as the last unit operation in the treatment plant. Typically 30 minutes of contact time is required to kill microorganisms in the water with a chlorine residual often remaining in the water. Unfortunately, this residual, if discharged into a lake or a river, could damage the natural aquatic ecosystem, and *dechlorination* of the effluent is necessary. Because of the potential problems associated with the use of chlorine, other methods of disinfection have gained favor in recent years, particularly nonionizing radiation when no residuals are apparent.

CONCLUSION

A typical wastewater treatment plant, shown schematically in Figure 8–9, includes primary, secondary, and tertiary treatment. The treatment and disposal of solids removed from the wastewater stream deserve special attention and are treated in the next chapter.

Figure 8–10 is an aerial view of a typical wastewater treatment plant. Well-operated plants produce effluents that are often much less polluted than the receiving waters into which they are discharged. However, not all plants perform that well. Many wastewater treatment plants are only marginally effective in controlling water pollution, and plant operation is often to blame.

Operation of a modern wastewater treatment plant is complex and demanding. Unfortunately, operators have historically been poorly compensated, so that recruitment of qualified operators is difficult. States now require licensing of operators, and operators' pay and status is improving. This is a welcome change, for it makes little sense to entrust unqualified operators with multimillion-dollar facilities, which require proper plant design and proper operation. One without the other is a waste of money.

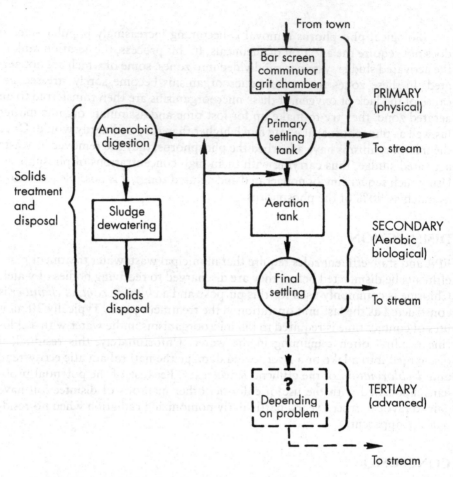

FIGURE 8–9. Block diagram of a complete wastewater treatment plant

FIGURE 8–10. Aerial view of a secondary wastewater treatment plant. [Courtesy Envirex.]

PROBLEMS

8.1 The following data were reported on the operation of a wastewater treatment plant:

	Influent (mg/L)	Effluent (mg/L)
BOD_5	200	20
SS	220	15
P	10	0.5

a. What percent removal was experienced for each of the individual processes?
b. What is the overall removal efficiency for BOD, SS, and P?
c. What type of treatment plant will produce such an effluent?

Draw a block diagram showing the treatment steps.

8.2 Describe the condition of a primary clarifier one day after the raw sludge pumps broke down. What might be happening to the raw sludge? Why?

8.3 Ponding, the excessive growth of slime on the rocks, is an operational problem with trickling filters. The excessive slime clogs spaces between the rocks so that water no longer flows through the filter. Suggest some cures for ponding.

8.4   Suppose you are an engineer hired to evaluate an industrial wastewater treatment plant for a specific wastewater with which you have no experience. What would you choose as the five most important wastewater parameters to be tested? Why would you want to know these values?

8.5   The influent and effluent data for a secondary treatment plant are

	Influent (mg/L)	Effluent (mg/L)
BOD_5	200	20
SS	200	100
P	10	8

Calculate the removal efficiency. Is the plant operating correctly. If not, what is wrong with the plant? What might be some actions to take to make it work better?

8.6   Draw block diagrams of the unit operations necessary to treat the following wastes to approximate effluent levels of BOD_5 = 20 mg/L, SS = 20 mg/L, P = 1 mg/L. Do no calculations but be prepared to argue that the proposed plants meet the required effluent limits.

Waste	BOD (mg/L)	Suspended Solids (SS) (mg/L)	Phosphorus (mg/L)
Domestic	200	200	10
Chemical industry	40,000	0	0
Pickle cannery	0	300	1
Fertilizer manufacturing	300	300	200

8.7   A percolation test shows that the measured water level drops 5 inches in 30 minutes. What size percolation field do you need for a two-bedroom house? What size would you need if the water level dropped only 0.5 inch?

8.8   A family of four wants to build a house on a lot for which the percolation rate is 1.00 mm/min. The county requires a septic tank hydraulic retention time of 24 hrs. Assume each person contributes 100 gal/day of wastewater. Find the volume of the tank required and the area of the drain field. Sketch the system, including all dimensions.

8.9   An activated sludge plant runs with a mixed liquor–solids concentration of 4000 mg/L. The operator finds that she is able to increase this to 6000 mg/L without any operational problems.

    a. How does she do this?
    b. What effect do you think this will have on the effluent BOD? Why?

8.10   A wastewater treatment plant operator runs a sludge volume index test in a liter cylinder and finds that at the end of 30 minutes' settling, the solids occupy 200 mL. He knows that the suspended solids concentration in his aeration tank from which he got the sample is 4000 mg/L.

a. What is the SVI?

b. Is this a good settling sludge, or is this operator is deep trouble?

8.11   The BOD_5 of the liquid from the primary clarifier is 100 mg/L at a flow rate of 0.05 mgd. The dimensions of the aeration tank are 20 ft × 10 ft × 20 ft and the MLSS = 4000 mg/L.

a. Calculate the F/M ratio.

b. What type of activated sludge system is this?

LIST OF SYMBOLS

BOD	biochemical oxygen demand, in mg/L
DO	dissolved oxygen
F/M	food-to-microorganism ratio
MLSS	mixed liquor suspended solids, in mg/L
P	phosphorus, in mg/L
SS	suspended solids, in mg/L
SVI	sludge volume index
t	time, in sec or days
$\bar{t}$	retention time, in sec or days

Chapter 9

Sludge Treatment, Utilization, and Disposal

When the wastewater is treated and discharged to a watercourse, the job is not over. Left behind are the solids, suspended in water, commonly called sludge. Currently sludge treatment and disposal accounts for over 50% of the treatment costs in a typical secondary plant, making this none-too-glamorous operation an essential aspect of wastewater treatment.

This chapter is devoted to the problem of sludge treatment and disposal. The sources and quantities of sludge from various types of wastewater treatment systems are examined first, followed by a definition of sludge characteristics. Solids concentration techniques, such as thickening and dewatering, are discussed next, concluding with considerations for ultimate disposal.

SOURCES OF SLUDGE

The first source of sludge is the suspended solids (SS) that enter the treatment plant and are partially removed in the primary settling tank or clarifier. Ordinarily about 60 percent of the SS becomes *raw primary sludge*, which is highly putrescent, contains pathogenic organisms, and is very wet (about 96% water).

The removal of BOD is basically a method of wasting energy, and secondary wastewater treatment plants are designed to reduce this high-energy material to low-energy chemicals, typically accomplished by biological means, using microorganisms (the "decomposers" in ecological terms) that use the energy for their own life and procreation. Secondary treatment processes such as the popular activated sludge system are almost perfect systems except that the microorganisms convert too little of the high-energy organics to CO_2 and H_2O and too much of it to new organisms. Thus the system operates with an excess of these microorganisms, or *waste activated sludge*. As defined in the previous chapter, the mass of waste activated sludge per mass of BOD removed in secondary treatment is known as the *yield*, expressed as mass of SS produced per mass of BOD removed. Typically, the yield of waste activated sludge is 0.5 pound of dry solids per pound of BOD reduced.

Phosphorus removal processes also invariably end up with excess solids. If lime is used, the calcium carbonates and calcium hydroxyapatites are formed and must be disposed of. Aluminum sulfate similarly produces solids, in the form of aluminum hydroxides and aluminum phosphates. Even the biological processes for phosphorus removal end up with solids. The use of an oxidation pond or marsh for phosphorus removal is possible only if some organics (algae, water hyacinths, fish, etc.) are periodically harvested.

SLUDGE TREATMENT

A great deal of money could be saved, and troubles averted, if sludge could be disposed of as it is drawn off the main process train. Unfortunately, the sludges have three characteristics that make such a simple solution unlikely: They are aesthetically displeasing, they are potentially harmful, and they have too much water.

The first two problems are often solved by *stabilization*, such as *anaerobic* or *aerobic digestion*. The third problem requires the removal of water by either thickening or dewatering. The next three sections cover the topics of stabilization, thickening, and dewatering, and then ultimate disposal of the sludge.

Sludge Stabilization

The objective of sludge stabilization is to reduce the problems associated with two detrimental characteristics—sludge odor and putrescence and the presence of pathogenic organisms. Sludge may be stabilized by use of lime, by aerobic digestion, or by anaerobic digestion.

Lime stabilization is achieved by adding lime (as hydrated lime, $Ca(OH)_2$, or as quicklime, CaO) to the sludge and thus raising the pH to 11 or above. This significantly reduces odor and helps in the destruction of pathogens. The major disadvantage of lime stabilization is that it is temporary. With time (days) the pH drops and the sludge once again becomes putrescent.

Aerobic digestion is a logical extension of the activated sludge system. Waste activated sludge is placed in dedicated tanks, and the concentrated solids are allowed to continue their decomposition. The food for the microorganisms is available only by the destruction of other viable organisms and both total and volatile solids are thereby reduced. However, aerobically digested sludges are more difficult to dewater than are anaerobic sludges and are not as effective in the reduction of pathogens as *anaerobic digestion*, a process illustrated in Figure 9–1. The biochemistry of anaerobic digestion is a staged process: Solution of organic compounds by extracellular enzymes is followed by the production of organic acids by a large and hearty group of anaerobic microorganisms known, appropriately enough, as the *acid formers*. The organic acids are in turn degraded further by a group of strict anaerobes called *methane formers*. These microorganisms are the prima donnas of wastewater treatment, becoming upset at the least change in their environment, and the success of anaerobic treatment depends on maintenance of suitable conditions for the methane formers. Since

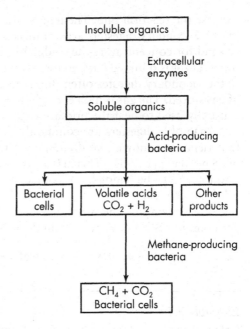

FIGURE 9–1
Generalized biochemical reactions
in anaerobic sludge digestion

they are strict anaerobes, they are unable to function in the presence of oxygen and are very sensitive to environmental conditions like pH, temperature, and the presence of toxins. A digester goes "sour" when the methane formers have been inhibited in some way and the acid formers keep chugging away, making more organic acids, further lowering the pH and making conditions even worse for the methane formers. Curing a sick digester requires suspension of feeding and, often, massive doses of lime or other antacids.

Most treatment plants have both a primary and a secondary anaerobic digester (Figure 9–2). The primary digester is covered, heated, and mixed to

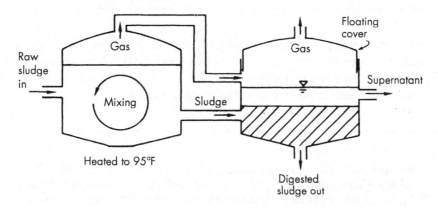

FIGURE 9–2. Anaerobic sludge digesters

increase the reaction rate. The temperature of the sludge is usually about 35°C (95°F). Secondary digesters are not mixed or heated and are used for storage of gas and for concentrating the sludge by settling. As the solids settle, the liquid supernatant is pumped back to the main plant for further treatment. The cover of the secondary digester often floats up and down, depending on the amount of gas stored. The gas is high enough in methane to be used as a fuel and in fact is usually used to heat the primary digester.

Anaerobic digesters are commonly designed on the basis of solids loading. Experience has shown that domestic wastewaters contain about 120 g (0.27 lb) of SS per day per capita. This may be translated, knowing the population served, into total SS to be handled. Of course, added to this must be the production of solids in secondary treatment, expressed as the yield of secondary solids. Experience has shown that the waste activated sludge yield is 0.5 pound of dry solids per pound of BOD destroyed. Pounds of BOD can be calculated as

$$\text{pounds per day} = \text{flow in mgd} \times \text{concentration in mg/L} \times 8.34$$

Example 9.1

Calculate the secondary (waste activated) solids production for a 10-mgd plant that achieves 60% BOD reduction in its biological treatment. Assume that the influent BOD is 200 mg/L and that the primary clarifier removes 30% of the BOD.

The BOD entering the secondary treatment step is $0.70 \times 200 = 140$ mg/L.

If 60% of this is destroyed in the secondary (aeration) step, the BOD destroyed is 140 mg/L $\times$ 0.60 = 84 mg/L.

The BOD reduction is then 84 mg/L $\times$ 10 mgd $\times$ 8.34 = 7005 lb/day.

Waste activated sludge produced = 0.5 lb/day $\times$ 7005 lb/day = 3502 lb/day.

Note that the units are taken care of by the conversion factor 8.34 as long as the flow is in million gallons per day and the concentration is in milligrams per liter.

Once the solids production is calculated, the digester volume is estimated by assuming a reasonable loading factor such as 4 kg of dry solids per cubic meter of digester volume per day (kg/m^3-day) (0.27 lb/ft^3 $\times$ day). This loading factor is decreased if a higher reduction of volatile solids is desired.

Example 9.2

Raw primary and waste activated sludge at 4% solids is to be anaerobically digested at a loading of 3 kg/m^3 $\times$ day. The total sludge produced in the plant is 1500 kg of dry solids per day. Calculate the required volume of the primary digester and the hydraulic retention time.

The production of sludge requires

$$\frac{1500 \text{ kg/day}}{3 \text{ kg/m}^3\text{-day}} = 500 \text{ m}^3 \text{ disgester volume} \qquad (9.1)$$

The total mass of wet sludge pumped to the digester is

$$\frac{1500 \text{ kg/day}}{0.04} = 37,500 \text{ kg/day} \qquad (9.2)$$

and since 1 L of sludge weighs about 1 kg, the volume of sludge is 37,500 L/day or 37.5 m³/day and the hydraulic residence time is

$$t = (500 \text{ m}^3)/(37.5 \text{ m}^3/\text{day}) = 13.3 \text{ days}$$

The production of gas from digestion varies with temperature, solids loading, solids volatility, and other factors. Typically, about 0.6 m³ of gas/kg of volatile solids added (10 ft³/lb) has been observed. This gas is about 60% methane and burns readily, usually being used to heat the digester and answer additional energy needs within a plant. It has been found that an active group of methane formers operates at 35°C (95°F) in common practice, and this process has become known as *mesophilic digestion*. As the temperature is increased to about 45°C (115°F), however, another group of methane formers predominates, and this process is tagged *thermophilic digestion*. Although the latter process is faster and produces more gas, the necessary elevated temperatures are more difficult and expensive to maintain.

All three stabilization processes reduce the concentration of pathogenic organisms, but to varying degrees. Lime stabilization achieves a high degree of sterilization, owing to the high pH. Further, if quicklime (CaO) is used, the reaction is exothermic and the elevated temperatures assist in the destruction of pathogens. Although aerobic digestion at ambient temperatures is not very effective in the destruction of pathogens, anaerobic digesters have been well studied from the standpoint of pathogen viability, since the elevated temperatures should result in substantial sterilization. Unfortunately, *Salmonella typhosa* organisms and many other pathogens can survive digestion, and polio viruses similarly survive with little reduction in virulence. Therefore, an anaerobic digester cannot be considered a method of sterilization.

Sludge Thickening

Sludge thickening is a process in which the solids concentration is increased and the total sludge volume is correspondingly decreased, but the sludge still behaves like a liquid instead of a solid. Thickening commonly produces sludge solids concentrations in the 3% to 5% range, whereas the point at which sludge begins to have the properties of a solid is between 15% and 20% solids. Thickening also implies that the process is gravitational, using the difference between particle and fluid densities to achieve the compaction of solids.

The advantages of sludge thickening in reducing the volume of sludge to be handled are substantial. With reference to Figure 9–3, a sludge with 1% solids thickened to 5% results in an 80% volume reduction (since 5% = 1/20). A concentration of 20% solids, which might be achieved by mechanical dewatering (discussed in the next section), results in a 95% reduction in volume, with resulting savings in treatment, handling, and disposal costs.

Two types of nonmechanical thickening operations are presently in use: the *gravity thickener* and the *flotation thickener*. The latter also uses gravity to separate the solids from the liquid, but for simplicity we continue to use both descriptive terms.

A typical gravity thickener is shown in Figure 9–4. The influent, or feed, enters in the middle, and the water moves to the outside, eventually leaving as the clear effluent over the weirs. The sludge solids settle as a blanket and are removed out the bottom.

A flotation thickener, shown in Figure 9–5, operates by forcing air under pressure to dissolve in the return flow and releasing the pressure as the return is mixed with the feed. As the air comes out of the solution, tiny bubbles attach themselves to the solids and carry them upward, to be scraped off.

Sludge Dewatering

Dewatering differs from thickening in that the sludge should behave as a solid after it has been dewatered. Dewatering is seldom used as an intermediate process unless the sludge is to be incinerated and most wastewater plants use dewatering as a final method of volume reduction before ultimate disposal.

In the United States, the usual dewatering techniques are sand beds, pressure filters, belt filters, and centrifuges. Each of these is discussed in the following paragraphs.

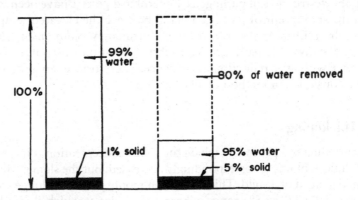

FIGURE 9-3. Volume reduction owing to sludge thickening

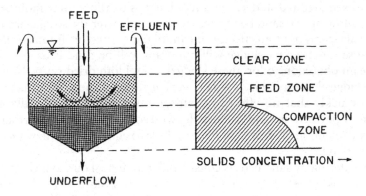

FIGURE 9-4. Gravity thickener

Sand beds have been used for a great many years and are still the most cost-effective means of dewatering when land is available. The beds consist of tile drains in sand and gravel, covered by about 26 cm (10 in) of sand. The sludge to be dewatered is poured on the sand. The water initially seeps into the sand and tile drains. Seepage into the sand and through the tile drains, although important in the total volume of water extracted, lasts only for a few days. The sand pores are quickly clogged, and all drainage into the sand ceases. The mechanism of evaporation takes over, and this process is actually responsible for the conversion of liquid sludge to solid. In some northern areas, sand beds are enclosed in greenhouses to promote evaporation as well as to prevent rain from falling into the beds.

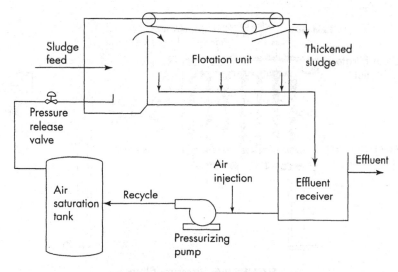

FIGURE 9-5. Flotation thickener

For mixed digested sludge, the usual design is to allow three months of drying time, allowing the sand bed to rest for a month after the sludge has been removed. This seems to be an effective means of increasing the drainage efficiency.

Because raw primary sludge will not drain well on sand beds and will usually have an obnoxious odor, these sludges are seldom dried on beds. Raw secondary sludges have a habit of either seeping through the sand or clogging the pores so quickly that no effective drainage takes place. Aerobically digested sludges may be dried on sand, but usually with some difficulty. In northern areas, sludges are intentionally frozen in *freezing beds* to enhance their dewatering after the spring thaw.

If dewatering by sand beds is considered impractical due to lack of land and high labor costs, mechanical dewatering techniques must be used. Three common mechanical dewatering processes are pressure and belt filtration and centrifugation.

The *pressure filter*, shown in Figure 9–6, uses positive pressure to force the water through a filter cloth. Typically, the pressure filters are built as plate-and-frame filters, in which the sludge solids are captured between the plates and frames, which are then pulled apart to allow for sludge cleanout.

The *belt filter*, shown in Figure 9–7, operates as both a pressure filter and a gravity drainage. As the sludge is introduced onto the moving belt, the free water drips through the belt but the solids are retained. The belt then moves into the dewatering zone, where the sludge is squeezed between two belts. These machines are quite effective in dewatering many different types of sludges and are being installed in many small wastewater treatment plants.

Centrifugation became popular in wastewater treatment only after organic polymers were available for sludge conditioning. Although the centrifuge will work on any sludge, most unconditioned sludges cannot be centrifuged with

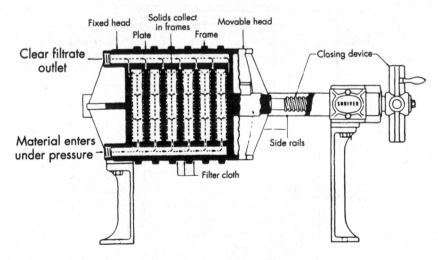

FIGURE 9-6. Pressure filters

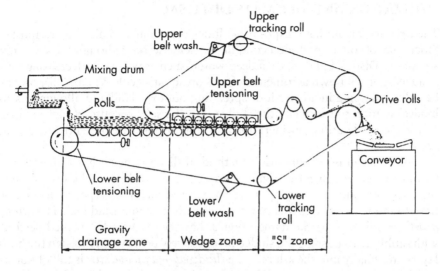

FIGURE 9-7. Belt filter

greater than 60% or 70% solids recovery. The centrifuge most widely used is the solid bowl decanter, which consists of a bullet-shaped body rotating on its axis. The sludge is placed in the bowl, and the solids settle out under about 500 to 1000 gravities (centrifugally applied) and are scraped out of the bowl by a screw conveyor (Figure 9–8). Although laboratory tests are of some value in estimating centrifuge applicability, tests with continuous models are considerably better and highly recommended whenever possible.

The solids concentration of the sludge from sand drying beds can be as high as 90% after evaporation. Mechanical devices, however, will produce sludge ranging from 15% to 35% solids.

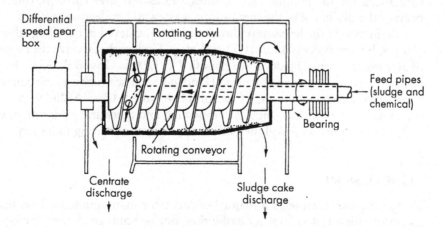

FIGURE 9-8. Solid bowl centrifuges

UTILIZATION AND ULTIMATE DISPOSAL

The options for ultimate disposal of sludge are limited to air, water, and land. Strict controls on air pollution complicate incineration, although this certainly is an option. Disposal of sludges in deep water (such as oceans) is decreasing owing to adverse or unknown detrimental effects on aquatic ecology. Land disposal may be either dumping in a landfill or spreading out over land and allowing natural biodegradation to assimilate the sludge into the soil. Because of environmental and cost considerations, incineration and land disposal are presently most widely used.

Incineration is actually not a method of disposal at all but rather a sludge treatment step in which the organics are converted to H_2O and CO_2 and the inorganics are oxidized to nonputrescent ash residue. Two types of incinerators have found use in sludge treatment: multiple hearth and fluid bed. The *multiple hearth incinerator*, as the name implies, has several hearths stacked vertically, with rabble arms pushing the sludge progressively downward through the hottest layers and finally into the ash pit. The *fluidized bed incinerator* is full of hot sand and is suspended by air injection; the sludge is incinerated within the moving sand. Owing to the violent motion within the fluid bed, scraper arms are unnecessary. The sand acts as a "thermal flywheel," allowing intermittent operation.

When sludge is destined for disposal on land and the beneficial aspects of such disposal are emphasized, sludge is often euphemistically referred to as *biosolids*. The sludge has nutrients (nitrogen and phosphorus), is high in organic content, and, as discussed, is full of water. Thus, its potential as a soil additive is often highlighted. However, both high levels of heavy metals, such as cadmium, lead, and zinc, as well as contamination by pathogens that may survive the stabilization process, can be troublesome.

Heavy metals entering the wastewater treatment plant tend to concentrate on the sludge solids, and thus far, we have found no effective means of removing these metals from the sludge prior to sludge disposal. Control must therefore focus on maintaining strict rules (industrial *pretreatment* rules) that prevent the discharge of the metals into wastewater collection systems.

Reduction in the levels of pathogens is often achieved in the sludge digestion process, but the process is not 100% effective. Sludges that receive the equivalent of 30 days anaerobic digestion are classified by EPA as Class B sludges which can be disposed of only on nonagricultural land (golf courses, highway median strips), but a 30-day delay in any use of the land is required. Class A sludges are disinfected by other processes, such as composting and quicklime addition, in which high temperatures act to kill the pathogens, or by nonionizing radiation.

CONCLUSION

Sludge disposal represents a major headache for many municipalities because its composition reflects our style of living, our technological development, and our ethical concerns. "Pouring things down the drain" is our way of getting rid of all manner of unwanted materials, not recognizing that these materials often become part of the sludge that must be disposed of in the environment. All of us need to become more sensitive to these problems and keep potentially harmful materials out of our sewage system and out of sludge.

PROBLEMS

9.1   Calculate the dry tons (1 ton = 2000 lb) per day of raw primary sludge that would be produced by a community of 100,000.

9.2   What sludge characteristics would be important if sludge from a wastewater treatment plant were to be

a. Placed on the White House lawn
b. Dumped into a trout stream
c. Sprayed on a playground
d. Sprayed on a vegetable garden

9.3   A sludge is thickened from 2000 mg/L to 17,000 mg/L. What is the reduction in volume, in percent?

9.4   Explain, in your own words, how a sludge dewatering centrifuge works.

9.5   A 50-mgd secondary wastewater treatment plant uses large aeration basins for BOD reduction. The influent BOD is 120 mg/L, and the primary clarifiers reduce this by 25%. The discharge permit requires that the effluent BOD be less than 20 mg/L. What will be the production of secondary sludge?

LIST OF SYMBOLS

BOD   biochemical oxygen demand, in mg/L
EPA   U.S. Environmental Protection Agency
SS   suspended solids

become part of the sludge that must be disposed of. If the emphasis shifts to medical care or more attention to these problems, the land, or open bodies of water, as opposed to a sewer system and treatment.

PROBLEMS

1. Calculate the dry weight in kg of sludge per capita of raw primary sludge that would be produced by a community of 100,000.

2. What sludge characteristics or sludge quantity or usage result from wastewaters that may result be:

   a. Present as waterborne solids
   b. Compacted to a dry slurry
   c. Stored on a playground
   d. Sprayed on a vacant land area

3. A sludge is thickened from 2000 to 15,000 mg/l. What is the reduction in volume in percent?

4. Explain in your own words how a sludge digester operates.

5. A mixing secondary wastewater treatment plant uses large aeration basins to 1000 volumes. The influent BOD is 230 mg/l, containing an slurry after removing the grit. The discharge permit require the treated effluent BOD to be only 20 mg/l. What would be the production of expanded sludge load per...

LIST OF SYMBOLS

BOD   Biochemical oxygen demand in mg/l
EPA   U.S. Environmental Protection Agency
SS   suspended solids

Chapter 10

Nonpoint Source Water Pollution

When the source of water pollution is an identifiable pipe discharge, such as the effluent from a wastewater treatment facility of an industrial plant, that source is labeled *point source pollution. Nonpoint source pollution* on the other hand comes from disperse overland flow associated with rain events.

As rain falls and strikes the ground, a complex runoff process begins, and nonpoint source water pollution is the unavoidable result. Even before people entered the picture, the rains came, raindrops picked up soil particles, muddy streams formed, and major watercourses became clogged with sediment. Witness the formation of the Mississippi River delta, which has been forming for tens of thousands of years. We can safely surmise that, even before humanity, rivers were contaminated by this natural series of events; sediment clogged fish gills and fish probably died. This natural runoff is classified as "background" nonpoint source runoff and is not generally labeled as "pollution."

Now view the world as it has been since the dawn of humankind—a busy place where human activities continue to influence the environment. For millennia, these activities have included farming, harvesting trees, constructing buildings and roadways, mining, and disposal of liquid and solid wastes. Each activity has led to disruptions in the surface of the earth's soil or has involved the application of chemicals to the soil. Increased transport of soil particles, with consequently increased sediment loading to watercourses and the application of chemicals to the soil, is generally labeled as pollution. In this chapter, we address:

- The runoff process
- Loading functions for sediment, a critical pollutant
- Control technologies applicable to nonpoint source pollution

Table 10–1 gives a list of nonpoint source categories, including sources ranging from agricultural practices to air pollution fallout to "natural" background. The focus of this chapter is the five major activities of concern: agriculture, urban stormwater, construction, silviculture, and urban stormwater runoff.

TABLE 10–1. Relative Importance of Pollutant Concentrations

NPS Category	Suspended Solids/Sediment	BOD	Nutrients	Toxic Metals	Pesticides	Pathogens	Salinity/TDS	Acids	Heat
Urban storm runoff	M	L-M	L	H	L	H	M	N	N
Construction	H	N	L	N-L	N	N	N	N	N
Highway de-icing	N	N	N	N	N	N	H	N	N
In-stream hydrologic modification	H	N	N	N-H	N	N	N-H	H	N
Non-coal mining	H	N	N	M-H	N	N	M-H	H	N
Agriculture									
Nonirrigated crop production	H	M	H	N-L	H	N-L	N	N	N
Irrigated crop production	L	L-M	H	N-L	M-H	N	H	N	N
Pasture and range	L-M	L-M	H	N·	N	N-L	N-L	N	N
Animal production	M	H	M	N-L	N-L	L-H	N-L	N	N
Forestry									
Growing	N	N	L	N	L	N-L	N	N	N
Harvesting	M-H	L-M	L-M	N	L	N	N	N	M
Residuals management	N-L	L-H	L-M	L-H	N	L-H	N-H	N-H	N
Onsite sewage disposal	L	M	H	L-M	L	H	N-L	N	N
Instream sludge accumulation	H	H	M-H	L-H	M	L	N	N	N
Direct precipitation	N	N	N-M	L	L	N-L	N	N-M	N
Air pollution fallout	M	L	L-M	L-H	L-M	N-L	N	L-M	N
"Natural" background	L-H	L-H	M	N-M	N	N-L	N-H	N-H	N-M

Key: N = Negligible; L = Low; M = Moderate; H = High; TDS = Total dissolved solids

THE RUNOFF PROCESS

The complex runoff process includes both the detachment and the transport of soil particles and chemical pollutants. For the remainder of this chapter, we simply include all human-induced soil erosion and chemical applications under the term *nonpoint source pollution*. Chemicals may be bound to soil particles or be soluble in rainwater; in either case, water movement is the prime mode of transport for solid and chemical pollutants. The characteristics of the rain indicate the ability of the rainwater to splash and detach the pollutants. This rain energy is defined by droplet size, velocity of fall, and the intensity characteristics of the particular storm.

Soil characteristics influence both the detachment and transport processes. Pollutant detachment is a function of an ill-defined motion of soil stability, since size, shape, composition, and strength of soil aggregates and soil clods all act to determine how readily the pollutants are detached from the soil to begin their movement to streams and lakes. Pollutant transport is influenced by the permeability of the soil to water, or the ease by which water passes through the soil, and this helps determine the infiltration capabilities and drainage characteristics of the surface receiving the rainfall. Pollutant transport is also a function of soil porosity, or the fraction of open space between the soil grains, which affects storage and movement of water, and soil surface roughness, which tends to create a potential for temporary and long-term detention of the pollutants.

Slope factors also influence pollutant transport. The slope gradient, as well as slope length, influences the flow and velocity of runoff, which in turn influences the quantity of pollutants that are moved from the soil to the watercourse.

Land cover conditions are another influence on the detachment and transport of pollutants. Vegetative cover helps to

- Provide protection from the impact of raindrops, thus reducing detachment.
- Make the soil aggregates less susceptible to detachment by protecting soil from evaporation and thus keeping the soil moist.
- Furnish roots, stems, and dead leaves that help slow overland flow and hold pollutant particles in place.

Only a portion of the pollution detached and transported from upland regions in a watershed is actually carried all the way to a stream or lake. In many cases, significant portions of the materials are deposited at the base of slopes or floodplains. The portion of the pollution detached, transported, and actually delivered from its source to the receiving waterway is defined as the *delivery ratio*. When chemical pollutants are involved, the whole spectrum of factors that determine reaction rates acts to limit the delivery ratio: temperature, times of transport, presence of other chemicals, and presence of sunlight, to name just a few. Whenever sediment or chemical pollutants become a problem, the list of physical factors becomes quite long, and includes:

- *Magnitude of pollutant sources.* Whenever the quantity of sediment pollution (sediment, phosphorus, nitrogen, pesticides, herbicides) available for transport is greater than the capacity of the runoff transport system, disposition will occur and the delivery ratio will be decreased.
- *Proximity of pollution sources to receiving waterways.* Pollutants trapped in runoff often move only short distances and, owing to factors such as surface roughness and slope, may be deposited far from the lake or stream. Areas close to a receiving waterway, or areas where channel-type erosion takes place, may be characterized with a relatively high delivery ratio.
- *Velocity and volume of water.* The characteristics of the pollutant transport system, particularly the velocity and volume of water from a given storm, affect the delivery ratio. A small storm may not supply enough water to carry a load of pollution to a lake or stream, resulting in a zero or very low delivery ratio. A large, lengthy rainfall may have the opposite effect and transport a very large portion of the pollution that is detached from its source to the receiving waterway.

By understanding rainfall characteristics, soil properties, slope factors, and vegetative covers, the loads of different nonpoint source pollutants to lakes and rivers can be predicted and possibly controlled.

CONTROL TECHNIQUES APPLICABLE TO NONPOINT SOURCE POLLUTION

Historically, control over municipal and industrial point sources of pollution has received considerable federal and corporate attention through construction grants and permit programs. However, public and private investment to significantly reduce point source pollution may be ill spent in cases in which water quality is governed instead by nonpoint source discharges. Two of the most important nonpoint sources of water pollution are runoff from construction sites and runoff from paved urban areas.

Construction

Erosion of soil at construction sites will not only cause water quality problems offsite but may be regarded as the loss of a valuable natural resource. Home buyers expect a landscaped yard, and lost topsoil is often costly for the contractor to replace. Builders of houses, highways, and other construction view soil erosion as a process that must be controlled in order to maximize economic return.

When construction is planned, controlled clearing of the proposed construction site is considered, so that the area to be disturbed during the construction and site restoration phases may be held to a minimum. Environmentally sensitive areas must be designated, and if any clearing is required in those areas, it should be limited as much as possible. Such critical areas include steep slopes, unaggregated soils like sand, natural sediment ponds, natural waterways, including intermittent streams, and floodplains. The planning phase of construction should also consider an erosion control system, including planned access as well as techniques for use in the operational and site restoration phases of the construction activity.

During construction, several pollution abatement techniques appear to be effective. Velocity regulation methods attempt to reduce the rate at which water moves over the construction site. Velocity reduction minimizes particle uptake by the water and may lead to particle deposition when the velocity reduction is sufficient. The result is an overall decrease in erosion. There are alternative methods of achieving velocity reduction, and all include the application of some material to the exposed soil, such as hay bales in ditches, plastic barriers, filter inlets, jute mesh, seeding, fertilizing, and mulching. The best method, and its corresponding cost, must be determined on a site-by-site basis.

Stormwater deflection methods attempt to reduce the amount of water passing over a construction site by diverting it. However, a diversion dike cannot limit or control runoff generated by rain falling directly on the site. Stormwater-channeling methods attempt to control the movement of falling rainwater through the site. Chutes, flumes, and flexible downdrains are effective in certain areas, but their costs are quite high and their outfall must be handled to ensure that it does not form a secondary pollution source.

Restoration of a site after construction is necessary if water pollution is to be controlled. Effective revegetation usually requires regrading as well as seeding, fertilizing, and mulching. Costs of revegetation depend on regrading costs, the type of mulch needed, and the slope (steeper slopes are more expensive to revegetate). Wood fiber mulch is effective only on relatively flat terrain, and more permanent excelsior mats and jute netting may be necessary.

Urban Stormwater Runoff

Rain falling on paved surfaces does not percolate into the ground, but runs off in storm sewers, carrying pollutants from these surfaces with it. The methods for controlling urban stormwater runoff include nonstructural housekeeping practices, such as litter control, as well as structural collection and treatment systems like settling tanks and possibly even secondary treatment. No single control can be used in all locations and situations since the factors affecting the choice of controls for a given site include

- Type of sewerage system (separate or combined)
- Status of development (planned, developing, or established urban area)
- Land use (residential, commercial, or industrial)

The number of pollution sources may be reduced by control at the planning stage. Street litter, often high in nitrogen and phosphorus, may be reduced by passage and enforcement of anti-littering regulation. Air pollution can be a source of water pollution when enough of it settles on surfaces in cities, particularly on streets and rooftops, so air pollution abatement planning can effectively reduce potential air pollution. Transportation residues like oil, gas, and grease from cars, and particulates from deteriorating road surfaces may be reduced by transportation planning, selection of road surfaces less susceptible to deterioration, and automobile exhaust inspection programs. Preventive actions may be taken as part of land use planning strategies to reduce potential runoff pollution, such as avoiding development in environmentally sensitive areas or in areas where urban runoff is an existing problem. Floodplain zoning, one type of land use regulation, often creates a buffer strip that is effective in reducing urban runoff pollution by filtering solids from overland flows and by stabilizing the soils of the floodplain.

Several control techniques prevent the buildup of pollutants on streets, parking lots, and other urban surfaces. If such buildup can be prevented, the total pollutant loading and the concentration of pollutants in the first rainfall flushing of an area can be reduced. The high concentration of pollutants in this first flush contributes heavily to the poor water quality of urban runoff. Street cleaning methods include sweeping, vacuuming, and flushing. Street sweeping, the oldest and least expensive technique and one still used in most cities, reduces soil loading in runoff but fails to pick up finer particles, which are often the more significant source of pollution (biodegradable and toxic substances, metals, and nutrients). Street vacuuming is more efficient in collecting the small particles, but is more expensive. Street flushing is effective in cleaning the street, and it flushes surface pollutants into storm sewers that empty into catch basins. Periodic catch basin cleaning removes refuse and other solids from catch basins. Significant reductions in biodegradables, nutrients, and other pollutants may result from regular cleaning. The basins may be cleaned by hand or by vacuum.

Urban stormwater runoff pollution may also be controlled after it enters the stormwater drainage system. Detention systems reduce runoff pollutant loading by retarding the rate of runoff and by encouraging the settling of suspended solids. These systems range from low-technology controls, such as rooftop storage, to intermediate technology controls, like small detention tanks interspersed in the collection network. In general, the size and number of units are directly proportional to the effectiveness of the system. Detention basins act as settling tanks and can be expected to remove 30% of the biodegradable compounds and 50% of the suspended solids in the stormwater.

In storage and treatment systems, the first flush of an area is retained in the collection network, in a storage unit, or in a flow equalization basin. The stormwater is then treated at a nearby wastewater treatment facility when the sanitary flow volume and the design capacity of the facilities allow. Several innovative storage methods exist, such as storing the stormwater in the drainage network, routing the flow by a computerized network of dams and drainage networks (in Seattle), or digging a subterranean storage tunnel (in Chicago). The effectiveness of these systems depends on the quantity of pollutants captured, the sizing of the storage units, and the extent of treatment received at the local wastewater facility. Storage capacity is the most critical factor. A debate exists on what size storm should be used as the design standard: a one-month, six-month, or less common storm. The larger the design storm, the more costly and (usually) more effective the system.

CONCLUSION

Nonpoint sources contribute major pollutant loadings to waterways. Control techniques are readily available, but vary considerably in both cost and effectiveness and are typically implemented through a generally confused institutional framework. This institutional setting sometimes poses an obstacle to nonpoint source abatement.

In addition, costs of control for abatement of nonpoint source pollution vary widely and are site-specific. Planning controls (preventive measures), however, are relatively inexpensive when compared with operational and site restoration controls (remedial measures) and thus may be the most efficient way to control nonpoint pollution.

PROBLEMS

10.1   Use Table 10–1 and rank the concentrations of the different non-point source pollutants coming from construction sites. How do you rank these same pollutants in terms of harm to the environment? Compare and discuss both rankings.

10.2   Onsite wastewater treatment and disposal systems are criticized for creating nonpoint source water pollution. Describe the pollutants that come from individual households, and identify alternatives to help control the problems.

10.3   Discuss the wastewater treatment technologies in Chapter 8 that are particularly applicable to control pollution from urban stormwater runoff.

10.4   Estimate the cost of controlling runoff from the construction of a five-mile highway link near your home town. Assume that bales of hay are sufficient if they are coupled with burlap barriers at key locations. No major collection and treatment facilities are required. Document your assumptions, including labor and materials charges.

10.5   Sediment pollution along the shorelines of rivers and lakes can be a particularly troublesome problem. Suggest a design for your grandparents' Lake Michigan home in northern Wisconsin.

Chapter 11

Water Pollution Law and Regulations

A complex system of laws requires industries and towns to treat wastewater flows before discharge to receiving waterways. In this system, *common law* and *statutory law* are intertwined to form the legal basis for pollution control.

The American legal tradition is based on common law, a body of law vastly different from statutory law as written by Congress and state governments. This common law is the aggregate body of decisions made in courtrooms as judges decide individual cases. An individual or group of individuals damaged by water pollution or any other wrong (the plaintiff) historically could seek relief in the courtroom in the form of an injunction to stop the polluter (the defendant) or in the form of payment for damages.

Court rulings in these cases were, and are, based on *precedents*. The underlying theory of precedents is that if a similar case or cases were brought before any court in the past, then the present-day judge violates the rules of fair play if the present-day case is not decided in the same manner and for the same party that the precedent cases dictated. Similar cases, defined to be so by the judge, theoretically have similar endings. If no precedent exists, then the plaintiff essentially rolls the dice in hopes of convincing the court to make a favorable precedent-setting decision.

Statutory law, on the other hand, is a set of rules mandated by a representative governing body, be it the Congress of the United States or a state legislature. Such legislation supplements or changes the effect of existing common law in areas in which Congress or state legislatures perceive shortcomings. For example, environmental quality in general, and public health in particular, were continually harmed under the common laws as they related to dirty water. Common law courts were taking years to reflect changed societal conditions because the courts were bound to precedents set during times in the nation's history when clean water was plentiful and essentially free. Finally, Congress decided to take the initiative with a series of laws aimed at abating water pollution and cleaning the surface waters of the nation.

In this chapter, we discuss the evolution of water law from the common law courtrooms, through the legislative chambers of Congress, to the administrative offices of the U.S. Environmental Protection Agency (EPA) and similar state agencies.

COMMON LAW

To date, common law has concerned itself with the disposition of surface water rather than groundwater. There are two major theories of common law as it applies to water. One theory, labeled the *riparian doctrine*, says that conflicts between plaintiffs and defendants must be decided by the ownership of the land underlying or adjoining a body of surface water (i.e., riparian land). The second theory, known as the *prior appropriations doctrine*, takes a different focus and simply states that water use is rationed on a first-come, first-served basis, regardless of land ownership. Note that in both of these doctrines, the focus is on water quantity—on deciding how to apportion a finite body of clean surface water. Common law is unclear about water quality considerations.

The principle underlying the riparian doctrine is that water is owned by the owner of the land underlying or adjoining the stream and that the owner generally is entitled to use the water as long as quantity is not depleted nor quality degraded.

The doctrine has a somewhat confused history. It was originally introduced to the New World by the French and was adopted by several colonies. The English court system eventually adopted the theory as common law in several court cases, and thus it eventually became the official law of the colonies. In colonial courtrooms, the riparian doctrine was a workable concept. The landowner was entitled to use the water for domestic purposes, such as washing and watering stock, but common law held that water could not be sold to nonriparian parties, simply because the water in the stream or river would be diminished in quantity and the downstream user would then not have access to the total flow. Even at the present time, in sparsely populated farming areas where water is plentiful, this system is still applicable. In urban and more densely populated areas, courts generally found that the riparian doctrine could not be applied in its purest form. Accordingly, several variations or ground rules were developed in the courts, based on the *principle of reasonable use* and the *concept of prescriptive rights*.

The *principle of reasonable use* holds that a riparian owner is entitled to reasonable use of the water, taking into account the needs of other riparians. Reasonable use is defined on a case-by-case basis by the courts. Obviously, this opens tremendous loopholes, which have been used in numerous litigations. Possibly the most famous example is the case of *New York City vs. the States of Pennsylvania and New Jersey*. In the 1920s, New York City began to pipe drinking water from the upper reaches of the Delaware River, and as the city grew the demand increased until the people downstream from these impoundments found that their rivers had disappeared. Many resort owners simply went out of business. After prolonged court battles it was finally determined that since the city did own the land around the impounded streams, and the use of this water was "reasonable," the city could continue to use the water. There was some monetary compensation for the downstream riparian owners, but in retrospect the failure of common law is quite evident.

The *concept of prescriptive rights* has evolved to the point where, if a riparian owner does not use the water and an upstream user "openly and notoriously" abuses the water quantity or quality, the upstream user is entitled to continue this practice. This concept holds that, through lack of use, the downstream riparian forfeits the water rights.

This concept was established in a famous case in 1886: *Pennsylvania Coal Co. vs. Sanderson.* Anthracite coal mines north of Scranton, Pennsylvania, at the headwaters of the Lackawanna River, were polluting the river and eventually made it unfit for aquatic life and human consumption. Mrs. Sanderson, a riparian landowner, built a house near the river before the polluted water quality conditions became noticeable. Her intent was to live there indefinitely, but the water quality soon deteriorated, eliminating her opportunities to benefit from the resource. She took the mining company to court and lost. The court basically held that the use of the river as a sewer was "reasonable," since the company had been in operation before Mrs. Sanderson built her house. Since water pollution was a necessary result of coal mining, the coal company could continue its open and notorious practice. This illustrates another example in which common water law broke down in its ability to serve the people.

Although historically important, the riparian doctrine is declining in use. It is, after all, only applicable to sparsely populated areas with no severe water supply and water quality problems. Most of its applicability is limited to areas east of the Mississippi River where there is sufficient rainfall to enable the system to work.

The other important water law concept is the *prior appropriations doctrine*, which states that water users "first in time" are necessarily "first in right." In other words, if one user puts surface water to some "beneficial use" before another user, the first user is guaranteed that quantity of water for as long as the use demands. Land ownership and user location, upstream or downstream, are irrelevant.

The doctrine began in the mid-1800s as gold miners and ranchers in the arid western United States sought to stake claims to water in the same manner that they staked mining claims. A farmer or rancher who irrigated had a particular concern that development of an upstream farm or ranch would reduce or eliminate stream flow to which he had a prior claim. The Reclamation Act of 1902, which provided cheap irrigation water to develop the West, made this need particularly critical.

Since the flow of most streams is highly variable, it is possible to own a water right and a dry stream bed simultaneously. This conflict is resolved under the appropriation doctrine by prior claim. For example, if a user has first claim to 1 million gallons per day (mgd), a second claimant has 3 mgd, and the third has 2 mgd, as long as the river flows at 6 mgd everyone is happy. If the flow drops to 4 mgd, the third claimant is completely out of business. If the third claimant happens to be upstream from claimants 1 and 2, the 4 mgd of water must be permitted to flow past that claimant's water intake. Many states that abide by the prior appropriations doctrine nonetheless permit withdrawal of water for personal use such as drinking, cooking, and washing by the holder of a junior right.

The Colorado River and its tributaries are completely appropriated. To set aside water for oil shale and uranium mining in the Colorado Basin, the developing corporations had to purchase water rights. The cities located along the Front Range of the Rocky Mountains—Denver, Laramie, Colorado Springs, Pueblo, and so on—have purchased water rights west of the Front Range of the Rocky Mountains and divert water through tunnels under the mountains. Albuquerque, Phoenix, and Tucson have also purchased water rights for urban development and divert water from the Rio Grande, Salt River, and lower Colorado drainages. Albuquerque arguably could have riparian rights as well to the Rio Grande, which flows through the city. The looming water shortage in the Colorado Basin has led to suggestions that water be diverted from the Columbia River to the Colorado, or even from the Yukon-Charlie river system in northern Canada to the Colorado. In 1968, Congress enacted a national water policy that prohibits these massive diversions, and this policy is still in force. Clearly, a doctrine of water conservation is needed as a supplement to the prior appropriations doctrine.

As the quantity of surface water decreases, the monetary value of water rights, which can be bought and sold, increases, leading to the mining of groundwater. Overappropriation and overuse of irrigation water result in concentration of pollutants in that water. The recycling of irrigation water in the Colorado River has resulted in an increase in dissolved solids and salinity so great that, at the Mexican border, Colorado River water is no longer fit for irrigation.

As is the case with the riparian doctrine, the prior appropriations doctrine says very little about water quality. Under the modified appropriations doctrine, the upstream user who is senior in time generally may pollute. If the downstream user is senior, the court in the past directed payments for losses. If the cost of cleanup is greater than the downstream benefits, however, courts have often found it "reasonable" to allow the pollution to continue. Under the appropriations doctrine, a downstream owner who is not actually using the water has no claim whatsoever.

The common law theories of public and private nuisance have been found to apply in certain cases. However, nuisance has found more application in air pollution control and is discussed in some detail in Chapter 22.

STATUTORY LAW

Citing the shortcomings in common law and continued water pollution problems, Congress and state governments have passed a series of laws designed to clean the surface waters across the nation. Although most states had some laws regulating water quality, it was not until 1965 that a concerted push was made to curb water pollution. In that year Congress passed the Water Quality Act, which among other provisions required each state to submit a list of water quality standards and to classify all streams by these standards.

Most states adopted a system similar to the *ambient water quality stream classifications* shown in Table 11–1. Streams were classified by their anticipated maximum beneficial use. This allowed some states to classify certain streams as low-quality waterways and others as virgin trout streams. The method of stream classification theoretically forces the states to limit industrial and municipal discharges, and prevents a stream from decaying further. As progress in pollution control is made, the classifications of various streams may be improved. However, lowering a stream classification is generally not allowed by state and federal regulatory agencies.

To attain the desired water quality, restrictions on wastewater discharges are necessary. Such restrictions, known as *effluent standards*, have been used by various levels of government for many years. For example, an effluent standard for all pulp and paper mills may require the discharge not to exceed 50 mg/L BOD. The total loading of the pollution (in pounds of BOD per unit time) or its effect on a specific stream is thus not considered. Using perfectly "reasonable" effluent standards, it is still possible that the effluent from a large mill, although it meets the effluent standards, completely destroys a stream. On the other hand, a small mill on a large river, which may in fact be able to discharge even untreated effluent without producing any appreciable detrimental effect on the water quality, must meet the same effluent standards.

This dilemma may be resolved by developing a system in which minimum effluent standards are first set for all discharges and then modified based on the actual effect the discharge would have on the receiving watercourse. For example, the large mill cited may have to meet a BOD effluent standard of 50 mg/L, but because of severe detrimental impact on the receiving water quality, this may be reduced to 5 mg/L BOD. This concept requires that each discharge be considered on an individual basis, a process spelled out in the 1972 Federal Water Pollution Control Act, which also established a nationwide policy of zero discharge by 1985. Congress mandated that EPA ensure that all waste be removed before discharge to a receiving waterway. Clearly, this and other goals have not been met. The control mechanism to achieve a reduction in pollution was EPA's prohibition of any discharge of pollutants into any public waterway unless authorized by a permit. The permit system, known as the National Pollutant Discharge Elimination System (NPDES), is administered by EPA, with direct permitting power transferred to states able to convince the agency that the state administering body has the authority and expertise to conduct the program. In Wisconsin, for example, the state government has been granted the authority to administer the Wisconsin Pollutant Discharge Elimination System (WPDES).

In situations in which an industry wishes to discharge into a municipal sewerage system, the industry must agree to contractual arrangements developed with the local governments to ensure compliance with federal industrial

TABLE 11–1. Typical State Stream Standards

| | | Fresh Waters | | | | Marine Waters | | | |
Class	Best Use	Minimum DO (mg/L)	Maximum Temperature (°C)	pH Range	Coliforms per 100 mL	Minimum DO (mg/L)	Maximum Temperature (°C)	pH Range	Coliforms per 100 mL
AA	All, fisheries	9.5	16	6.5–8.5	50	7.0	13	7.0–8.5	14
A	All, fisheries	8.0	18	6.5–8.5	100	6.0	16	7.0–8.5	14
B	No fish spawning	6.5	21	6.5–8.6	200	5.0	19	7.0–8.5	100
C	Fish passage, boating	—	24	6.5–9.0	—	4.0	22	6.5–9.0	200
Lake	All	Natural conditions			50				

pretreatment requirements. For selected industries, pretreatment guidelines are being developed that require facilities to treat their wastewater flows before discharge to municipal sewer systems as discussed later in this chapter.

The 1977 amendments to the Clean Water Act, in recognition of the limits of technology and management of wastewater treatment systems to achieve a zero discharge, proposed that eventually all discharges be treated by "best conventional pollutant control technology," even though this would not be 100 percent removal. In addition, EPA is now responsible for setting effluent limits to a list of about 100 toxic pollutants, which must be controlled by "best available technology" (BAT).

EPA also promulgated NPDES regulations requiring permits for selected discharges of stormwater. Cities, counties, and industries are identified that must complete stormwater permit applications. BCTs—best conventional control technologies—and receiving water quality based controls will be necessary, depending on the pollutants found in the stormwater. Combined sewer overflows (CSOs) also require permits under NPDES.

An industrial facility has two choices in the disposal of wastewater:

- Discharge to a watercourse—in which case an NPDES permit is required and the discharge will have to be continually monitored.
- Discharge to a public sewer.

The latter method may be preferable if the local publicly owned treatment works (POTW) have the capacity to handle the industrial discharge and if the discharge contains nothing that will poison the biological treatment processes in the POTW secondary treatment system. However, some industrial discharges may cause severe treatment problems in the POTWs, and so tighter restrictions on what industries may and may not discharge into public sewers have become necessary. This has evolved into what is now known as the *pretreatment* program.

Pretreatment Guidelines

Under pretreatment regulations developed by EPA any municipal facility or combination of facilities operated by the same authority, with a total design flow greater than 5 mgd and receiving pollutants from industrial users, is required to establish a pretreatment program. Discharge, reporting, and permitting obligations are imposed on all significant individual wastes (SIW) discharging into POTWs. The EPA regional administrator may require that a municipal facility with a design flow of 5 mgd or less develop a pretreatment program if it is found that the nature or the volume of the industrial effluent disrupts the treatment process, causes violations of effluent limitations, or results in the contamination of municipal sludge.

In addition to these general pretreatment regulations, EPA is developing specific regulations for the 34 major industries listed in Table 11–2. The regulations for each industry are designed to limit the concentration of certain pollutants

TABLE 11–2. Industries Whose Effluents Require Pretreatment

Adhesives	Mining
Aluminum working	Paint and ink manufacture
Batteries	Paving and roofing
Coated coil fabrication	Petroleum refining
Copper working	Pharmaceutical manufacture
Electroplating	Plastics
Enameled products	Plastics/synthetics manufacture
Explosives fabrication	Printing
Foundries	Pulp and paper
Gum and wood	Rubber manufacture
Iron and steel working	Soaps and detergents
Laundries	Textiles
Leather tanning	Thermal electric generation
Machinery	Timber processing

that may be introduced into sewerage systems by the respective industries. The standards require limitations on the discharge of pollutants that are toxic to human beings as well as to aquatic organisms. Examples include cadmium, lead, chromium, copper, nickel, zinc, and cyanide.

Table 11–3 summarizes the effect of pretreatment on the sludge quality at the Northeast Water Pollution Control Plant in Philadelphia. Note that although the *identified* industrial contribution of cadmium is only 25%, the effect of the pretreatment regulations reduced the cadmium concentration in the sludge by 90%.

Sludge Disposal

The problem of disposing of residuals (sludges) presents a totally different problem for regulators. In this case there is no discharge that can be readily sampled, nor are there clear-cut parameters that can cause problems. In response to this

TABLE 11–3. Reduction in Heavy Metal Concentration in Sludge

Metallic	Industrial Contribution (%)	Sludge Reduction (%)
Cd	25	90
Cr	47	89
Cu	24	64
Ni	23	73
Pb	12	88
Zn	22	76

problem, EPA has promulgated sludge disposal standards that basically divide sludges into three classifications:

"A" Sludge—sludge that has been completely disinfected and has low metal concentrations.

"B" Sludge—sludge having been treated to a point where the level of pathogens equals that typically achieved by 30 days of anaerobic digestion.

"C" Sludge—sludge that has not been treated in any way.

Actually, the regulations do not even mention "C" sludge because it is illegal to dispose of such sludges in any manner. The most likely disposal of "A" sludge is land disposal. In fact, it is an excellent fertilizer and soil conditioner, and many farms, golf courses, and other large land areas will gladly accept it.

The disposal of "B" sludges is more problematical. There are severe restrictions on the placement of such sludges on pasture land or farm land, such as not being able to use the pasture for 30 days after application. The most likely disposal method for "B" sludges is by dedicated land—that is, land dedicated solely to sludge disposal. Sludge is either sprayed on or injected into the land and allowed to assimilate into the soil. The process can be repeated forever, and the soil will get better and better. Communities that have available land often find this method the most economical for their sludge disposal.

Finally, it is possible to burn sludge in sludge incinerators and produce an ash that obviously meets all of the "A" sludge pathogen standards. Since the metals are concentrated, however, there may be concern about the disposal of this ash on farmland. Typically, sludge incinerator residue is placed in solid waste landfills.

Drinking Water Standards

Drinking water standards are equal if not more important to public health than stream standards. These standards have a long history. In 1914, faced with the questionable quality of potable water in the towns along their routes, the railroad industry asked the U.S. Public Health Service (USPHS) to suggest standards that characterize drinking water. As a result, the first USPHS drinking water standards were born. There was no law passed to require that all towns abide by these standards, but it was established that interstate transportation would not be allowed to stop at towns that could not provide water of adequate quality. Over the years most water supplies in the United States have not been closely regulated, and the high-quality water provided by municipal systems has been as much the result of the professional pride of water industry personnel as of any governmental restrictions.

Because of a growing concern with the quality of some urban water supplies and reports that not all waters are as pure and safe as people have always assumed, the federal government passed the Safe Drinking Water Act in 1974. This law authorizes EPA to set minimum national drinking water standards.

Some of the EPA published standards, which are quite similar to the USPHS water standards, are shown in Table 11–4. Potable water standards used to describe these contaminants may be divided into three categories: physical, bacteriological, and chemical.

Physical standards include color, turbidity, and odor, all of which are not dangerous in themselves but could, if present in excessive amounts, drive people to drink other, perhaps less safe, water.

Bacteriological standards are in terms of coliform, the indicator of pollution by wastes from warm-blooded animals. Present EPA standards call for a concentration of coliform of less than 1/100 mL of water. This is a classic example of how the *principle of expediency* is used to set standards. Before modern water treatment plants were commonplace, the bacteriological standard stood at 10 coliform/100 mL. In 1946, this was changed to the present level of 1/100 mL. In reality, with modern methods we can attain about 0.01 coliform/100 mL, and this will doubtless be a future standard.

Chemical standards include a long list of chemical contaminants, beginning with arsenic and ending with zinc. Two classifications exist, the first a suggested limit, the latter a maximum allowable limit. Arsenic, for example, has a suggested limit of 0.01 mg/L. From experience, this concentration has been shown to be a safe level even when ingested over an extended period. The maximum allowable arsenic level is 0.05 mg/L, which is still under the toxic threshold but close enough to create public health concern. On the other hand, some chemicals such as chlorides have no maximum allowable limits since at concentrations above the suggested limits the water becomes unfit to drink on the basis of taste or odor.

TABLE 11–4. Selected EPA Drinking Water Standards

Standard	Suggested Standard (mg/L)	Maximum Allowable (mg/L)
Physical		
Turbidity	5 units	
Color	15 units	
Odor	3 (threshold)	
Bacteriological		
Coliform	1 coliform/100 mL	
Chemical		
Arsenic	0.01	0.05
Chloride	250.	
Copper	1.	
Cyanide	0.01	0.2
Iron	0.3	
Phenols	0.001	
Sulfate	250.	
Zinc	5.	

At present, the only legislation that directly protects groundwater quality is the Safe Drinking Water Act. Increasing pollution of groundwater from landfill leachate and inadequately stabilized waste sites is a matter for public concern. Products of the anaerobic degradation of synthetic materials are found in groundwater in increasing concentration. Some provisions of the Resource Conservation and Recovery Act (RCRA), particularly the provision prohibiting landfill disposal of organic liquids and pyrophoric substances, also provide groundwater protection.

CONCLUSION

Over the years, the battles for clean water have moved from the courtroom, through the chambers of Congress, to the administrative offices of EPA and state departments of natural resources. The strengths and weaknesses of water pollution law are not unique to the United States. Throughout central and eastern Europe, for example, massive problems exist because of (1) pollution from agricultural runoff, including soil, nitrates, pesticides, and industrial contamination by toxic organic compounds and metals; and (2) discharge of untreated or poorly treated wastewater having high levels of BOD, nutrients, and suspended solids. Governments worldwide are both successful and unsuccessful with different legal and economic systems and address similar problems differently.

In the United States, permitting systems have replaced inconsistent, one-case-at-a-time judicial proceedings as ambient water quality standards and effluent standards are sought. Tough decisions lie ahead as current water programs are administered, particularly NPDES permitting for polluters discharging to waterways and the pretreatment guidelines for polluters discharging to municipal sewer systems. Even tougher decisions must be faced in the future as regulations are developed for the control of toxic substances.

PROBLEMS

11.1   Describe the NPDES reporting requirements for the local wastewater treatment facility in your home town. What data are required, how often are summary forms completed, and what agency reviews the data on the forms?

11.2   Health departments often require that chlorine be added to water as it enters municipal distribution systems. Discuss the benefits and risks associated with this requirement, and describe alternative ways to ensure potable water at the household tap.

11.3   Many industrial processes are water intensive. That is, to produce a product that will sell in the marketplace, many gallons of water must flow into the factory. Develop a sample listing of such industries, and discuss the legal and administrative problems generally associated with securing this water for new factories. Compare and contrast these problems with respect to the generally wet eastern states and the dry western states.

11.4   Federal regulations are designed to achieve "zero discharge" of pollutants from point sources located along surface waterways. Land application of liquid waste is an option often proposed in many sections of the nation. Discuss the advantages and disadvantages of such systems, particularly in terms of heavy metal pollutants, and outline possible restrictions on land where such wastes have been applied.

11.5   Assume you work for EPA and are assigned to propose a standard for the allowable levels of arsenic for household drinking water. What data do you collect, where do you go to get the data (literature or laboratory), and in what professions do you seek experts to help guide you?

LIST OF SYMBOLS

BAT	best available technology
BCT	best conventional control technology
BOD	biochemical oxygen demand
CSO	combined sewer overflow
EPA	U.S. Environmental Protection Agency
NPDES	National Pollutant Discharge Elimination System
POTW	publicly owned (wastewater) treatment works
RCRA	Resource Conservation and Recovery Act
USPHS	U.S. Public Health Service
WPDES	Wisconsin Pollutant Discharge Elimination System

Chapter 12

Solid Waste

Solid wastes other than hazardous and radioactive materials are considered in this chapter. Such solid wastes are often called *municipal solid waste* (MSW) and consist of all the solid and semisolid materials discarded by a community. The fraction of MSW produced in domestic households is called *refuse*. The composition of refuse has been changing over the past decades. Much of the material historically has been food wastes, but new materials such as plastics and aluminum cans have been added to refuse, and the use of kitchen garbage grinders has decreased the food waste component. Most of the 2000 new products created each year by American industry eventually find their way into MSW and contribute to individual disposal problems.

The components of refuse are *garbage*, or food wastes; *rubbish*, including glass, tin cans, and paper; and *trash*, including larger items like tree limbs, old appliances, pallets, and so forth, that are not usually deposited in garbage cans.

The relationship between solid waste and human disease is intuitively obvious but difficult to prove. If a rat is sustained by an open dump, and that rat sustains a flea that transmits murine typhus to a human, the absolute proof of the pathway requires finding the particular rat and flea—an obviously impossible task. Nonetheless, we have observed more than twenty human diseases that are associated with solid waste disposal sites, and there is little doubt that improper solid waste disposal is a health hazard.

Disease vectors are the means by which disease organisms are transmitted, such as water, air, and food. The two most important disease vectors related to solid waste are rats and flies. Flies are such prolific breeders that 70,000 flies can be produced in 1 ft³ of garbage, and they carry many diseases like bacillary dysentery. Rats not only destroy property and infect by direct bite, but carry insects like fleas and ticks that may also act as vectors. The plagues of the Middle Ages were directly associated with the rat populations.

Public health is also threatened by infiltration of leachate from MSW disposal into groundwater, particularly drinking water supplies. Leachate is formed when rainwater collects in landfills, pits, waste ponds, or waste lagoons, and stays in contact with waste material long enough to leach out and dissolve some of its chemical and biochemical constituents. Leachate may be a major groundwater and surface water contaminant, particularly where there is heavy rainfall and rapid percolation through the soil.

QUANTITIES AND CHARACTERISTICS OF MUNICIPAL SOLID WASTE

The quantities of MSW generated in a community may be estimated by one of three techniques: input analysis, secondary data analysis, and output analysis. Input analysis estimates MSW based on use of a number of products. For example, if 100,000 cans of beer are sold each week in a particular community, the MSW, including litter, might be expected to include 100,000 aluminum cans per week. But obtaining waste characteristics' data from such information is often difficult and inaccurate. When possible, solid waste generation should be measured by output analysis—that is, by weighing the refuse deposited at the disposal site. Refuse must generally be weighed in any case, because fees for use of the facility (called *tipping fees*) depend on the weight of the refuse. Daily weight of refuse varies with the day of the week and the week of the year. Weather conditions also affect refuse weight, since moisture content can vary widely depending on how much rainwater enters the waste. If every truckload cannot be weighed, statistical methods must be used to estimate the total quantity from sample truckload weights.

Characteristics of Municipal Solid Waste

Refuse management depends on both the characteristics of the site and the characteristics of the MSW itself: gross composition, moisture content, particle size, chemical composition, and density.

Gross composition may be the most important characteristic affecting MSW disposal, or the recovery of materials and energy from refuse. Composition varies from one community to another, as well as with time in any one community. Refuse composition is expressed "as generated" or "as disposed," since moisture transfer takes place during the disposal process and thereby changes the weights of the various fractions of refuse. Table 12–1 shows typical components of average U.S. refuse. The numbers in the table are useful only as guidelines, as each community has characteristics that influence its solid waste production and composition.

The moisture content of MSW may vary between 15% and 30% and is usually about 20%. Moisture is measured by drying a sample at 77°C (170°F) for 24 hours, weighing, and calculating as follows:

$$M = \frac{w - d}{w} \times 100 \qquad (12.1)$$

where M = moisture content, in percent
        w = initial, wet weight of sample
        d = final, dry weight of sample

The chemical composition of typical refuse is shown in Table 12–2. The use of both proximate and ultimate analysis in the combustion of MSW and its various fractions is discussed further in Chapters 13 and 14.

TABLE 12–1. Average Annual Composition of MSW in the United States

Category	As Generated		As Disposed	
	Millions of tons	*%*	*Millions of tons*	*%*
Paper	37.2	29.2	44.9	35.3
Glass	13.3	10.4	13.5	10.6
Metal				
Ferrous	8.8	6.9	8.8	6.9
Aluminum	0.9	0.7	0.9	0.7
Other, nonferrous	0.4	0.3	0.4	0.3
Plastics	6.4	5.0	6.4	5.0
Rubber and leather	2.6	2.0	3.4	2.7
Textiles	2.1	1.6	2.2	1.7
Wood	4.9	3.8	4.9	3.8
Food waste	22.8	20.4	20.0	15.7
Miscellaneous	1.9	1.5	2.8	2.1
Total	127.3	100.0	127.3	100.0

COLLECTION

In the United States, and in most other industrialized countries, solid waste is collected by trucks. These may be open-bed trucks that carry trash or bagged refuse, but they are usually trucks that carry hydraulic rams to compact the refuse to reduce its volume so the trucks can carry larger loads (Figure 12–1). Commercial and industrial collections are facilitated by the use of containers, which are either emptied into the truck with a hydraulic mechanism or carried by truck to the disposal site. Collection is an expensive part of waste management, and many new devices and methods have been proposed in order to cut costs.

TABLE 12–2. Proximate and Ultimate Chemical Analysis of MSW

	Proximate Analysis (%)	Ultimate Analysis (%)
Moisture	15–35	
Volatile matter	50–60	
Fixed carbon	3–9	
Noncombustibles	15–25	
Higher heat value	3000–6000 Btu/lb	
Carbon		15–30
Hydrogen		2–5
Oxygen		12–24
Nitrogen		0.2–1.0
Sulfur		0.02–0.1

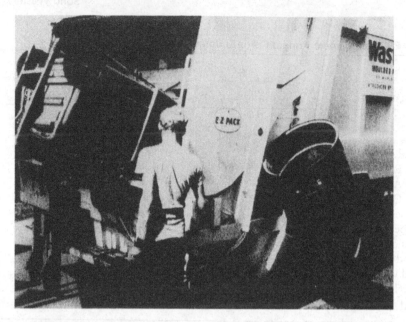

FIGURE 12–1. Packer truck used for residential refuse collection

Garbage grinders reduce the amount of garbage in refuse. If all homes had garbage grinders, the frequency of collection could be reduced. Twice-a-week collection is only needed in warm weather when garbage decomposes rapidly. Garbage grinders do put an extra load on the wastewater treatment plant, but sewage is relatively dilute and ground garbage can be accommodated in both sewers and treatment plants. The increased burden may be problematic in water-short communities.

Pneumatic pipes have been installed in some small communities, mostly in Sweden and Japan. The refuse is ground at the residence and sucked through underground lines.

Kitchen garbage compactors can reduce collection and MSW disposal costs and thus reduce local taxes, but only if every household has one. A compactor costs about as much as other large kitchen appliances, but uses special high-strength bags, so that the operating cost is also a consideration. At present they are beyond the means of many households, but stationary compactors for commercial establishments and apartment houses have already had significant influence on collection practices.

Transfer stations are part of many urban refuse collection systems. A typical system, as shown in Figure 12–2, includes several stations, located at various points in a city, to which collection trucks bring the refuse. The drive to each transfer station is relatively short, so that workers spend more time collecting and less time traveling. At the transfer station, bulldozers pack the refuse

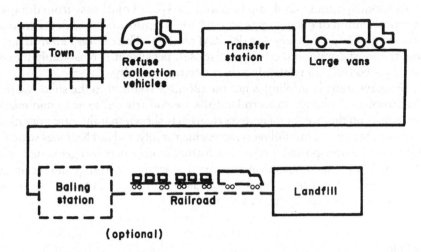

FIGURE 12–2. Transfer station method of solid waste collection

into large containers that are trucked to the landfill or other disposal facility. Alternatively, the refuse may be baled before disposal.

Green cans on wheels are widely used for transfer of refuse from the household to the collection truck. The cans are pushed to the curb by the householder and emptied into the truck by a hydraulic lift. This system saves money and has reduced occupational injuries dramatically. Garbage collection workers traditionally suffer higher lost-time accident rates than other municipal or industrial workers but the use of hydraulically lifted green cans on wheels has reduced such injuries.

Route optimization may result in significant cost saving as well as increased effectiveness. An optimal route is one in which collection takes place without wasted travel.

DISPOSAL OPTIONS

Ever since the Romans invented city dumps, municipal refuse has been disposed of outside the city walls. As cities and suburbs grew, as metropolitan areas grew contiguous, and as the use of "throwaway" packages and containers increased, finding a place for MSW disposal became a critical problem. Many cities in the United States encouraged "backyard burning" of trash in order to reduce MSW volume and disposal cost. Building codes in many cities mandated the installation of garbage grinders in new homes. Cities like Miami, Florida, that have no landfill sites at all built MSW incinerators.

Increasing urban air pollution has resulted in the prohibition of backyard burning, even of leaves and grass clippings, and the de-emphasis of municipal incineration. Spontaneous dump fires and the spread of disease from dumps led to the prohibition of open dumps after 1980, in conformance with the Resource Conservation and Recovery Act (RCRA) of 1976. The *sanitary landfill* has become the most common method of disposal, because it is reasonably inexpensive and is considered relatively sound environmentally.

Unfortunately, landfilling is not the ultimate solution to the solid waste disposal problem. Although modern landfills are constructed so as to minimize adverse effects on the environment, experience has shown that they are not fail-safe. Moreover, the cost of landfilling is increasing rapidly, as land becomes scarce and refuse must be transported further and further from where it is generated. Rising public environmental consciousness is making waste processing and reclamation of waste material and energy appear increasingly attractive. Options for resource recovery are discussed further in Chapter 14.

LITTER

Litter is unsightly, a breeding ground for rats and other rodents, and hazardous to wildlife. Plastic sandwich bags are mistaken for jellyfish by tortoises, and birds strangle themselves in the plastic rings from six-packs.

Anti-litter campaigns and attempts to increase public awareness have been ongoing for many years. Bottle manufacturers and bottlers encourage voluntary bottle return. The popularity of "Adopt-a-road" programs has also sharply increased litter awareness and has the potential to reduce roadside litter.

Restrictive beverage container legislation is a more drastic assault on litter. The Oregon "Bottle Law" prohibits pop-top cans and discourages the use of non-returnable glass beverage bottles. The law operates by placing an artificial deposit value on all carbonated beverage containers so that it is in the user's interest to bring them back to the retailer for a deposit return. The retailer in turn must recover the money from the manufacturer and sends all of the bottles back to the bottling company. The bottling company must now either discard these bottles, send them back to the bottle manufacturer, or refill them. In any case, it becomes more efficient for the manufacturer to either refill or recover the bottles rather than to throw them away. The beverage industry is thus forced to rely more heavily on returnable containers, reducing the one-way containers such as steel cans or plastic bottles. Such a process saves money, materials, and energy, and has the added effect of reducing litter.

POLLUTION PREVENTION

One means of getting a handle on questions of material and product use is not to produce the materials in the first place that end up as waste. This principle has become known as *pollution prevention*, and it is probably the wave of the future in solid waste management.

Pollution prevention actually began as recognition by some industries (e.g., 3M and DuPont) that if they produced less waste, they might actually save money. Indeed, through material inventories within a manufacturing plant, it was discovered that small and low-cost changes in the processes would save large amounts of money (and be beneficial to the environment). In the United States, virtually every large corporation is now actively engaged in pollution prevention, and the effects are beginning to show in the total production of solid waste (as well as other forms of waste).

Pollution prevention when applied to the consumer is more difficult to apply. Consumers have the power of deciding what to purchase, and by this means they affect the solid waste produced. But this decision is often difficult to make.

In order to begin estimating the effect of a material or product on the environment, it is necessary to conduct a *life-cycle analysis*. Such an analysis is a holistic approach to pollution prevention that analyzes the entire life of a product, process, or activity, encompassing raw materials, manufacturing, transportation, distribution, use, maintenance, recycling, and final disposal. In other words, life-cycle analysis should yield a complete picture of the environmental impact of a product.

Life-cycle analyses are done for several reasons, including the comparison of products for purchase and the comparison of products by industry. In the former case, it should be possible to establish the total environmental effect of returnable bottles compared to the environmental effect of nonrecyclable bottles. If all of the factors going into the manufacture, distribution, and disposal of both types of bottles is considered, one should be clearly superior.

One problem with such studies is that they are often conducted by industry groups or individual corporations, and the results often promote their own product. For example, Proctor & Gamble, the manufacturer of a popular brand of disposable baby diapers, found in a study done for them that cloth diapers consume three times more energy than the disposable kind. But a study by the National Association of Diaper Services found that disposable diapers consume 70 percent more energy than cloth diapers. The difference was in the accounting procedure. If one uses the energy contained in the disposable diaper as recoverable in a waste-to-energy facility, then the disposable diaper is more energy efficient.[1]

Life-cycle analyses also suffer from a dearth of data. It is virtually impossible to obtain some of the information critical to the calculations. For example, something as simple as the tonnage of solid waste collected in the United States is not readily calculable or measurable. And even if the data *were* there, the procedure would suffer from the unavailability of a single accounting system. Is there an optimal level of pollution, or must all pollutants be removed 100 percent (a virtual impossibility)? If there is air pollution and water pollution, how must these be compared?

---

[1]"Life Cycle Analysis Measures Greenness, But Results May Not Be Black and White," *Wall Street Journal* (28 February 1991).

A simple example of the difficulties in life-cycle analysis would be in finding the solution to the great coffee cup debate—whether to use paper or polystyrene. The answer most people would give is not to use either but instead to rely on the permanent mug. Nevertheless, there are times when disposable cups are necessary, and a decision must be made as to which type to choose. So let's use life-cycle analysis to make a decision.

The paper cup comes from trees, but cutting trees and producing paper result in environmental degradation. The foam cup comes from hydrocarbons such as oil and gas, which also results in adverse environmental impact, including the use of nonrenewable resources. The manufacture of the paper cup results in significant water pollution, with 30 to 50 kg of BOD per cup produced, while that of the foam cup contributes essentially no BOD. The paper cup's manufacture also results in the emission of chlorine, chlorine dioxide, reduced sulfides, and particulates, while that of the foam cup results in none of these. The paper cup does not require chlorofluorocarbons, but neither do the newer foam cups since the CFCs in polystyrene were phased out. However, the foam cup contributes from 35 to 50 kg per cup of pentane emissions, while the paper cup contributes none. The recyclability of the foam cup is much higher than that of the paper cup since the latter is made from several materials, including the plastic coating on the paper. They both burn well, although the foam cup produces 40,000 kJ/kg, while the paper cup produces only 20,000 kJ/kg. In the landfill, the paper cup degrades into CO_2 and CH_4, both greenhouse gases, while the foam cup is inert. Since it is inert, it will remain in the landfill for a very long time whereas the paper cup will eventually (but very slowly!) decompose. If the landfill is considered a waste storage receptacle, the foam cup is superior, since it does not participate in the reaction while the paper cup produces gases and probably leachate. If, on the other hand, the landfill is thought of as a treatment facility, then the foam cup is less desirable.

It should be obvious that though pollution prevention is a great idea, a great deal of work needs to be done in order to make decisions as to how we are going to change our lifestyle to produce the least waste, or to produce the waste that least affects the environment.

CONCLUSION

The solid waste problem has three facets: source, collection, and disposal. The first is perhaps the most difficult. A "new economy" of reduced waste, increased longevity instead of planned obsolescence, and thriftier use of natural resources is needed. Collection and disposal of refuse are discussed in the next chapter.

PROBLEMS

12.1   Walk along a stretch of road and collect the litter in two bags, one for beverage containers only and one for everything else. Calculate (a) the number of items per mile, (b) the number of beverage containers per mile, (c) the weight of litter per mile, (d) the weight of beverage containers per mile, (e) the percent of beverage containers by weight, and (f) the percent of beverage containers by count. If you are working for the bottle manufacturers, will you report your data as (e) or (f)? Why?

12.2   How would a tax on natural resource withdrawal affect the economy of solid waste management?

12.3   What effect do the following have on the quantity and composition of MSW: (a) garbage grinders, (b) home compactors, (c) nonreturnable beverage containers, (d) a newspaper strike. Make quantitative estimates of the effects.

12.4   Drive along a low-traffic measured stretch of road or highway and count the pieces of litter visible from the car. (Do this with one person driving and another counting!) Then walk along the same stretch and pick up the litter, counting the pieces and weighing the full bags. What percent of the litter by piece (and by weight if you have enough information) is visible from the car?

12.5   On a map of your campus or your neighborhood develop an efficient route for refuse collection, assuming that the truck has to travel down each street.

12.6   Using a study hall, lecture hall, or student lounge as a laboratory, study the prevalence of litter by counting the items in the waste receptacles vs. the items improperly disposed of. Vary the conditions of your laboratory in the following way (you may need cooperation from the maintenance crew):

- Day 1: normal conditions (baseline)
- Day 2: all waste receptacles removed except one
- Day 3: additional receptacles added (more than normal)

If possible, do several experiments with different numbers of receptacles. Plot the percent of material properly disposed of vs. the number of receptacles, and discuss the implications.

12.7   Argue one side of the "great coffee cup debate"—should we use disposable cups made of paper or those made of foam polystyrene? Then reflect on the use of nondisposable coffee mugs.

LIST OF SYMBOLS

CFC	chlorofluorocarbons
MSW	municipal solid waste
RCRA	Resource Conservation and Recovery Act

Chapter 13

Solid Waste Disposal

Disposal of solid wastes is defined as placement of the waste so that it no longer impacts society or the environment. The wastes are either assimilated so that they can no longer be identified in the environment, as by incineration to ash, or they are hidden well enough so that they cannot be readily found. Solid waste may also be processed so that some of its components may be recovered and used again for a beneficial purpose. Collection, disposal, and recovery are all part of the total solid waste management system, and this chapter is devoted to disposal.

DISPOSAL OF UNPROCESSED REFUSE IN SANITARY LANDFILLS

The only two realistic options for disposal are in the oceans and on land. Because the environmental damage done by ocean disposal is now understood, the United States prohibits such disposal by federal law, and many developed nations are following suit. This chapter is therefore devoted to a discussion of land disposal.

Until the mid-1970s, a solid waste disposal facility was usually a *dump* in the United States and a *tip* (as in "tipping") in Great Britain. The operation of a dump was simple and inexpensive: Trucks were directed to empty loads at the proper spot on the dump site. The piled-up volume was often reduced by setting the refuse on fire, thereby prolonging the life of the dump. Rodents, odor, insects, air pollution, and the dangers posed by open fires all became recognized as serious public health and aesthetic problems, and an alternative method of refuse disposal was sought. Larger communities frequently selected incineration as the alternative, but smaller towns could not afford the capital investment required and opted for land disposal.

The term *sanitary landfill* was first used for the method of disposal employed in the burial of waste ammunition and other material after World War II, and the concept of burying refuse was used by several Midwestern communities. The sanitary landfill differs markedly from open dumps: Open dumps are simply places to deposit wastes, but sanitary landfills are engineered operations, designed and operated according to acceptable standards (Figure 13–1).

Sanitary landfilling is the compaction of refuse in a lined pit and the covering of the compacted refuse with an earthen cover. The liner is made of plastic

FIGURE 13–1. The sanitary landfill

(typically PVC) and a layer of clay that further reduces the chance of leakage into the groundwater of the liquid produced by the landfill during the decomposition of the waste. The liquid produced is collected by pipes laid into the landfill as it is constructed. Gases produced by the decomposing waste must be collected and either vented or collected and burned. When the landfill is full, a cover must be placed on it such that the seepage of rainwater into the landfill is minimized. Vegetation must then be established on the landfill, and its effect on groundwater must be monitored by wells sunk around it. In effect, the landfill will continue to cost the community many years after the last waste is deposited.

Typically, refuse is unloaded, compacted with bulldozers, and covered with compacted soil. The landfill is built up in units called *cells* (Figure 13–2). The daily cover is between 6 and 12 inches thick depending on soil composition (Figure 13–3), and a final cover at least 2 feet thick is used to close the landfill. A land-

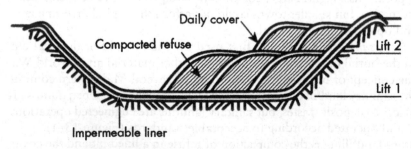

FIGURE 13–2. Arrangement of cells in an area-method landfill

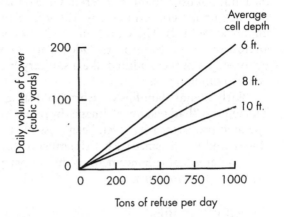

FIGURE 13–3
Daily volume of cover versus
refuse disposal rate

fill continues to subside after closure, so that permanent structures cannot be built onsite without special foundations. Closed landfills have potential uses as golf courses, playgrounds, tennis courts, winter recreation, or parks and greenbelts.

The sanitary landfilling operation involves numerous stages, including siting, design, operation, and closing.

Siting Landfills

Siting of landfills is rapidly becoming the most difficult stage of the process, since few people wish to have landfills in their neighborhoods. In addition to public acceptability, considerations include

- *Drainage:* Rapid runoff will lessen mosquito problems, but proximity to streams or well supplies may result in water pollution.
- *Wind:* It is preferable that the landfill be downwind from any nearby community.
- *Distance from collection sites.*
- *Size:* A small site with limited capacity is generally not acceptable since finding a new site entails considerable difficulty.
- *Rainfall patterns* influence the production of leachate from the landfill.
- *Soil type:* Can the soil be excavated and used as cover?
- *Depth of the water table:* The bottom of the landfill must be substantially above the highest expected groundwater elevation.
- *Treatment of leachate* requires proximity to wastewater treatment facilities.
- *Proximity to airports:* All landfills attract birds to some extent and are therefore not compatible with airport siting.
- *Ultimate use:* Can the area be used for private or public use after the landfilling operation is complete?

Although daily cover helps to limit disease vectors, a working landfill still has a marked and widespread odor during the working day. The working face of the landfill must remain uncovered while refuse is added and compacted. Wind can pick up material from the working face, and the open refuse attracts feeding flocks of birds. These birds are both a nuisance and a hazard to low-flying aircraft using nearby airports. Odor from the working face and the truck traffic to and from the landfill make a sanitary landfill an undesirable neighbor to nearby communities.

Early sanitary landfills were often indistinguishable from dumps, thereby enhancing the "bad neighbor" image. In recent years, as more landfills have been operated properly, it has even been possible to enhance property values with a closed landfill site, since such a site must remain open space. Acceptable operation and eventual enhancement of the property are understandably difficult to explain to a community.

Design of Landfills

Modern landfills are designed facilities, much like water or wastewater treatment plants. The landfill design must include methods for the recovery and treatment of the leachate produced by the decomposing refuse and for the venting or use of the landfill gas. Full plans for landfill operation must be approved by the appropriate state agencies before construction can begin.

Since landfills are generally in pits, the soil characteristics are important. Areas with high groundwater are not acceptable. The management of rainwater during landfilling operations as well as when the landfill is closed must be part of the design.

Operation of Landfills

The landfill operation is actually a biological method of waste treatment. Municipal refuse deposited as a fill is anything but inert. In the absence of oxygen, anaerobic decomposition steadily degrades the organic material to more stable forms. This process is very slow and may still be going on as long as 25 years after the landfill closes.

The liquid produced during decomposition, as well as water that seeps through the groundcover and works its way out of the refuse, is known as *leachate*. Though relatively small in volume, this liquid contains pollutants in high concentration. Table 13-1 shows typical leachate composition. Should leachate escape the landfill, its effects on the environment may be severe. In a number of instances, leachate has polluted nearby wells to a degree that they have ceased to be sources of potable water.

The amount of leachate produced by a landfill is difficult to predict. The only available method is water balance: The water entering a landfill must equal the water flowing out of the landfill—the leachate. The total water entering the top soil layer is

$$C = P(I - R) - S - E \qquad (13.1)$$

where   C = total percolation into the top soil layer, in mm
        P = precipitation, in mm
        R = runoff coefficient (fraction of precipitation that runs off)
        S = storage, in mm
        E = evapotranspiration, in mm

The percolation for three typical landfills is shown in Table 13–2.

Using these figures it is possible to predict when landfills will produce leachate. Clearly, Los Angeles landfills may virtually never produce leachate, but leaching through a 7.5-m (25-ft) deep landfill in Orlando, Florida, might take 15 years, while a 20-m (65-ft) deep landfill in Cincinnati can produce leachate after only 11 years. Leachate production depends on rainfall patterns as well as on total amount of precipitation. The figures given for Cincinnati and Orlando are typical of the "summer thunderstorm" climate that exists in most of the United States. The Pacific Northwest (west of the Pacific Coast Range) has a maritime climate, in which rainfall is spread more evenly through the year. Seattle landfills produce leachate at approximately twice the rate of that of Cincinnati landfills, although the annual rainfall amount is approximately the same.

TABLE 13–1. Typical Sanitary Landfill
Leachate Composition

Component	Typical Value
BOD_5	20,000 mg/L
COD	30,000 mg/L
Ammonia nitrogen	500 mg/L
Chloride	2000 mg/L
Total iron	500 mg/L
Zinc	50 mg/L
Lead	2 mg/L
Total polychlorinated biphenyl (PCB) residue	1.5 µg/L
pH	6.0

TABLE 13–2. Percolation in Three Landfills

Percolation Location	Precipitation P (mm)	Runoff Coefficient R	Evapotranspiration E (mm)	C (mm)
Cincinnati	1025	0.15	568	213
Orlando	1342	0.07	1172	70
Los Angeles	378	0.12	334	0

From Tenn, D.G., Haney, K.J., and Degeare, T.V., *Use of the Water Balance Method for Predicting Leachate Generation from Solid Waste Disposal Sites*, Washington, DC: U.S. Environmental Protection Agency, OSWMP, SW-168 (1975).

Gas is a second by-product of a landfill. Since landfills are anaerobic biological reactors, they produce CH_4 and CO_2. Gas production occurs in four distinct stages, as illustrated in Figure 13–4. The first stage is aerobic and may last from a few days to several months, during which time aerobic organisms are active and affect the decomposition. As the organisms use up the available oxygen, the landfill enters the second stage, when anaerobic decomposition begins but methane-forming organisms have not yet become productive. During the second stage, the acid formers cause a buildup of CO_2. The length of this stage varies with environmental conditions. The third stage is the anaerobic methane production stage, during which the percentage of CH_4 progressively increases and the landfill interior temperature rises to about 55°C (130°F). The final, steady-state condition occurs when the fractions of CO_2 and CH_4 are about equal and microbial activity has stabilized. The amount of methane produced from a landfill may be estimated using the following semi-empirical relationship:

$$+ \frac{1}{4}(4 - a - 2b + 3c)H_2O \rightarrow \frac{1}{8}[(4 - a + 2b + 3c)CO_2 \\ + (4 + a - 2b - 3c)CH_4]$$

(13.2)

Equation 13.2 is useful only if the chemical composition of the waste is known. This is not usually the case with municipal solid waste.

The rate of gas production from sanitary landfills may be controlled by varying the particle size of the refuse by shredding before it is placed in the landfill, and by changing the moisture content. Gas production may be minimized with the combination of low moisture, large particle size, and high density. Unwanted gas migration may be prevented by installing escape vents in the land-

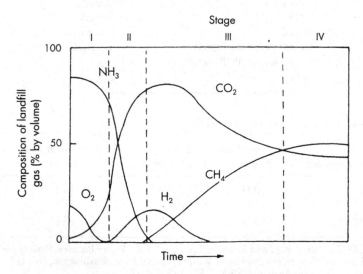

FIGURE 13–4. States in the decomposition of organic matter in landfills

fill. These vents, called "tiki torches," are kept lit and the gas is burned off as it is formed. Improper venting may lead to dangerous accumulation of methane. In 1986, a dozen homes near the Midway Landfill in Seattle were evacuated because potentially explosive quantities of methane had leaked through underground fissures into the basements. Venting of the accumulated gas, so that the occupants could return to their homes, took three years.

Since landfills produce considerable quantities of methane, landfill gas can be burned to produce electric power. Alternatively, the gas can be cleaned of CO_2 and other contaminants and used as pipeline gas. Such cleaning is both expensive and troublesome, and the most reasonable use of landfill gas is to burn it in some industrial application like brickmaking.

Closure and Ultimate Use of Landfills

Municipal landfills must be closed according to state and federal regulations. Such closure includes the permanent control of leachate and gas and the placement of an impermeable cap. The cost of closure is very high and must be incorporated in the tipping fee during the life of the landfill. This is one of the primary factors responsible for the dramatic increase in landfill tipping fees.

Biological aspects of landfills as well as the structural properties of compacted refuse limit the ultimate uses of landfills. Landfills settle unevenly, and it is generally suggested that nothing at all be constructed on a landfill for at least two years after closure, and that no large permanent structures ever be built. With poor initial compaction, about 50% settling can be expected in the first five years. The owners of the motel shown in Figure 13–5 learned this the hard way.

FIGURE 13–5. A motel built on a landfill that experienced differential settling

Landfills should never be disturbed. Disturbance may cause structural problems, and trapped gases can present a hazard. Buildings constructed on landfills should have spread footings (large concrete slabs) as foundations, although some have been constructed on pilings that extend through the fill onto rock or some other strong material.

VOLUME REDUCTION BEFORE DISPOSAL

Refuse is bulky and does not compact easily, so volume requirements of landfills are significant. Where land is expensive, the costs of landfilling may be high. Accordingly, various ways to reduce refuse volume have been found effective.

Under the right circumstances, incineration is an effective treatment of municipal solid waste. It reduces the volume of waste by a factor of 10 to 20, and incinerator ash is both more stable and more compactible than the refuse itself. Disposal of ash that concentrates heavy metal oxides may be problematic, however, and capital costs of incinerator construction are also high.

Figure 13–6 is a diagram of a large incinerator. The grapple bucket lifts the refuse from a storage pit and drops it into the charging chute. The stoker, a traveling grate in this case, moves the refuse to the furnace area. Combustion occurs both on the stoker and in the furnace. Air is fed under and over the burning refuse.

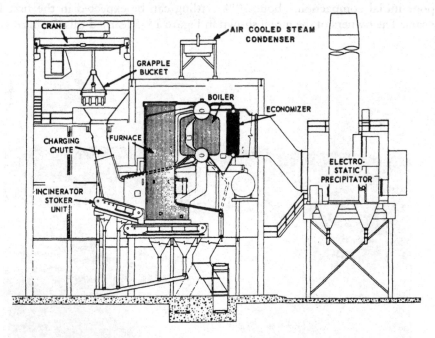

FIGURE 13–6. Schematic of a typical solid waste incinerator

The walls of the furnace are cooled by pipes filled with water for waste-to-energy production. The flue gases exit through an electrostatic precipitator (or some other device for controlling airborne pollutants) and then up the stack. In many cases, hazardous pollutants, particularly dioxin, can be emitted as discussed in Chapter 15.

CONCLUSION

This chapter began by defining the objective of solid waste disposal as the placement of solid waste so that it no longer impacts society or the environment. At one time, this was fairly easy to achieve: Dumping solid waste over city walls was quite adequate. In modern civilization, however, this is no longer possible and adequate disposal is becoming increasingly difficult.

The disposal methods discussed in this chapter are only partial solutions to the solid waste problem. Another would be to redefine solid waste as a resource and use it to produce usable goods. This is explored in the next chapter.

PROBLEMS

13.1   Suppose that the municipal garbage collectors in a town of 100,000 go on strike, and as a gesture to the community your college or university decides to accept all city refuse temporarily and pile it on the football field. If all the people dump refuse into the stadium, how many days must the strike continue before the stadium is filled to 1 yard deep? Assume the density of the refuse as 300 lb/yd^3, and assume the dimensions of the stadium as 120 yards long and 100 yards wide.

13.2   If a town has a population of 100,000, what is the approximate daily production of wastepaper?

13.3   What are some environmental impacts and effects of depositing dewatered (but sloppy wet) sludge from a wastewater treatment plant into a sanitary landfill?

13.4   Choose a place for a 25-acre landfill on the map shown in Figure 13–7. What other information do you need? Justify your selection of the site.

LIST OF SYMBOLS

A	area, in m^2
C	total percolation of rain into the soil, in mm
E	evapotranspiration, in mm
P	precipitation, in mm
PCB	polychlorinated biphenyl
R	runoff coefficient
S	storage, in mm

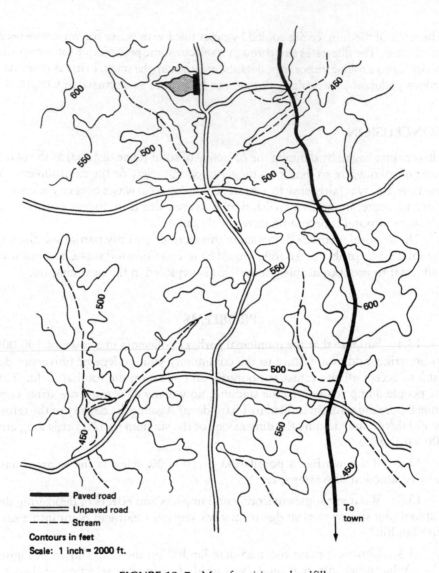

FIGURE 13–7. Map for siting a landfill

Chapter 14

Reuse, Recycling, and Recovery

Finding new sources of energy and materials is becoming increasingly difficult. Concurrently, we are finding it more and more difficult to locate solid waste disposal sites, and the cost of disposal is escalating exponentially. As a result, society's interest in reuse, recycling, and recovery of materials from refuse has grown.

Reuse of materials involves either the voluntary continued use of a product for a purpose for which it may not have been originally intended, such as the reuse of coffee cans for holding nails, or the extended use of a product, such as retreaded automobile tires. In materials reuse the product does not return to the industrial sector, but remains within the public or consumer sector.

Recycling is the collection of a product by the public and the return of this material to the industrial sector. This is very different from reuse, where the materials do not return for remanufacturing. Examples of recycling are the collection of newspapers and aluminum cans by individuals and their eventual return to paper manufacturers or aluminum companies, and the remanufacture and sale of recycled papers and aluminum cans. The recycling process requires the participation of the public, since the public must perform the separation step.

Recovery differs from recycling in that the waste is collected as mixed refuse, and then the materials are removed by various processing steps. For example, refuse can be processed by running it under a magnet that removes the steel cans and other ferrous materials. This material is then sold to the ferrous metals industry for remanufacturing. Recovery of materials is commonly conducted in a *Materials Recovery Facility* (MRF, pronounced "murph"). The difference between recycling and recovery is that in the latter the user of the product is not asked to do any separation, while in the former that crucial separation step is done voluntarily by a person who gains very little personal benefit from going to the trouble of separating out waste materials. Recycling and recovery, the two primary methods of returning waste materials to industry for remanufacturing and subsequent use, are discussed in more detail in the next section.

RECYCLING

Two incentives could be used to increase public participation in recycling. The first is regulatory, wherein the government *dictates* that only separated material will be picked up. This type of approach has had only limited success in democracies like the United States because dictation engenders public resentment.

A more democratic approach to achieve cooperation in recycling programs is to appeal to the sense of community and to growing concern about environmental quality. Householders usually respond very positively to surveys about prospective recycling programs, but the *active* response, or participation in materials separation, has been less enthusiastic.

Participation can be increased by making separation easy. The city of Seattle has a very high participation in its household recycling program because the separate containers for paper, cans, and glass are provided and the householder only needs to put the containers out on the curb. The city of Albuquerque sells, for ten cents each, large plastic bags to hold aluminum and plastic containers for recycling. The bags of recyclables, and bundled newspapers, are picked up at curbside along with garbage. Municipal initiatives like this are costly, however.

A major factor in the success or failure of recycling programs is the availability of a market for the pure materials. Recycling can be thought of as a chain, which can be pulled by the need for post-consumer materials but cannot be pushed by the collection of such materials by the public. A recycling program therefore includes, by necessity, a market for the materials collected; otherwise, the separated materials will end up in the landfill along with the mixed unseparated refuse.

In recent years there has been a strong indication that the public is willing to spend the time and effort to separate materials for subsequent recycling. What has been lacking is the markets. How can these be created? Simply put, markets for recycled materials can be created by public demand. If the public insists, for example, on buying only newspapers that have been printed on recycled newsprint, then the newspapers will be forced, in their own interest, to use recycled newsprint and this will drive up and stabilize the price of used newsprint.

Knowing this, and sensing the mood of the public, industry has been quick to produce products that are touted as being from "recycled this" and "recycled that." Most often, the term "recycled" is incorrect in such claims, since the material used has never been in the public sector. Paper, for example, has for years included fibers produced during the production of envelopes and other products. This waste paper never enters the public sector, but is an industrial waste that gets immediately used by the same industry. Although such use of materials is efficient, this is not "recycling," and such products will not drive the markets for truly recycled materials. The public has to become more knowledgeable about what are and are not legitimate recycled products, and the government may force industries to adopt standards for the use of such terms as "recycled."

RECOVERY

Most processes for separation of the various materials in refuse rely on a characteristic or property of the specific materials, and this characteristic is used to separate the material from the rest of the mixed refuse. Before such separation can be achieved, however, the material must be in separate and discrete pieces, a condition clearly not met by most mixed refuse components. An ordinary "tin can" contains steel in its body, zinc on the seam, a paper wrapper on the outside, and perhaps an aluminum top. Other common items in refuse provide equally or more challenging problems in separation.

The separation process can be facilitated by decreasing the particle size of refuse, thus increasing the number of particles and achieving a greater number of "clean" ones. Size reduction, although not strictly materials separation, is commonly a first step in a solid waste processing facility.

Size Reduction

Size reduction, or *shredding*, is brute force breaking of particles of refuse by swinging hammers in an enclosure. Two types of shredder are used in solid waste processing: vertical and horizontal hammermills, as shown in Figure 14–1. In vertical hammermills, the refuse enters at the top and must work its way past the rapidly swinging hammers, clearing the space between the hammer tips and the enclosure. Particle size is controlled by adjusting this clearance. In horizontal hammermills, the hammers swing over a grate that may be changed depending on the size of product required.

The solid waste processing facility in Figure 14–2 has a conveyor belt leading up to a vertical shredder, with a control room above and to the left. The

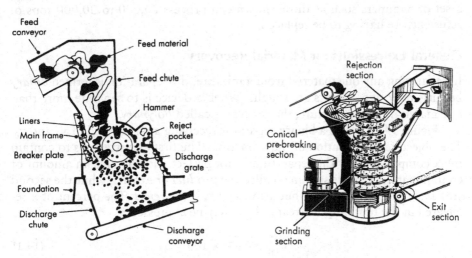

FIGURE 14–1. Vertical and horizontal hammermills

FIGURE 14–2
A shredding facility showing
the conveyor belt leading
to a vertical shredder

hammers inside the shredder are shown in Figure 14–3. As the hammers reduce the size of the refuse components, they are themselves worn down. Typically, a set of hammers such as those shown can process 20,000 to 30,000 tons of refuse before having to be replaced.

General Expressions for Material Recovery

In separating any one material from a mixture, the separation is termed *binary* because only two outputs are sought. When a device is to separate more than one material from a mixture, the process is called *polynary*.

Figure 14–4 shows a binary separator receiving a mixed feed of x_0 and y_0. The objective is separation of the x fraction: The first exit stream is to contain the x component, but the separation is not perfect and contains an amount of contamination y_1. This stream is called the *product* or *extract*, while the second stream, containing mostly y but also some x, is the *reject*. The percent of x recovered in the first output stream, $R_{(x_1)}$, may be expressed as

$$R_{(x_1)} = \frac{x_1}{x_0} \times 100 \tag{14.1}$$

FIGURE 14–3
Inside a vertical
hammermill. The
hammers have been
worn down by the
shredding process.
[Courtesy W. A.
Worrell.]

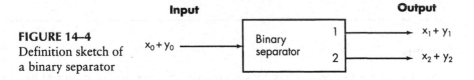

FIGURE 14–4
Definition sketch of
a binary separator

$R_{(x_1)}$ alone does not describe the performance of the binary separator adequately. If the separator were turned *off*, all of the feed would go to the first output; the extract would be $x_0 = x_1$, making $R_{(x_1)} = 100\%$. However, in this case there would have been no separation. Accordingly, the *purity* of the extract stream as percent must be considered and can be defined as

$$P_{(x_1)} = \left(\frac{x_1}{x_1 + y_1} \right) 100 \qquad (14.2)$$

A separator might extract only a small amount of pure x, so that the recovery, $R_{(x_1)}$, would also be very small. The performance of a materials separator is assessed by both recovery and purity and may thus be characterized by an additional parameter, the separator efficiency $E_{(x,y)}$, as

$$E_{(x,y)} = \left(\frac{x_1}{x_0} \times \frac{y_2}{y_0} \right)^{1/2} \times 100 \tag{14.3}$$

Example 14.1

A binary separator, a magnet, is to separate a product, ferrous materials, from a feed stream of shredded refuse. The feed rate to the magnet is 1000 kg/hr and contains 50 kg of ferrous materials. The product stream weighs 40 kg, of which 35 kg are ferrous materials. What is the percent recovery of ferrous materials, their purity, and the overall efficiency?

The variables in Equations 14.1, 14.2, and 14.3 are

$x_0 = 50$ kg	$y_0 = 1000 - 50 = 950$ kg
$x_1 = 35$ kg	$y_1 = 40 - 35 = 5$ kg
$x_2 = 50 - 35 = 15$ kg	$y_2 = 950 - 5 = 945$ kg

Then

$$R_{(x_1)} = \left(\frac{35}{50} \right) 100 = 70\%$$

$$P_{(x_1)} = \left(\frac{35}{35 + 5} \right) 100 = 88\% \tag{14.4}$$

$$E_{(x,y)} = \left(\frac{35}{50} \right) \left(\frac{945}{950} \right) 100 = 70\%$$

Screens

Screens separate material solely by size and do not identify the material by any other property. Consequently, screens are most often used in materials recovery as a classification step before a materials separation process. For example, glass can be sorted (technically but perhaps not economically) into clear and colored fractions by optical coding. However, this process requires that the glass be of a given size, and screens may be used for the necessary separation.

The *trommel*, shown in Figure 14–5, is the most widely used screen in materials recovery. The charge inside the trommel behaves in three different ways depending on the speed of rotation. At slow speeds, the trommel material is *cascading*—not being lifted but simply rolling back. At higher speed, *cataracting* occurs, in which centrifugal force carries the material up to the side and then

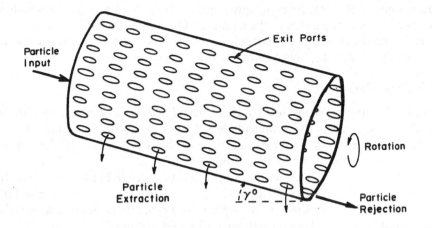

FIGURE 14–5. Trommel screen

the material falls back. At even higher speeds, *centrifuging* occurs, in which material adheres to the inside of the trommel. Obviously, the efficiency of a trommel is enhanced when the particles have the greatest opportunity to drop through the holes, and this occurs during cataracting.

Air Classifiers

Materials may be separated by their aerodynamic properties. In shredded MSW, most of the aerodynamically less dense materials are organic, and most of the denser materials are inorganic; thus air classification can produce an RDF superior to unclassified shredded refuse.

Most air classifiers are similar to the unit pictured in Figure 14–6. The fraction escaping with the air stream is the *extract* or *overflow*; the fraction falling out the bottom is the *reject* or *underflow*. The recovery of organic materials by air classification is adversely influenced by two factors:

- Not all organic materials are aerodynamically less dense, nor are all inorganic materials more dense.
- Perfect classification of more and less dense materials is difficult because of the unpredictable nature of material movement in the classifier.

Complete separation of organic from inorganic material can never occur, regardless of the chosen air velocity.

Magnets

Ferrous material may be removed from refuse with magnets, which continually extract the ferrous material and reject the remainder. Figure 14–7 shows two types of magnets.

With the belt magnet, recovery of ferrous material is enhanced by placing the belt close to the refuse, but such placement decreases the purity of the product. The depth of refuse on the belt can also pose difficulties, since the heavy

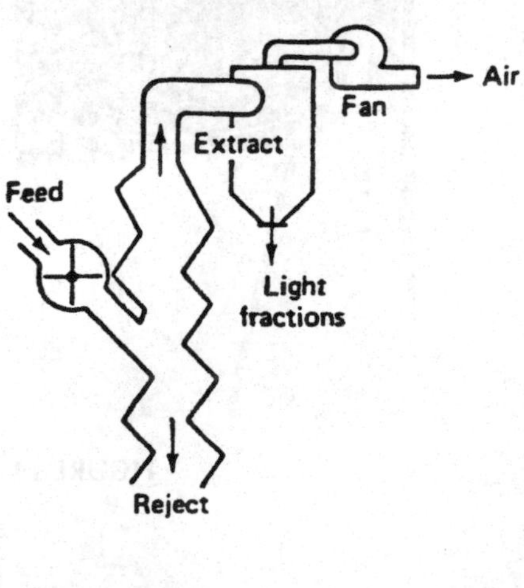

FIGURE 14–6
Air classifier

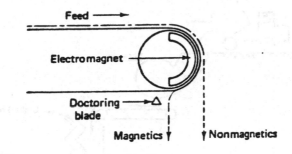

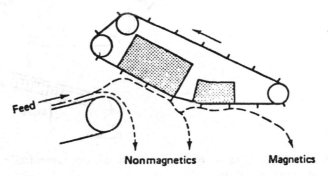

FIGURE 14–7. Two types of magnet used for ferrous recovery

ferrous particles tend to settle to the bottom of the refuse on the conveyor and are then further from the magnet than other refuse components.

Separation Equipment

Countless other unit operations for materials handling and storage have been tried. Jigs have been used for removing glass; froth flotation has been success-fully employed to separate ceramics from glass; eddy current devices have re-covered aluminum in commercial quantities; and so on. As recovery operations evolve, more and better materials separation and handling equipment will be introduced. Figure 14–8 is a diagram of a typical MRF.

ENERGY RECOVERY FROM THE ORGANIC FRACTION OF MSW

As it comes off the truck refuse is a useful fuel, and it is combusted routinely in many communities. Such facilities used to be known as incinerators, but since old incinerators were highly inefficient and grossly polluting, the refuse com-bustion industry has avoided that name. The modern facilities are known as *waste-to-energy* plants, since they not only combust the refuse but also use the heat to produce steam, which is then used to power turbines that produce

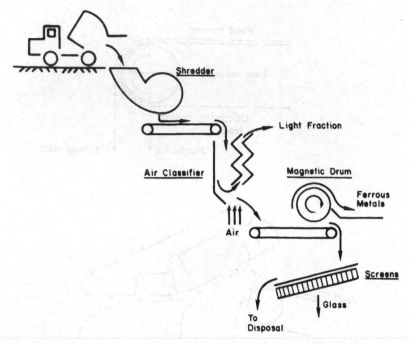

FIGURE 14–8. Diagram of a typical materials separation facility for refuse processing

electricity. Theoretically, a community should be able to produce about 20% of its total electricity needs by burning its refuse in such facilities.

An alternative method of refuse combustion is to use the shredded and separated organic fraction of refuse from materials recovery facilities. Such a fuel, called *refuse derived fuel* or RDF, may be used in existing electric generating plants as a supplement to coal or as the sole fuel in separate boilers. Figure 13–7 showed a cross-section of a boiler that could be used to recover energy.

Combustion of organic material is assumed to proceed by the reaction

$$(HC)_x + O_2 \rightarrow CO_2 + H_2O + heat \tag{14.5}$$

However, not all hydrocarbons are oxidized completely to carbon dioxide and water, and other components of the fuel, like nitrogen and sulfur, are also oxidized, by the reactions

$$N_2 + O_2 \rightarrow 2NO$$
$$2NO + O_2 \rightarrow 2NO_2 \tag{14.6}$$

$$S + O_2 \rightarrow SO_2$$
$$2SO_2 + O_2 \rightarrow 2SO_3 \tag{14.7}$$

As discussed in Chapter 18, NO_2 is an important component in the formation of photochemical smog. SO_2 is damaging to health and vegetation, and the reaction product, SO_3, forms "acid rain" by the reaction

$$SO_3 + H_2O \rightarrow H_2SO_4 \qquad (14.8)$$

Stoichiometric oxygen is the theoretical amount of oxygen required for combustion (in terms of air it is *stoichiometric air*) and is calculated from the chemical reaction, as in the following example.

Example 14.2

If carbon is combusted as

$$C + O_2 \rightarrow CO_2 + heat \qquad (14.9)$$

how much air is required? One mole of oxygen is required for each mole of carbon used. The atomic weight of carbon is 12 g/g-atom, and the molecular weight of O_2 is $2 \times 16 = 32$ g/mole. Hence, 1 gram of C requires

$$\frac{32}{12} = 2.28 \ gO_2/gC \qquad (14.10)$$

Air is 23.15% O_2 by weight. The total amount of air required to combust 1 gram of C is

$$\frac{2.28}{0.2315} = 9.87 \ g \ air \qquad (14.11)$$

The yield of energy from combustion is measured as the calories of heat liberated per unit weight of material burned. This is the *heat of combustion* or, in engineering terms, the *heat value*. Heat value is measured using a calorimeter, in which a small sample of fuel is placed in a water-jacketed stainless steel bomb under high pressure of pure oxygen and then fired. The heat generated is transferred to the water in the water jacket, and the rise in water temperature is measured. Knowing the mass of the water, the energy liberated during combustion can be calculated. In SI units the heat value is expressed as kilojoules (kJ) per kg; in British units, as British thermal units (Btu) per pound. Table 14–1 lists some heats of combustion for common hydrocarbons and gives some typical values for refuse and RDF. The rate at which heat goes into a boiler is sometimes called the *heat rate*.

Combustion of the organic fraction of refuse is not the only means of extracting useful energy. Extraction may also be by chemical or biochemical means. The cellulose fraction of RDF may be treated by *acid hydrolysis* to produce methane gas. Other chemical processes are presently being developed for

TABLE 14–1. Typical Values of Heats of Combustion

	Heat of Combustion	
Fuel	kJ/kg	Btu/lb
Carbon (to CO_2)	32,800	14,100
Hydrogen	142,000	61,100
Sulfur (to SO_2)	9,300	3,980
Methane	55,500	23,875
Residual oil	41,850	18,000
Raw refuse	9,300	4,000
RDF (air classified)	18,600	8,000

producing alcohol from RDF. Ten percent (by volume) alcohol is now required as a gasoline additive in many U.S. cities during the winter months.

COMPOSTING

Both aerobic and anaerobic decomposition can extract useful products biochemically from RDF. In the anaerobic system, refuse is mixed with sewage sludge and the mixture is digested. Operational problems have made this process impractical on a large scale, although single household units that combine human excreta with refuse have been used.

Aerobic decomposition of refuse is better known as *composting* and results in the production of a useful soil conditioner that has moderate fertilizer value. The process is exothermic and has been used at the household level as a means of producing hot water for home heating. On a community scale, composting may be a mechanized operation, using an aerobic digester (Figure 14–9), or a low-technology operation using long rows of shredded refuse known as *windrows*. Windrows are usually about 3 m (10 ft) wide at the base and 1.5 m (4 to 6 ft) high. Under these conditions, known as *static pile composting* (Figure 14–10), sufficient moisture and oxygen are available to support aerobic life. The piles must be turned periodically to allow sufficient oxygen to penetrate all parts of the pile; alternatively, air can be blown into the piles.

Temperatures within a windrow approach 50°C, entirely because of biological activity. The pH will approach neutrality after an initial drop. With most wastes, additional nutrients are not needed, but the composting of bark and other materials is successful only with the addition of nitrogen and phosphorus. Moisture must usually be controlled because excessive moisture makes maintenance of aerobic conditions difficult, while a dearth of moisture inhibits biological life. A moisture content of 40% to 60% is considered desirable.

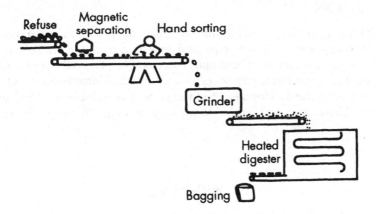

FIGURE 14–9. Mechanical composting operations

There has been some controversy over the use of inoculants, that is, freeze-dried cultures used to speed up the process. Once the composting pile is established, requiring about two weeks, the inoculants have not proved to be of any significant value. Most MSW contains sufficient organisms for successful composting, and "mystery cultures" are not needed.

The endpoint of a composting operation is reached when the temperature drops. The compost should have an earthy smell, similar to peat moss, and a dark brown color. Compost is an excellent soil conditioner, but is not yet widely used by U.S. farmers. Inorganic fertilizers are cheap and easy to apply, and most farms are located where soil conditions are good. As yet, plentiful food supplies in developed countries do not dictate the use of marginal cropland where compost would be of real value.

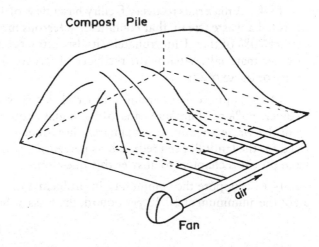

FIGURE 14–10
Static pile composting

CONCLUSION

The solid waste problem must be addressed from the point of view of source control as well as disposal. Many reuse and recycling methods are still in the exploratory stage, but they need development as land for disposal grows scarcer and more expensive, and refuse continues to accumulate. Unfortunately, we are still years away from the development and use of fully recyclable and biodegradable materials. The only truly disposable package available today is the ice-cream cone.

PROBLEMS

14.1   A power plant burns 100 ton/hr of coal. How much air is needed if 50% excess air is used? Assume coal is all carbon (C). Express your answer in lb air/lb coal.

14.2   An air classifier performance is

	Organics (kg/hr)	Inorganics (kg/hr)
Feed	80	20
Product	60	10
Reject	20	10

Calculate the recovery, purity, and efficiency.

14.3   Suppose a materials recovery facility can recover 100% of newsprint for a community (a totally unrealistic assumption, of course). Approximately what fraction of the total solid waste stream is then diverted from the landfill? Discuss your assumptions.

14.4   A materials recovery facility has a flow of 100 tons per hour. A magnet is fed a waste stream that contains 2% ferrous materials, and it manages to extract 70% of that. Unfortunately, it also extracts an (unwanted) 3 tons/hour of other materials. What is the recovery of ferrous materials, and what is the purity of the extract?

14.5   Suppose a community undertakes a recycling program and is able to divert 90% of aluminum cans, 50% of glass bottles, and 30% of newsprint from the waste stream. Approximately how much has their total waste (destined for the landfill) been reduced (as percent)? (You will have to use numbers from previous chapters to answer this question.)

14.6   Suppose the community in problem 14.5 is 100,000 people. It sells all of the aluminum at $0.20 per pound, the glass at $0.005 per pound, and the newsprint at $0.05 per pound. What will the total income be as the result of the recycling operation? If the collection of these materials cost $50 per year per household (4 people per household), what will the recycling program cost the community per capita? That is, how much will each person have to pay in additional taxes to have the recycling program?

LIST OF SYMBOLS

Btu	British thermal units
Kj	kilojoules
MRF	materials recovery facility
MSF	materials separation facililty
MSW	municipal solid waste
RDF	refuse derived fuel
SI	International System of Units

now given at $0.05 per pound. What will the cost to the city be as a result of the recycling operation? If the collection of these materials costs $20 per year per household of 4 people per household, what will the recycling program cost the community per month? That is, how much will each person have to pay in an amount they desire to have the new recycling program?

LIST OF SYMBOLS

Btu	British thermal units
KJ	kilojoules
MRF	materials recovery facility
MSW	municipal solid waste
RDF	refuse-derived fuel
Σ	summation (Greek letter of Sigma)

Chapter 15

Hazardous Waste

For centuries, chemical wastes have been the by-products of developing societies. Disposal sites were selected for convenience and placed with little or no attention to potential impacts on groundwater quality, runoff to streams and lakes, and skin contact as children played hide-and-seek in a forest of abandoned 55-gallon drums. Engineering decisions here were made by default—lack of planning for handling or processing or disposal at the corporate or plant level necessitated "quick and dirty" decisions by mid- and entry-level engineers at the end of production processes. These production engineers solved disposal problems by simply piling or dumping these waste products "out back."

Attitudes eventually began to change and air, water, and land were no longer viewed as commodities to be polluted, with the problems of cleanup freely passed to neighboring towns or future site users. Individuals responded with court actions against polluters, and governments responded with revised local zoning ordinances, updated public health laws, and new major federal clean air and water acts. In 1976, the Federal Resource Conservation and Recovery Act (RCRA) was enacted to give EPA specific authority to regulate the generation and disposal of dangerous and hazardous materials. This chapter discusses the state of knowledge in the field of hazardous waste engineering, tracing the quantities of wastes generated, their handling and processing options through transportation controls, resource recovery, and ultimate disposal alternatives.

THE MAGNITUDE OF THE PROBLEM

Over the years, the term *hazardous* has evolved in a confusing setting as different groups have advocated many criteria for classifying a waste as such. Within the federal government, different agencies use descriptions such as toxic, explosive, and radioactive to label a waste as hazardous.

The federal government has developed a nationwide classification system under the implementation of RCRA, in which a hazardous waste is defined by the degree of flammability, corrosivity, reactivity, or toxicity. This definition includes acids, toxic chemicals, explosives, and other harmful or potentially harmful waste. In this chapter, this is the applicable definition of hazardous waste.

Radioactive wastes are excluded because, although they obviously are hazardous, their generation, handling, processing, and disposal differ from those of nonnuclear hazards. The radioactive waste problem is addressed separately in Chapter 16.

The four criteria for defining hazardous waste—*flammability, corrosivity, reactivity*, and *toxicity*—must be quantified if specific materials are to be included or excluded in the list of hazardous waste. Tests have been developed for rating the flammability of a material by measuring its kindling temperature. Such materials as gasoline are obviously included on this list. Corrosivity is the ability of a chemical to react with common materials, such as sulfuric acid with steel. Reactivity is the propensity of the material to react under normal conditions, such as the reaction of sodium with water. The most difficult criterion to define is toxicity. To define toxicity we need to specify a toxicity to some organism, and how fast, and what effect is considered a toxic effect? Usually it is defined on the basis of effect on humans, although phytotoxicity (damage to plants) and bioconcentration (the ability of the chemical to be concentrated as it moves up the food chain) are equally important. Toxicity to humans is often determined on the basis of experiments on animals, and these results are expressed in terms of the death of the organisms resulting from some high dose of the toxicant. Specifically, two measures have been adopted:

- LD_{50} *(lethal dose 50)*—a calculated dose of a chemical substance that is expected to kill 50% of a population exposed through a route other than respiration (mg/kg of body weight).
- LC_{50} *(lethal concentration 50)*—a calculated concentration of a chemical substance that, when following the respiratory route, will kill 50 percent of a population during a 4-hour exposure period (ambient concentration in ppm).

LD_{50} and LC_{50} values are measured using laboratory animals to determine the relative toxicity (death) caused by selected chemicals. Those chemicals with low LD_{50} or low LC_{50} values are then considered to be toxic to humans and are *listed* as toxic chemicals and hazardous wastes. The assumption is, of course, that the toxic effect on a fish or laboratory rat, expressed as mg of toxin per kg of body weight, corresponds to the same effect on humans. In the absence of human experiments (thankfully!), this assumption will never be tested. A second way a chemical can be listed as a hazardous chemical is if it is determined to be a carcinogen. Many volatile organics, while not being identifiably toxic on contact or ingestion, are highly carcinogenic, thus they are considered hazardous.

Given this somewhat limited definition of hazardous waste, more than 60 million metric tons, by wet weight, of hazardous waste are generated annually throughout the United States. More than 60% is generated by the chemical and allied products industry. The machinery, primary metals, paper, and glass products industries each generate between 3% and 10% of the nation's total. Approximately 60% of the hazardous waste is liquid or sludge. Major generating states, including New Jersey, Illinois, Ohio, California, Pennsylvania, Texas, New York, Michigan, Tennessee, and Indiana, contribute more than 80% of the nation's total production of hazardous waste, and most of it is disposed of on the generator's property.

Most hazardous waste is generated and inadequately disposed of in the eastern portion of the United States. In this region, the climate is wet with patterns of rainfall that permit infiltration or runoff to occur. Infiltration permits the transport of hazardous waste into groundwater supplies, and surface runoff leads to the contamination of streams and lakes. Moreover, most hazardous waste is generated and disposed of in areas where people rely on aquifers for drinking water. Major aquifers and well withdrawals underlie areas where the wastes are generated. Thus, the hazardous waste problem is compounded by two considerations: The wastes are generated and disposed of in areas where it rains and in areas where people rely on aquifers for supplies of drinking water.

WASTE PROCESSING AND HANDLING

Waste processing and handling are key concerns as a hazardous waste begins its journey from the generator site to a secure long-term storage facility. Ideally, the waste can be stabilized, detoxified, or somehow rendered harmless in a treatment process similar to the following.

Chemical Stabilization/Fixation. In these processes, chemicals are mixed with waste sludge, the mixture is pumped onto land, and solidification occurs in several days or weeks. The result is a chemical nest that entraps the waste, and pollutants such as heavy metals may be chemically bound in insoluble complexes. Asphalt-like compounds form "cages" around the waste molecules, while grout and cement form actual chemical bonds with the trapped substances. Chemical stabilization offers an alternative to digging up and moving large quantities of hazardous waste, and it is particularly suitable for treating large volumes of dilute waste. Proponents of these processes have argued for building roadways, dams, and bridges with a selected cement as the fixing agent. The adequacy of the containment offered by these processes has not been documented, however, as long-term leaching and defixation potentials are not well understood.

Volume Reduction. Volume reduction is usually achieved by incineration, which takes advantage of the large organic fraction of waste being generated by many industries but which may lead to secondary problems for hazardous waste engineers: air emissions in the stack of the incinerator and ash production in the base. Both by-products of incineration must be addressed in terms of risk as well as legal and economic constraints (as must all hazardous waste treatment, for that matter). Because incineration is often considered a very good method for the ultimate disposal of hazardous waste, we discuss it in some detail later in this chapter.

Waste Segregation. Before shipment to a processing or long-term storage facility, wastes are segregated by type and chemical characteristics. Similar wastes are grouped in a 55-gallon drum or group of drums, segregating liquids like acids from solids such as contaminated laboratory clothing and equipment. Waste segregation is generally practiced to prevent undesirable reactions at disposal sites and may lead to economies of scale in the design of detoxification or resource recovery facilities.

Detoxification. Many thermal, chemical, and biological processes are available to detoxify chemical wastes. Options include:

- Neutralization
- Ion exchange
- Incineration
- Aerated lagoons
- Waste stabilization ponds

These techniques are specific; ion exchange obviously does not work for every chemical, and some forms of heat treatment may be prohibitively expensive for sludge that has a high water content.

Degradation. Methods exist that chemically degrade some hazardous wastes and render them less hazardous. Chemical degradation is a form of chemical detoxification. Waste-specific degradation processes include hydrolysis, which destroys organophosphorus and carbonate pesticides, and chemical dechlorination, which destroys some polychlorinated pesticides. Biological degradation generally involves incorporating the waste into the soil. *Landfarming*, as it has been termed, relies on healthy soil microorganisms to metabolize the waste components. Landfarming sites must be strictly controlled for possible water and air pollution that results from overactive or underactive organism populations.

Encapsulation. A wide range of material is available to encapsulate hazardous waste. Options include the basic 55-gallon steel drum (the primary container for liquids), clay, plastics, and asphalt; these materials may also serve to solidify the waste. Several layers of different materials are often recommended for the outside of the drum, such as an inch or more of polyurethane foam to prevent corrosion.

TRANSPORTATION OF HAZARDOUS WASTES

Hazardous wastes are transported across the nation on trucks, rail flatcars, and barges. Truck transportation, particularly in small trucks, is a highly visible and constant threat to public safety and the environment. There are four basic elements in the control strategy for the movement of hazardous waste from a generator—a strategy that forms the basis of U.S. Department of Transportation

(USDOT) regulation of hazardous materials transportation as set forth in Volume 49, Parts 170–180 of the Code of Federal Regulations.

Haulers. Major concerns over hazardous waste haulers include operator training, insurance coverage, and special registration of transport vehicles. Handling precautions include workers wearing gloves, face masks, and coveralls, as well as registration of handling equipment to control its future use so that, for example, hazardous waste trucks today are not used to carry produce to market tomorrow. Schedules for relicensing haulers and checking equipment are part of an overall program for ensuring proper transport of hazardous wastes. The Chemical Manufacturer's Association and USDOT operate a training program for operators of long-distance vehicles hauling hazardous materials.

Hazardous Waste Manifest. A cradle-to-grave tracking system has long been considered key to proper management of hazardous waste. A "bill of lading" or "trip ticket" ideally accompanies each barrel of waste and describes its content to its recipient. Copies of the manifest are submitted to generators and state officials so all parties know that each waste has reached its desired destination in a timely manner. This system serves four major purposes: (1) it provides the government with a means of tracking waste within a given state and of determining quantities, types, and locations where the waste originates and is ultimately disposed of; (2) it certifies that wastes being hauled are accurately described to the manager of the processing/disposing facility; (3) it provides information for recommended emergency response if a copy of the manifest is not returned to the generator; and (4) it provides a database for future planning within a state. Figure 15–1 illustrates one possible routing of copies of a selected manifest. In this example, the original manifest and five copies are passed from

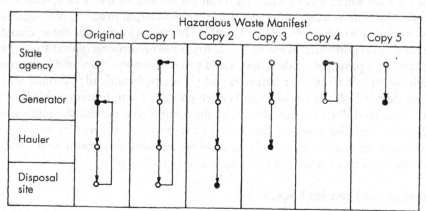

KEY: O = Transshipment point: for signature and relay with waste shipment to next location
● = Final destination: to file

FIGURE 15–1. Possible routing of copies of a hazardous waste manifest the state regulatory agency to the generator of the waste. Copies accompany each barrel of waste that leaves the generating site, and they are signed and mailed to the respective locations to indicate the transfer of the waste from one location to another.

Labeling and Placarding. Before a waste is transported from a generating site, each container is labeled and the transportation vehicle is placarded. Appropriate announcements include warnings for explosives, flammable liquids, corrosive material, strong oxidizers, compressed gases, and poisonous or toxic substances. Multiple labeling is desirable if, for example, a waste is both explosive and flammable. These labels and placards warn the general public of possible dangers and assist emergency response teams as they react in the event of a spill or accident along a transportation route.

Accident and Incident Reporting. Accidents involving hazardous wastes must be reported immediately to state regulatory agencies and local health officials. Accident reports that are submitted immediately and indicate the amount of materials released, the hazards of these materials, and the nature of the failure that caused the accident may be instrumental in containing the spilled waste and in cleaning the site. For example, if liquid waste can be contained, groundwater and surface water pollution may be avoided.

RECOVERY ALTERNATIVES

Recovery alternatives are based on the premise that one person's waste is another person's prize. What may be a worthless drum of electroplating sludge to the plating engineer may be a silver mine to an engineer skilled in metals recovery. In hazardous waste management, two types of system exist for transferring this waste to a location where it is viewed as a resource: *hazardous waste materials transfers* and *hazardous waste information clearinghouses*; in practice, one organization may display characteristics of both. The rationale behind these transfer mechanisms is illustrated in Figure 15–2. An industrial process typically has three outputs: (1) a principal product that is sold to a consumer; (2) a useful by-product available for sale to another industry; and (3) waste, historically destined for ultimate disposal. Waste transfers and clearinghouses act to minimize this flow of waste to a landfill or to ocean burial by directing it to a previously unidentified industry or firm that perceives the waste as a resource. As the regulatory and economic climate of the nation evolves, these perceptions may continue to change and more and more waste may be economically recovered.

Information Clearinghouses

The pure clearinghouse has a limited function. It offers a central point for collecting and displaying information about industrial wastes. The goal is to introduce interested potential trading partners to each other through anonymous advertisements and contacts. Clearinghouses generally do not seek customers,

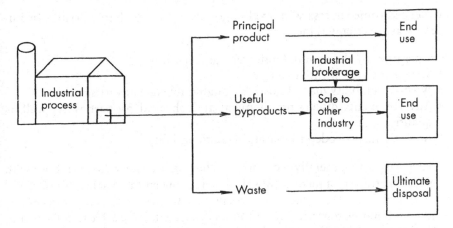

FIGURE 15–2. Rationale for hazardous waste clearinghouses and exchanges

negotiate transfers, set prices, process materials, or provide legal advice to interested parties. One major function of a clearinghouse is to keep all data and transactions confidential so trade secrets are not compromised.

Clearinghouses are also generally subsidized by sponsors, either trade or governmental. Small clerical staffs are organized in a single office or in offices spread throughout a region. Little capital is required to get these operations off the ground, and annual operation expenses are relatively low.

The value of clearinghouses should not be overemphasized. Often they are only able to operate in the short term; they evolve from an organization with many listings and active trading to a business with minimal activity as plant managers make their contacts directly with waste suppliers and short-circuit the system by eliminating the clearinghouse.

Materials Exchanges

In comparison with the clearinghouse concept, a pure materials exchange has many complex functions. A transfer agent within the exchange typically identifies both generators and potential users of the waste. The exchange will buy or accept waste, analyze its chemical and physical properties, identify buyers, reprocess the waste as needed, and sell it at a profit.

The success of an exchange depends on several factors. Initially, a highly competent technical staff is required to analyze waste flows and design and prescribe methods for processing the waste into a marketable resource. The ability to diversify is critical to the success of an exchange. Its management must be able to identify local suppliers and buyers of their products. Additionally, an exchange may even enter the disposal business and incinerate or landfill waste.

Although exchanges have been attempted with some success in the United States, they have a longer track record in Europe. Belgium, Switzerland, Germany, most of the Scandinavian countries, and the United Kingdom all have experienced some success with exchanges. The general characteristics of European waste exchanges include:

- Operation by the national industrial associations
- Services offered without charge
- Waste availability made known through published advertisements
- Advertisements discussing chemical and physical properties, as well as quantities, of waste
- Advertisements coded to maintain confidentiality

Five wastes are generally recognized as having transfer value: (1) those with a high concentration of metals, (2) solvents, (3) concentrated acids, (4) oils, and (5) combustibles for fuel. That is not to say these wastes are the only transferable items. Four hundred tons per year of foundry slag containing 50% to 60% metallic Al, 150 m³/yr of 90% methanol with trace mineral acids, and 4 tons of deep-frozen cherries were transformed from waste to resource in one European exchange. One person's waste may truly be another person's valued resource.

HAZARDOUS WASTE MANAGEMENT FACILITIES

Siting Considerations

A wide range of factors must be considered in siting hazardous waste management facilities. Some of these are determined by law: For example, RCRA prohibits the landfilling of flammable liquids. Socioeconomic factors are often the key to siting. In selecting a site, all of the relevant "-ologies" must be considered: hydrology, climatology, geology, and ecology, as well as current land use, environmental health, and transportation.

Hydrology. Hazardous waste landfills should be located well above historically high groundwater tables. Care should be taken to ensure that a location has no surface or subsurface connection, such as a crack in confining strata, between the site and a watercourse. Hydrologic considerations limit direct discharge of wastes into groundwater or surface water supplies.

Climatology. Hazardous waste management facilities should be located outside the paths of recurring severe storms, since hurricanes and tornadoes disrupt the integrity of landfills and incinerators, and cause immediate catastrophic effects on the surrounding environment and public health in the region of the facility. In addition, areas of high air pollution potential should be avoided, such as valleys where winds or inversions act to hold pollutants close to the surface of the earth and areas on the windward side of mountain ranges, that is, areas similar to the Los Angeles area, where long-term inversions are prevalent.

Geology. A disposal or processing facility should be located only on stable geologic formations. Impervious rock, which is not littered with cracks and fissures, is an ideal final liner for hazardous waste landfills.

Ecology. The ecological balance must be considered when hazardous waste management facilities are located in a region. Ideal sites in this respect include areas of low fauna and flora density, and efforts should be made to avoid wilderness areas, wildlife refuges, and animal migration routes. Areas with unique plants and animals, especially endangered species and their habitat, should also be avoided.

Alternative Land Use. Areas with low alternate land use should receive prime consideration. Areas with high recreational use potential should be avoided because of the increased possibility of direct human contact with the wastes.

Transportation. Transportation routes to facilities are a major consideration in siting hazardous waste management facilities. USDOT guidelines suggest the use of interstate and limited-access highways whenever possible. Other roads to the facilities should be accessible by all-weather highways to minimize spills and accidents during periods of rain and snowfall. Ideally, the facility should be close to the generation of the waste in order to reduce the probability of spills and accidents as wastes are transported.

Socioeconomic Factors. Factors that could make or break an effort to site a hazardous waste management facility fall under this major heading. Such factors, which range from public acceptance to long-term care and monitoring of the facility, are:

1. *Public control over the opening, operation, and closure of the facility.* Who will make policy for the facility?
2. *Public acceptance and public education programs.* Will local townspeople permit it?
3. *Land use changes and industrial development trends.* Does the region wish to experience the industrial growth that is induced by such facilities?
4. *User fee structures and recovery of project costs.* Who will pay for the facility? Can user charges induce industry to reuse, reduce, or recover the resources materials in the waste?
5. *Long-term care and monitoring.* How will postclosure maintenance be guaranteed and who will pay?

All are critical concerns in a hazardous waste management scheme.

The term *mixed waste* refers to mixtures of hazardous and radioactive wastes; organic solvents used in liquid scintillation counting are an excellent example. Siting a mixed waste facility is difficult because the laws and regulations governing the handling of chemically hazardous waste overlap and sometimes conflict with those governing the handling of radioactive waste.

Incinerators

Incineration is a controlled process that uses combustion to convert a waste to a less bulky, less toxic, or less noxious material. The principal products of incineration from a volume standpoint are carbon dioxide, water, and ash, but the products of primary concern because of their environmental effects are compounds containing sulfur, nitrogen, and halogens. When the gaseous combustion products from an incineration process contain undesirable compounds, a secondary treatment such as afterburning, scrubbing, or filtration is required to lower concentrations to acceptable levels before atmospheric release. The solid ash products from the incineration process are also a major concern and must reach adequate ultimate disposal.

The advantages of incineration as a means of disposal for hazardous waste follow:

1. Burning wastes and fuels in a controlled manner has been carried on for many years, and the basic process technology is available and reasonably well developed. This is not the case for some of the more exotic chemical degradation processes.
2. Incineration is broadly applicable to most organic wastes and can be scaled to handle large volumes of liquid waste.
3. Incineration is the best known method for disposal of "mixed waste" (see previous description).
4. Incineration is an excellent disposal method for biologically hazardous ("biohazard") wastes, like hospital waste.
5. Large expensive land areas are not required.

The disadvantages of incineration include the following:

1. The equipment tends to be more costly to operate than many other alternatives, and the process must meet the stringent regulatory requirements of air pollution control.
2. It is not always a means of ultimate disposal in that normally an ash remains that may or may not be toxic but that in any case must be disposed of properly and with minimal environmental contamination.
3. Unless controlled by air pollution control technology, the gaseous and particulate products of combustion may be hazardous to health or damaging to property.

The decision to incinerate a specific waste depends on the environmental adequacy of incineration as compared with other alternatives and on the relative costs of incineration and other environmentally sound alternatives.

The variables that have the greatest effect on the completion of the oxidation of wastes are waste combustibility, residence time in the combustor, flame temperature, and the turbulence present in the reaction zone of the incinerator. The combustibility is a measure of the ease with which a material may be oxidized in a combustion environment. Materials with a low flammability limit, a low flash point, and low ignition and autoignition temperatures may be combusted in a less severe oxidation environment, that is, at a lower temperature and with less excess oxygen.

Of the three "T's" of good combustion—time, temperature, and turbulence—only the *temperature* may be readily controlled after the incinerator unit is constructed, by varying the air-to-fuel ratio. If solid carbonaceous waste is to be burned without smoke, a minimum temperature of 760°C (1400°C) must be maintained in the combustion chamber. Upper temperature limits in the incinerator are dictated by the refractory materials available to line the inner wall of the burn chamber. Above 1300°C (2400°F) special refractories are needed.

The degree of *turbulence* of the air for oxidation with the waste fuel will affect the incinerator performance significantly. In general, both mechanical and aerodynamic means are utilized to achieve mixing of the air and fuel. The completeness of combustion and the time required for complete combustion are significantly affected by the amount and the effectiveness of the turbulence.

The third major requirement for good combustion is *time*. Sufficient time must be provided to the combustion process to allow slow-burning particles or droplets to burn completely before they are chilled by contact with cold surfaces or the atmosphere. The amount of time required depends on the temperature, fuel size, and degree of turbulence achieved.

If the waste gas contains organic materials that are combustible, incineration should be considered as a final method of disposal. When the amount of combustible material in the mixture is below the lower flammable limit, it may be necessary to add small quantities of natural gas or other auxiliary fuel to sustain combustion in the burner. Economic considerations are critical in the selection of incinerator systems because of the high costs of these additional fuels.

Boilers for some high-temperature industrial processes may serve as incinerators for toxic or hazardous carbonaceous waste. Cement kilns, which must operate at temperatures in excess of 1400°C (2500°F), can use organic solvents as fuel, providing an acceptable method of waste solvent and waste oil disposal.

Incineration is also a possibility for the destruction of liquid wastes. Liquid wastes are of two types from a combustion standpoint: combustible liquids and partially combustible liquids. Combustible liquids include all materials having sufficient calorific value to support combustion in a conventional combustor or burner. Noncombustible liquids cannot be treated by incineration and include materials that would not support combustion without the addition of auxiliary fuel and would have a high percentage of noncombustible constituents such as water. To support combustion in air without the assistance of an auxiliary fuel, the waste must generally have a heat content of 18,500 kJ/kg to 23,000 kJ/kg (8000–10,000 Btu/lb) or higher. Liquid waste having a heating value below 18,500 kJ/kg (8000 Btu/lb) is considered a partially combustible material and requires special treatment.

When starting with a waste in liquid form, it is necessary to supply sufficient heat for vaporization in addition to raising it to its ignition temperature. For a waste to be considered combustible, several rules of thumb should be used. The waste should be pumpable at ambient temperature or capable of being pumped after heating to some reasonable temperature level. Since liquids vaporize and react more rapidly when finely divided in the form of a spray, atomizing nozzles are usually used to inject waste liquids into incineration equipment whenever the viscosity of the waste permits atomization. If the waste cannot be pumped or atomized, it cannot be burned as a liquid but must be handled as a sludge or solid.

The design of an incinerator for a partially combustible waste should ensure that the waste material is atomized as finely as possible, to present the greatest surface area for mixing with combustion air, and that adequate combustion air is provided to supply the oxygen required for oxidation or incineration of the organic present. In addition, the heat from the auxiliary fuel must be sufficient to raise the temperature of the waste and the combustion air to a point above the ignition temperature of the organic material in the waste.

Incineration of wastes that are not pure liquids but might be considered sludge or slurries is also an important waste disposal problem. Incinerator types applicable for this kind of waste would be fluidized bed incinerators, rotary kiln incinerators, and multiple hearth incinerators.

Incineration is not a total disposal method for many solids and sludge because most of these materials contain noncombustibles and have residual ash. Complications develop with the wide variety of materials that must be burned. Controlling the proper amount of air to allow for the combustion of both solids and sludge is difficult, and incinerator designs must incorporate all important considerations.

Closed incinerators such as rotary kilns and multiple hearth incinerators are also used to burn solid wastes. Generally, the incinerator design does not have to be limited to a single combustible or partially combustible waste. Often it is both economical and feasible to use a combustible waste, either liquid or gas, as the heat source for the incineration of a partially combustible waste that may be either liquid or gas.

Experience indicates that wastes containing only carbon, hydrogen, and oxygen and that may be handled in power generation systems may be destroyed in a way that reclaims some of their energy content. These types of waste may also be judiciously blended with wastes having low energy content, such as the highly chlorinated organics, to minimize the use of purchased fossil fuel. On the other hand, rising energy costs are not necessarily a deterrent to the use of thermal destruction methods when they are clearly indicated to be the most desirable method on an environmental basis.

Air emissions from hazardous waste incineration systems illustrated in Figure 15–3 include the common air pollutants, discussed in Chapter 18. In addition, inadequate incineration may result in emission of some of the hazardous materials that the incineration was intended to destroy. Incomplete combustion, particularly at relatively low temperatures, may also result in production of a class of compounds known collectively as dioxin, including both polychlorinated dibenzodioxins (PCDD) and polychlorinated dibenzofurans (PCDF). The com-

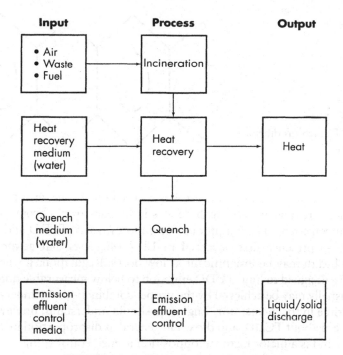

FIGURE 15–3. Waste incineration system

pound in this class that has been identified as a carcinogen and teratogen is 2,3,7,8-tetrachlorodibenzo-*p*-dioxin (2,3,7,8-TCDD), shown in Figure 15–4.

TCDD was first recognized as an oxidation product of trichlorophenol herbicides (2,4-D and 2,4,5-T, one of the ingredients of Agent Orange). In 1977, it was one of the PCDDs found present in municipal incinerator fly ash and air emissions, and it has subsequently been found to be a constituent of gaseous emissions from virtually all combustion processes, including wood stove fires, trash fires, and barbecues. Forest fires and brush fires are the major source of environmental PCDD—Canadian forest fires produce about 130 pounds of it per year. In addition, marine organisms, terrestrial plants, fungi, and mammalian thyroid glands chlorinate organic compounds and produce PCDD.[1] TCDD is degraded by sunlight in the presence of water.

The acute toxicity of TCDD in animals is extremely high (LD_{50} in hamsters of 3.0 µg/kg); carcinogenesis and genetic effects (teratogenesis) have also been observed in chronic exposure to high doses in experimental animals. In humans, the evidence for these adverse effects is mixed. Although acute effects such as skin rashes and digestive difficulties have been observed on high accidental exposure,

---

[1]Gribble, Gordon W., "The Natural Production of Chlorinated Compounds," *Environment, Science, & Technology* 28 (1994): 310–319.

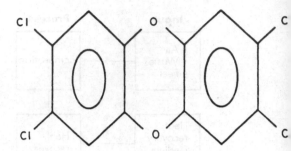

FIGURE 15–4
2,3,7,8-Tetrachlorodibenzo-
p-dioxin

these are transitory. Public concern has focused on chronic effects, but evidence for either carcinogenesis or birth defects in humans from chronic TCDD exposure is inconsistent. The ubiquitous natural presence of TCDD and related compounds—a presence first observed in 1877—suggests that adverse effects on human health may be insignificant at low doses. Regulations governing incineration are designed to limit TCDD emission to below measurable quantities; these limits usually may be achieved by the proper combination of temperature and residence time in the incinerator. Engineers should understand, however, that public concern about TCDD, and dioxin in general, is disproportionate to the known hazards and is a major factor in opposition to incinerator siting.

Hazardous Waste Landfills

Hazardous waste landfills must be adequately designed and operated if public health and the environment are to be protected.

Design. The three levels of safeguard that must be incorporated into the design of a hazardous landfill are displayed in Figure 15–5. The primary system is an impermeable liner, either of clay or synthetic material, coupled with a leachate collection and treatment system. Infiltration may be minimized with a cap of impervious material overlaying the landfill, sloped to permit adequate runoff and discourage pooling of water. The objectives are to prevent rainwater and snow melt from entering the soil and percolating to the waste containers and, if water does enter the disposal cells, to collect and treat it as quickly as possible. Side slopes of the landfill should be a maximum of 3:1 to reduce stress on the liner material. Research and testing of the range of synthetic liners must be viewed with respect to a liner's strength, compatibility with wastes, costs, and life expectancy. Rubber, asphalt, concrete, and a variety of plastics are available, and combinations such as polyvinyl chloride overlaying clay may prove useful on a site-specific basis.

A leachate collection system must be designed by contours to promote movement of the waste to pumps for extraction to the surface and subsequent treatment. Plastic pipes, or sand and gravel, similar to systems in municipal landfills and used on golf courses around the country, are adequate to channel the leachate to a pumping station below the landfill. One or more pumps direct

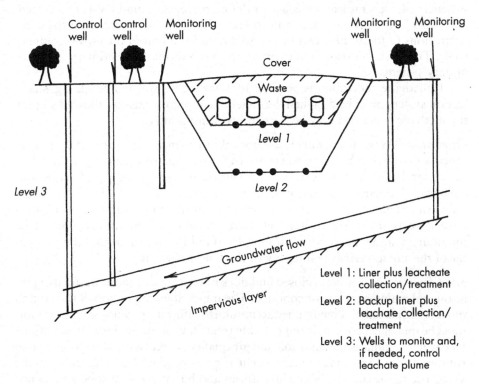

FIGURE 15–5. Three levels of safeguard in hazardous waste landfills

the collected leachate to the surface, where a wide range of waste-specific treatment technologies are available, including:

- *Sorbent material*: carbon and fly ash arranged in a column through which the leachate is passed.
- *Packaged physical-chemical units*, including chemical addition and flash mixing, controlled flocculation, sedimentation, pressure filtration, pH adjustment, and reverse osmosis.

The effectiveness of each method is highly waste specific, and tests must be conducted on a site-by-site basis before a reliable leachate treatment system can be designed. All methods produce waste sludge that must reach ultimate disposal.

A secondary safeguard system consists of another barrier contoured to provide a backup leachate collection system. In the event of failure of the primary system, the secondary collection system conveys the leachate to a pumping station, which in turn relays the wastewater to the surface for treatment.

A final safeguard system is also advisable. This system consists of a series of discharge wells up-gradient and down-gradient to monitor groundwater quality in the area and to control leachate plumes if the primary and secondary systems fail. Up-gradient wells act to define the background levels of selected chemicals in the groundwater and to serve as a basis for comparing the concentrations of these chemicals in the discharge from that of the down-gradient wells. This system thus provides an alarm mechanism if the primary and secondary systems fail.

If methane generation is possible in a hazardous waste landfill, a gas collection system must be designed. Sufficient vent points must be allowed so that the methane generated may be burned off continuously.

Operation.   As waste containers are brought to a landfill site for burial, specific precautions should be taken to ensure the protection of public health, worker safety, and the environment. Wastes should be segregated by physical and chemical characteristics, and buried in the same cells of the landfill. Three-dimensional mapping of the site is useful for future mining of these cells for recovery purposes. Observation wells with continuous monitoring should be maintained, and regular core soil samples should be taken around the perimeter of the site to verify the integrity of the liner materials.

Site Closure.   Once a site is closed and does not accept any more waste, its operation and maintenance must continue. The impervious cap on top of the landfill must be inspected and maintained to minimize infiltration. Surface water runoff must be managed, collected, and possibly treated. Continuous monitoring of surface water, groundwater, and soil and air quality is necessary, as ballooning and rupture of the cover material may occur if gases produced or released from the waste rise to the surface. Waste inventories and burial maps must be maintained for future land use and waste reclamation. A major component of postclosure management is maintaining limited access to the area.

POLLUTION PREVENTION

Hazardous wastes historically have been thought of as an "end of the pipeline" problem—the unwanted chemicals and other materials left over from various industrial operations. The notorious Love Canal, for example, was the waste from a large chemical facility that operated in the Niagara Falls area. Not having anything else to do with the waste, the facility operators piled it into an open canal. This operation, which was perfectly legal at that time, exposed the individuals who bought homes in the area and the children who went to a school that was constructed on the site to noxious and potentially very hazardous substances. The chemical firm that deposited the chemicals in the canal in the first place incurred high costs for remediation and long-term monitoring.

Because some chemical wastes are difficult to control or dispose of, it makes sense to ask whether these wastes are needed in the first place. As discussed in Chapter 13, this is the essence of pollution prevention.

In no other aspect of waste management is pollution prevention more effective and efficient than in the elimination of hazardous waste materials. It also turns out to be fairly easy, requiring only that someone analyze the plant operations and identify how certain chemicals end up in the waste stream. For example, if the waste stream is high in heavy metals such as chromium and zinc, is it possible that these metals are being wasted during the production operation, and, if so, how can this waste be prevented? In an electroplating plant, for example, the simple expedient of not dripping the solutions on the floor (from where it flowed via the floor drain to the waste treatment facility) saved the plant from having to construct an expensive metals recovery operation. A number of manufacturing plants, including the Boeing Corporation, now use citric-acid-based solvents instead of organic solvents. Life-cycle analysis is important in pollution prevention also. For example, detergent and water may be substituted for an acetone-based cleaner, but items that are water-washed usually need heat drying, while acetone evaporates with air drying. Pollution prevention in hazardous waste is a major option for industries, and all major firms are studying ways to reduce the amount of hazardous waste generated by the facility.

CONCLUSION

For years, the necessary by-products of an industrialized society were piled "out back" on land that had little value. As time passed and the rains came and went, the migration of harmful chemicals moved hazardous waste to the front page of the newspaper and into the classroom. Environmental scientists and engineers employed in all public and private sectors must now face head-on the processing, transportation, and disposal of these wastes.

PROBLEMS

15.1   Assume you are an engineer working for a hazardous waste processing firm. Your vice president thinks it would be profitable to locate a new regional facility near your home town. Given what you know about that region, rank the factors that distinguish a good site from a bad site. Discuss the reasons for this ranking. Why, for example, are hydrologic considerations more critical in that region than, say, the region's geology?

15.2   You are a town engineer just informed of a chemical spill on Main Street. Sequence your responses. List and describe the actions your town should take for the next 48 hours if the spill is relatively small (100–500 gallons) and confined to a small plot of land.

15.3   The manifest system, through which hazardous waste must be tracked from generator to disposal site, is expensive for industry. Make these assumptions about a simple electroplating operation: 50 barrels of a waste per day, 1 "trip ticket" per barrel, and a $25-per-hour labor charge. Assume the generator's technician can identify the contents of each barrel at no additional time or cost to the company because she has done that routinely for years. What is the cost to the generator in person-hours and dollars to comply with the manifest system shown in Figure 15–1? Document assumptions about the time required to complete each step of each trip ticket.

15.4 Compare and contrast the design considerations of the hazardous waste landfill with the design considerations of a conventional municipal refuse landfill.

15.5 Design a system to detect and stop the movement of hazardous wastes into your municipal refuse landfill. Consider all of the alternatives, including pollution preventions.

LIST OF SYMBOLS

LC_{50} lethal dose concentration, at which 50 percent of the subjects are killed

LD_{50} lethal dose, at which 50 percent of the subjects are killed

PCDD   polychlorinated dibenzodroxins

PCDF   polychlorinated dibenzofurans

RCRA   Resource Conservation and Recovery Act

TCDD   tetrachlorodibenzo-*p*-dioxin

USDOT   U.S. Department of Transportation

Chapter 16

Radioactive Waste

This chapter presents a general background discussion of the interaction of ionizing radiation with matter, as well as a discussion of the environmental effects of nuclear generation of electricity and of radionuclides that are in the accessible environment. The chapter focuses on radioactive waste as an environmental pollutant, discusses the impact of ionizing radiation on environmental and public health, and summarizes options available today for the management and disposal of radioactive waste.

RADIATION

X-rays were discovered by Wilhelm Roentgen in 1895. The following year Henri Becquerel observed radiation similar to x-rays emanating from certain uranium salts. In 1898, Marie and Pierre Curie studied radiation from two uranium ores, pitchblende and chalcolite, and isolated two additional elements that exhibited radiation similar to uranium but considerably stronger. These two elements were named radium and polonium. The discovery and isolation of these radioactive elements marks the beginning of the "atomic age."

The Curies classified the radiation from radium and polonium into three types, according to the direction of deflection in a magnetic field. These three types of radiation were called alpha (α), beta (β), and gamma (γ). Becquerel's observation correlated gamma radiation with Roentgen's x-rays. In 1905, Ernest Rutherford identified alpha particles emanating from uranium as ionized helium atoms, and in 1932 Sir James Chadwick characterized as neutrons the highly penetrating radiation that results when beryllium is bombarded with alpha particles. Modern physics has subsequently identified other subatomic particles, including positrons, muons, and pions, but not all of these are of equal concern. Management of radioactive waste requires an understanding of the sources and effects of alpha, beta, gamma, and neutron emissions.

Radioactive Decay

A radioactive atom has an unstable nucleus. The nucleus moves to a more stable condition by emitting an alpha or beta particle; this emission is frequently accompanied by emission of additional energy in the form of gamma radiation, although gamma radiation may also be emitted by itself. Collectively these emissions are known as *radioactive decay*. The rate of radioactive decay, or rate of decrease in the number of radioactive nuclei, can be expressed by an equation:

$$\ln \frac{N}{N_0} = -K_b t \tag{16.1}$$

$$\frac{N}{N_0} = e^{-K_b t} \tag{16.2}$$

where   N = number of radioactive nuclei
K_b = a factor called the *disintegration constant*; time^{-1}
N_0 = N at time t = 0

The product $K_b \cdot N$ is sometimes called the *activity*. The data points shown in Figure 16–1 can be calculated using this equation.

After a specific time period $t = t_{1/2}$, the value of N is equal to one-half of N_0, and after each succeeding period of time $t_{1/2}$, the value of N is one-half of the preceding N. That is, one-half of the radioactive atoms have decayed (or disintegrated) during each time period $t_{1/2}$, called the *radiological half-life*, or sometimes simply the half-life. Looking at Figure 16–1, we see that at $t = 2t_{1/2}$, N becomes 1/4 N_0; at $t = 3t_{1/2}$, N becomes 1/8 N_0; and so on. Equation 16.2 is so constructed that N never becomes zero in any finite time period; for every

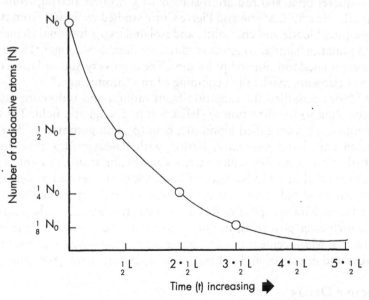

FIGURE 16–1. General description of radioactive decay

half-life that passes, the number of atoms is halved. The half-life may be determined from Equation 16.3:

$$t_{1/2} = \frac{\ln 2}{K_b} = \frac{0.693}{K_b} \tag{16.3}$$

We note that specific activity (radioactivity per unit weight) is inversely proportional to half-life. Long-lived radionuclides are considerably less radioactive than short-lived ones. The radiological half-lives of selected radionuclides are presented in Table 16–1.

TABLE 16–1. Some Important Radionuclides

Radionuclide	Type of Radiation	Half-life
Americium-241	Alpha	432 years
Carbon-14	Beta	5770 years
Cesium-137	Beta and gamma	30 years
Cobalt-60	Beta and gamma	5 years
Iodine-131	Beta and gamma	8.3 days
Krypton-85	Beta and gamma	10 years
Plutonium-239	Alpha	24,600 years
Strontium-90	Beta	29.8 years
Tritium (Hydrogen-3)	Beta	12 years
Uranium-238	Alpha	4.9×10^9 years

Example 16.1
10.0 grams of pure $_6C^{11}$ is prepared. The equation for this nuclear reaction is

$$_6C^{11} \rightarrow {}_1e^0 + {}_5B^{11} \tag{16.4}$$

The half-life of C-11 is 21 minutes. How many grams of C-11 will be left 24 hours after the preparation? (Note that one atomic mass unit (amu) = 1.66×10^{-24} g.)

Equation 16.2 refers to the number of atoms, so we must calculate the number of atoms in 1.0 gram of Carbon-11:

$$\frac{1 \text{ atom } _6C^{11}}{11.0 \text{ amu}} \times \frac{1 \text{ amu}}{1.66 \times 10^{-24} \text{ g } _6C^{11}} \times 10.0\text{g } _6C^{11} = 55 \times 10^{22} \text{ atoms } _6C^{11} \tag{16.5}$$

Applying Equation 16.2,

$$K_b = \frac{0.693}{t_{1/2}} = \frac{0.693}{21 \text{ min}} = 33 \times 10^{-3} \text{min}^{-1} \tag{16.6}$$

Since there are 1440 minutes in a day,

$$K_b t = (33 \times 10^{-3} \, min^{-1})(1440 \, min) = 47.5 \qquad (16.7)$$

and

$$\frac{N_0}{N} = \exp(47.5) = 4.3 \times 10^{20} \qquad (16.8)$$

$$N = \frac{55 \times 10^{22}}{4.3 \times 10^{20}} = 1300 \text{ atoms of } C^{11} = 240 \times 10^{-22} \text{ grams of } C^{11} \qquad (16.9)$$

remaining after a day.

Useful figures of merit for radioactive decay follow:

- After 10 half-lives, 10^{-3} (or 0.1%) of the original quantity of radioactive material is left.
- After 20 half-lives, 10^{-6} of the original quantity of radioactive material is left.

Alpha, Beta, and Gamma Radiation

Emissions from radioactive nuclei are called, collectively, *ionizing radiation* because collision between these emissions and an atom or molecule ionizes that atom or molecule. Ionizing radiation may be characterized further as alpha, beta, or gamma radiation by its behavior in a magnetic field. Apparatus for such characterization is shown in Figure 16–2. A beam of radioactively disintegrating atoms is aimed with a lead barrel at a fluorescent screen that is designed to glow when hit by the radiation. Alternately charged probes direct the α and β radiation accordingly. The γ radiation is seen to be "invisible light," a stream of neutral particles that passes undeflected through the electromagnetic field. α and β emissions have some mass and are considered particles, while γ emissions are photons of electromagnetic radiation.

Alpha radiation has been identified as helium nuclei that have been stripped of their planetary electrons, and each consists of two protons and two neutrons. α particles thus have a mass of about 4 amu (6.642×10^{-4} g) each and a positive charge of 2.[1] External radiation by α particles presents no direct health hazard because even the most energetic are stopped by the epidermal layer of skin and rarely reach more sensitive layers. A health hazard occurs when material contaminated with α-emitting radionuclides is eaten or inhaled, or otherwise absorbed inside the body, so that organs and tissues more sensitive than skin are

---

[1]This electric charge is expressed in units relative to an electronic charge of −1.

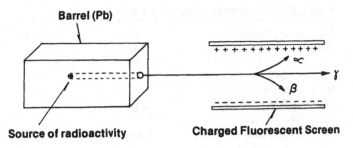

FIGURE 16–2. Controlled measurement of alpha (α), beta (β), and gamma (γ) radiation

exposed to α radiation. Collisions between α particles and the atoms and molecules of human tissue may cause disorder of the chemical or biological structure of the tissue.

Beta radiation is a stream of electrons emitted at a velocity approaching the speed of light, with kinetic energy between 0.2 MeV and 3.2 MeV. Given their lower mass of approximately 5.5×10^{-4} amu (9.130×10^{-24} g), interactions between β particles and the atoms of pass-through materials are much less frequent than α particle interactions: fewer than 200 ion pairs are typically formed in each centimeter of passage through air. The slower rate of energy loss enables β particles to travel several meters through air and several centimeters through human tissue. Internal organs are generally protected from external β radiation, but exposed organs such as eyes are sensitive to damage. Damage may also be caused by incorporation of β emitters into the body and resulting exposure of internal organs and tissue.

Gamma radiation is invisible electromagnetic radiation, composed of photons, much like medical x-rays. γ photons are electrically neutral and collide randomly with the atoms of the material as they pass through. The considerably longer distance that γ rays travel in all media is defined by the *relaxation length*, the distance that the γ photon travels before its energy is decreased very quickly. A typical 0.7-MeV γ photon has a relaxation length of 5 cm, 50 cm, and 10,000 cm in lead, water, and air, respectively—much longer than an α or β particle of the same energy. External doses of γ radiation may have significant human health consequences because the dose is not greatly affected by passage of the radiation through air. The properties of the more common radioactive emissions are summarized in Table 16–2.

When ionizing radiation is emitted from a nucleus, the nature of that nucleus changes: Another element is formed, and there is a change in nuclear mass. This process may be written as a nuclear reaction in which both mass and charge must balance for reactants and products. For example, the beta decay of Carbon-14 may be written as

$$_6C^{14} = {_{-1}}\beta^0 + {_7}N^{14} \qquad (16.10)$$

TABLE 16–2. Properties of Ionizing Radiation

Particle or Photon (Wave)	Mass (amu)	Electric Charge
Alpha ($_2$He^4)	4	+2
Beta (electron)	5.5×10^{-4}	−1
Gamma (x-ray)	Approx. 0	0
Neutron	1	0
Positron (positive electron)	5.5×10^{-4}	+1

That is, C-14 decays to ordinary stable nitrogen (N-14) with emission of a beta particle. The mass balance for this equation is

$$14 = 0 + 14 \tag{16.11}$$

and the charge balance is

$$6 = -1 + 7 \tag{16.12}$$

A typical reaction for α decay, the first step in the U-238 decay chain, is

$$_{92}U^{238} = {_2}He^4 + {_{90}}Th^{234} \tag{16.13}$$

When a radionuclide emits a β, the mass number remains unchanged and the atomic number increases by 1 (β decay is thought to be the decay of a neutron in the nucleus to a proton and a β, with subsequent emission of the β). When a nuclide emits an α, the atomic mass decreases by 4 and the atomic number decreases by 2. γ emission does not result in a change of either atomic mass or atomic number.

Nuclear reactions may also be written for bombardment of nuclei with subatomic particles. For example, tritium (H-3) is produced by bombarding a lithium target with neutrons:

$$_0n^1 + {_3}Li^6 = {_1}H^3 + {_2}He^{+4} \tag{16.14}$$

These reactions tell us nothing about the energy with which ionizing radiation is emitted, however, or the relative biological damage that can result from transfer of this energy in collisions.

Units for Measuring Ionizing Radiation

Damage to living organisms is directly related to the amounts of energy transferred to tissue by collisions with α and β particles, neutrons, and γ radiation. This energy, in the form of ionization and excitation of molecules, results in heat damage to the tissue. Many of the units discussed in this section are thus related to energy transfer.

The International System of Units (SI units) for measuring ionizing radiation, based on the meter-kilogram-second (mks) system, was defined by the General Conference on Weights and Measures in 1960, and adoption of these units has been recommended by the International Atomic Energy Agency. SI units replaced the units that had been in use since about 1930. They are used in this chapter, and their relationship to the historical units is discussed.

A *becquerel* (Bq) is the SI measure of source strength or total radioactivity, and is defined as one disintegration per second; the units of becquerels are sec^{-1}. The decay rate, $\Delta N/\Delta t$, is measured in Bq. The historical unit of source strength, the *curie* (Ci), is the radioactivity of one gram of the element radium and is equal to 3.7×10^{10} disintegrations per second.[2] The source strength in Bq is not sufficient for a complete characterization of a source; the nature of the radionuclide (e.g., Pu-239, Sr-90) and the energy and type of emission (e.g., 0.7 MeV γ) are also necessary.

The relationship between activity and mass of a radionuclide is given by

$$Q = \frac{K_b M N^0}{W} \tag{16.15}$$

where Q = number of becquerels
        K_b = disintegration constant = $0.693 t_{1/2}$; the fraction of atoms that decay each second
        $t_{1/2}$ = half-life of the radionuclide, in seconds
        M = mass of the radionuclide, in grams
        N^0 = Avogadro's number, 6.02×10^{23} atoms/gram-atom
        W = atomic weight of the radionuclide, in grams/gram-atom

The *gray* (Gy) is the SI measure of the quantity of ionizing radiation that results in absorption of one joule of energy per kg of absorbing material. One Gy is the equivalent of 100 *rad*, the historical unit of "radiation absorbed dose" equal to 100 ergs/gm of absorbing material. The gray is a unit of *absorbed radiation dose*.

The *sievert* (Sv) is the SI measure of *absorbed radiation dose equivalent*. That is, one Sv of absorbed ionizing radiation does the same amount of biological damage to living tissue as one Gy of absorbed x-ray or γ radiation. The standard for comparison is γ radiation having a linear energy transfer (LET) in water of 3.5 keV/γm and a dose rate of 0.1 Gy/min. As previously stated, all ionizing radiation does not produce the same biological effect, even for a given amount of energy delivered to human tissue. The dose equivalent is the product of the absorbed dose in Gy and a *quality factor* (QF) (sometimes called the *relative biological effectiveness*, or RBE). The historical dose-equivalent unit is

---

[2]Marie Curie is said to have rejected a unit of 1 disintegration/sec as too small and insignificant. The much larger unit of the curie was then established.

the *rem*, for "roentgen equivalent man." One sievert is equivalent to 100 rem. These units are related as follows:

$$\text{Quality Factor} = \text{Sv/Gy} = \text{rem/rad} \qquad (16.16)$$

Quality factors take into account chronic low-level exposures, as distinct from acute high-level exposures, and the pathway of the radionuclide into the human body (inhalation, ingestion, dermal absorption, external radiation) as well as the nature of the ionizing radiation. Table 16–3 gives sample quality factors for the internal dose from radionuclides incorporated into human tissue.

Dose equivalents are often expressed in terms of *population dose equivalent*, which is measured in *person-Sv*. The population dose equivalent is the product of the number of people affected and the average dose equivalent in Sv. That is, if a population of 100,000 persons receives an average dose equivalent of 0.05 Sv, the population dose equivalent will be 5000 person-Sv.

Table 16–4 gives current (1990) estimated average radiation dose equivalents in the United States. The dose equivalent is the product of the absorbed dose and the QF. The *effective dose equivalent* is the risk-weighted sum of the dose equivalents to the individually irradiated tissues, and it is the figure usually cited by EPA and other regulatory agencies. Note that 55% of the effective dose equivalent is from radon exposure and 82% is from natural (nonanthropogenic) sources.

Measuring Ionizing Radiation

The particle counter, the ionization chamber, photographic film, and the thermoluminescent detector are four methods widely used to measure radiation dose, dose rate, and the quantity of radioactive material present.

Particle counters detect the movement of single particles through a defined volume. Gas-filled counters collect the ionization produced by the radiation as

TABLE 16–3. Sample Quality Factors for Internal Radiation

Type of Radiation	Quality Factor
Internal alpha	10
Plutonium alpha in bone	20
Neutrons (atom bomb survivor dose)	20
Fission spectrum neutrons	2–100
Beta and gamma	
[a]$E_{max} > 0.03$ MeV	1.0
[a]$E_{max} < 0.03$ MeV	1.7

[a]E_{max} refers to the maximum energy of emissions from the β or γ source.

TABLE 16–4. Average Annual Dose Equivalent of Ionizing Radiation to a Member of the U.S. Population

Source of Radiation	Dose Equivalent (mSv)	(mrem)	Effective Dose Equivalent mSv
Natural			
Radon	24	(2400)	2.0
Cosmic radiation	0.27	(27)	0.27
Terrestrial	0.28	(28)	0.28
Internal	0.39	(39)	0.39
Total natural	24.94		3.0
Anthropogenic			
Medical: diagnostic x-ray	0.39	(39)	0.39
Medical: nuclear medicine	0.14	(14)	0.14
Medical: consumer products	0.10	(10)	0.1
Occupational	0.009	(0.9)	< 0.1
Nuclear fuel cycle	< 0.01	(< 1.0)	< 0.1
Fallout	< 0.01	(< 1.0)	< 0.1
Miscellaneous	< 0.01	(< 1.0)	< 0.1
Total anthropogenic	0.64		0.63
Total			3.63[a]

[a]The number usually cited is 3.6 mSv, or 360 mrem effective dose equivalent.
From National Academy of Sciences, *Health Effects of Exposure to Low Levels of Ionizing Radiation (BEIR V)*, Upton, Arthur, ed., National Academy Press (1990).

it passes through the gas and amplify it to produce an audible pulse or other signal. Counters are used to determine radioactivity by measuring the number of particles emitted by radioactive material in a given time.

Ionization chambers consist of a pair of charged electrodes that collect ions formed within their respective electric fields. They can measure dose or dose rate because they provide an indirect representation of the energy deposited in the chamber.

Photographic film darkens on exposure to ionizing radiation and is an indicator of the presence of radioactivity. Film is often used for determining personnel exposure and making other dose measurements for which a record of dose accumulated over a period of time is necessary, or for which a permanent record is required.

Thermoluminescent detectors (TLD) are crystals, such as NaI, that can be excited to high electronic energy levels by ionizing radiation. The excitation energy is then released as a short burst or flash of light that can be detected by a photocell or photomultiplier. TLD systems are replacing photographic film because they are more sensitive and consistent. Liquid scintillators—organic phosphors operating on the same principle as TLDs—are used in biochemical applications.

HEALTH EFFECTS

When α and β particles and γ radiation interact with living tissue they transfer energy to the receiving material through a series of collisions with the material's atoms or nuclei. Molecules along the path of the ionizing radiation are damaged in the process as chemical bonds are broken and electrons are ejected (ionization). Resulting biological effects are due mainly to the interactions of these electrons with molecules of tissue. The energy transferred through these collisions and interactions per unit path length through the tissue is called *linear energy transfer* (LET) of the radiation. The more ionization observed along the particle's path, the more intense the biological damage. LET can serve as a qualitative index for ranking ionizing radiation with respect to biological effect. Table 16–5 gives some typical LET values.

Biological effects of ionizing radiation may be grouped as *somatic* and *genetic*. Somatic effects are impacts on individuals who are directly exposed to the radiation. Radiation sickness (circulatory system breakdown, nausea, hair loss, and sometimes death) is an *acute* somatic effect occurring after very high exposure, as from a nuclear bomb, intense radiation therapy, or a catastrophic nuclear accident. Such an accident occurred at the Chernobyl nuclear electric generating plant (located near Kiev, Ukraine) in April 1986. Thirty-one people received whole body doses between 4 Gy and 16 Gy and died during the 50 days following the accident. An additional 158 people who received doses between 0.8 Gy and 4 Gy suffered acute radiation sickness; all but one of these individuals recovered after treatment with red blood cell replacement.

Chronic effects resulting from long-term exposure to low doses of ionizing radiation can potentially include both somatic and genetic effects that occur because ionizing radiation damages the genetic material of the cell. Our knowledge of both somatic and genetic effects of low-dose ionizing radiation is based

TABLE 16–5. Representative LET Values

Radiation	Kinetic Energy (MeV)	Average LET (keV/μm)	Dose Equivalent
X-ray	0.01–0.2	3.0	1.00[a]
γ	1.25	0.3	0.7
β	0.1	0.42	1.0
	1.0	0.25	1.4
	0.1	260	
α	5.0	95	10
Neutrons	Thermal		4–5
	1.0	20	2–10
Protons	2.0	16	2
	5.0	8	2

[a]These equivalents are in terms of x-ray dose, which is defined as 1.00.

on animal studies and a very limited number of human epidemiological studies: studies of occupational exposure, the Japanese Atomic Bomb Survivors' Life Study, and studies of effects of therapeutic radiation treatment. Dose-response analysis based on these studies indicates that the dose-response relationship is linear at low doses and quadratic at high doses and may be linear at low doses (Figure 16–3). Much of the low-dose part of the curve is below any range of observation and is extrapolated. As with all carcinogens, the absence of a threshold is assumed. Data collected since 1980 indicate the possible existence of a threshold dose below which there is no evident health effect. Although we do not know the precise mechanism by which ionizing radiation produces these somatic and genetic effects, we understand that it involves damage to the DNA of the cell nucleus.

Somatic effects include decrease in organ function and carcinogenesis. Genetic effects result from radiation damage to chromosomes and have been demonstrated to be inheritable in animals but not in humans. The human genetic risk is estimated to be between 1 and 45 additional genetic abnormalities per 10 mSv (1 rem) per million liveborn offspring in the first generation affected, and between 10 and 200 per 10 mSv per million liveborn at equilibrium. The spontaneous rate of human genetic abnormality is presently estimated to be about 50,000 per million liveborn in the first generation.

Health risk from ionizing radiation may be summarized as follows: There is a documented risk from ionizing radiation, but it is apparently small, clearly uncertain, and depends on a number of factors. These include

- The magnitude of the absorbed dose
- The type of ionizing radiation
- The penetrating power of the radiation
- The sensitivity of the receiving cells and organs

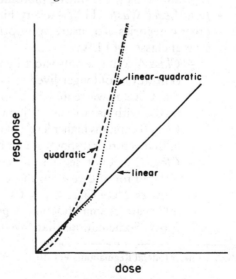

FIGURE 16–3
Linear-quadratic dose-response
relationship

- The rate at which the dose is delivered
- The proportion of the target organ or organism exposed

Although the risk of damage due to exposure is uncertain, prudence indicates that we minimize unnecessary exposure to ionizing radiation.

SOURCES OF RADIOACTIVE WASTE

The nuclear fuel cycle, nuclear weapons manufacture, radiopharmaceutical manufacture and use, biomedical research and application, and a number of industrial uses generate radioactive waste. The behavior of the radionuclides in waste (as in any other form) is determined by their physical and chemical properties; they may exist as gases, liquids, or solids and may be soluble or insoluble in water or other solvents.

The U.S. Nuclear Regulatory Commission (NRC)[3] has classified radioactive wastes and other materials into the following categories:

- *High-Level Waste* (HLW)—spent nuclear fuel from commercial nuclear reactors and the solid and liquid waste from solvent, the first cycle of extraction in reprocessing spent or irradiated fuel. In the United States, only military irradiated fuel is reprocessed, to recover plutonium and fissile uranium. NRC reserves the right to classify additional materials as HLW as necessary.
- *Uranium mining and mill tailings*—The pulverized rock and leachate from uranium mining and milling operations.
- *Transuranic Waste* (TRU)—radioactive waste that is not HLW but contains more than 3700 Bq (100 nCi) per gram of elements with an atomic number higher than 92. Most TRU waste in the United States is the product of plutonium purification and weapons fabrication.
- *By-product material*—Any radioactive material, except fissile nuclides, that is produced as waste during plutonium production or fabrication.
- *Low-Level Waste* (LLW)—everything that is not included in one of the other categories. To ensure appropriate disposal, the NRC has designated several classes of LLW:
  - *Class A* contains only short-lived radionuclides or extremely low concentrations of longer-lived radionuclides, and must be chemically stable. Class A waste may be disposed of in landfills as long as it is not mixed with hazardous or flammable waste.
  - *Class B* contains higher levels of radioactivity and must be physically stabilized before transportation or disposal. It must not contain free liquid.
  - *Class C* will not decay to acceptable levels in 100 years and must be isolated for 300 years or more. Power plant LLW is in this category.
  - *Greater Than Class C (GTCC)* will not decay to acceptable levels in 300 years. A small fraction of power plant Class C waste is in this category. Some nations, like Sweden, treat GTCC waste like HLW.

---

[3]U.S. Code of Federal Regulations, Vol. 10, Part 60.

- *Formerly Used Sites Remedial Action Program* (FUSRAP) waste is contaminated soil from radium and World War II uranium refining and atomic bomb development. Little was known at that time about the long-term effects of ionizing radiation. Consequently, there was widespread contamination of the structures in which this work was done and of the land surrounding those structures. FUSRAP waste contains very low concentrations of radionuclides, but there is a large amount of it.
- *Naturally Occurring Radioactive Material (NORM),* waste from several mining and milling processes, is contaminated with NORM. NORM is not presently regulated.

The Nuclear Fuel Cycle

The nuclear fuel cycle (shown in Figure 16–4) generates radioactive waste at every stage. Uranium mining and milling generate the same sort of waste that any mining and milling operations generate, including acid mine drainage and radioactive uranium daughter elements, including a considerable amount of Rn-222. Mining and milling dust must be stabilized to prevent both windborne dispersion and leaching into ground- and surface water.

Partially refined uranium ore, called "yellowcake" because of its bright yellow color, must be enriched in the fissile isotope U-235 before nuclear fuel can be fabricated from it. Mined uranium is more than 99% U-238, which is not fissile, and approximately 0.7% U-235. The concentration of U-235 is increased to about 3% by converting to UF_6; the lighter isotope is then converted to UO_2 or U_3O_8 and fabricated into fuel. Enrichment and fabrication both produce radioactive waste.

Plutonium is produced by neutron irradiation of U-238. In the United States, this plutonium is purified and fabricated into nuclear weapons. TRU and

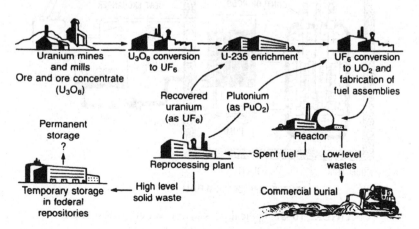

FIGURE 16–4. The nuclear fuel cycle

some HLW waste, now stored at Department of Energy plutonium-handling sites, is the result of this process. Although plutonium is no longer produced, TRU and HLW waste from past production still exist and more is being generated by environmental remediation.

In electric power generation, nuclear fuel is inserted into a reactor core, where this same (controlled) fission reaction produces heat that in turn produces pressurized steam. The steam system, turbines, and generators in a nuclear power plant are essentially the same as those in any thermal fossil fuel burning electric generating plant. The difference between nuclear and fossil fuel electric generation is in the evolution of the heat that drives the plant.

Figure 16–5 is a diagram of a typical pressurized water nuclear reactor. Commercial reactors in the United States are either pressurized water reactors, in which the water that removes heat from the nuclear reactor core (the "primary coolant") is under pressure and does not boil, or boiling water reactors, in which the primary coolant is permitted to boil. The primary coolant transfers heat from the core to the steam system (the "secondary coolant") by a heat exchange system that ensures complete physical isolation of the primary from the secondary coolant. A third cooling system provides water from external sources to condense the spent steam in the steam system.

All thermal electric power generation produces large quantities of waste heat. Fossil fuel electric generating plants typically have a thermal efficiency of about 42%; that is, 42% of the heat generated by combustion of the fuel is converted to electricity and 58% is dissipated in the environment. By comparison, nuclear plants are about 33% thermally efficient.

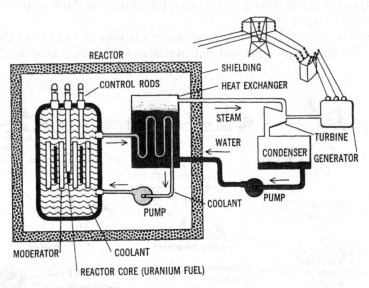

FIGURE 16–5. Typical pressurized water nuclear reactor

The nuclear reactor is perhaps the key radioactive waste producer in the nuclear fuel cycle. The production of HLW is a direct consequence of fission reactions in the core. One such reaction is

$$_{92}U^{235} + _{0}n^1 = _{42}Mo^{95} + _{57}La^{139} + 2_{0}n^1 + 204 \text{ MeV} \tag{16.17}$$

Several features of this reaction merit discussion:

- A large amount of energy is released in the reaction: 204 MeV per uranium atom, or 80 million Btu per gram of uranium. Uranium fission releases about 100,000 times as much heat per gram of fuel as natural gas combustion. The development of commercial nuclear power is based on this phenomenon.
- One neutron is required for the fission reaction, but the reaction itself produces two neutrons, each of which can initiate another fission reaction that will produce two neutrons, and so on, resulting in a *fission chain reaction*. However, the concentration of fissile material (U-235 in commercial power reactors) must be high enough so that neutrons produced are likely to collide with U-235 nuclei. The mass of fissile material needed to sustain a fission chain reaction is called the *critical mass*.
- Neutron flow in a reactor may be interrupted, and the fission reaction stopped, by insertion of control rods that absorb neutrons (Figure 16–5). Although control rods stop the fission reaction, they do not stop heat generation in the core since the many radioactive fission products continue to emit energy. Therefore, continued coolant flow is critical.
- In this particular reaction, Mo-95 and La-139 are the products of the fission. However, a fissile nucleus can split apart in about 40 different ways, yielding approximately 80 different fission fragments. Although many of these have very short half-lives and decay very quickly, some have long half-lives. These long half-life fission products make up most of the radioactivity of HLW.

This mixture of fission products has a very high specific radioactivity (radioactivity per unit mass or volume). Fission products are beta or gamma emitters because they are too small to emit alpha particles. The longest-lived fission products, Cs-137 and Sr-90, have half-lives of 30 years and 29 years, respectively. A large quantity of fission products containing these two radionuclides could thus be a significant source of radioactivity for over 600 years.

About one fission reaction in 10,000 yields three fission fragments instead of two; the third fragment is tritium (H-3), which has a 12.3-year half-life. Since tritium is virtually identical to hydrogen chemically, it exchanges freely with nonradioactive hydrogen gas in a reactor and with H^+ ions in the reactor cooling water. Containment of tritium is thus difficult.

In a fission reaction, not all of the neutrons will collide with a fissile nucleus to initiate fission. Some neutrons will react with the fuel container and the reactor vessel itself, producing neutron activation products. Some of these are relatively long-lived, notably Co-60 (half-life, 5.2 years) and Fe-59 (half-life, 45 days). Other neutrons react with nonfissile isotopes, like U-238, as in the reaction

$$_{92}U^{238} + _0n^1 = _{92}U^{239} = _{93}Np^{239} + _{-1}e^0 \tag{16.18}$$

$$_{93}Np^{239} = _{94}Pu^{239} + _{-1}e^0 \tag{16.19}$$

Pu-239, highly refined "weapons-grade" plutonium, has a half-life of 24,600 years. Other isotopes of plutonium are formed also. Radionuclides like U-238, from which fissile material like plutonium can be produced by nuclear reaction, are known as *fertile* isotopes. Another isotope of plutonium, Pu-241, decays by beta emission to Am-241 (half-life, 433 years). Am-241 is commonly used as an ion producer in smoke detectors.

Each uranium isotope is part of a decay series; when U-238 decays, the daughter element is also radioactive and decays, producing another radioactive daughter, and so on, until a stable element (usually a lead isotope) is reached. The plutonium isotopes also decay in a series of radionuclides. Spent nuclear fuel (HLW) thus contains fission products, tritium, neutron activation products, plutonium, and plutonium and uranium daughter elements in a very radioactive, very long-lived mixture that is chemically difficult to segregate.

Reprocessing Waste and Other Reactor Waste

In the United States, plutonium is used only for nuclear warheads, which were manufactured from the end of World War II until 1989. Plutonium was produced by irradiating U-238 with neutrons (in the reaction given above) in military breeder reactors. The irradiated fuel was then completely dissolved, and plutonium, along with fissile uranium and neptunium, was extracted. Further precipitation resulted in recovery of plutonium, uranium, neptunium, strontium, and cesium. The fissile isotopes of plutonium and uranium are categorized as *special nuclear material*; the other nuclides are considered by-product material. Although the production and extraction of special nuclear material have ceased, the sludge and organic extraction solvents used contain high concentrations of radionuclides and are classified as HLW, TRU, or LLW.

The primary and secondary coolants in a nuclear generating plant pick up considerable radioactive contamination through controlled leaks. Contaminants are removed from the cooling water by ion exchange, and the loaded ion-exchange columns are Class C or GTCC waste. Class A and Class B wastes are also produced in routine clean-up activities in nuclear reactors.

After thirty to forty years of operation, the reactor core and the structures immediately surrounding it will have become very radioactive, primarily by neutron activation, and the reactor must be shut down and decommissioned. Except for spent fuel, most decommissioning waste is LLW.

Additional Sources of Radioactive Waste

The increasingly widespread use of radionuclides in research, medicine, and industry has created radioactive waste. Sources include laboratories using a few isotopes, medical and research laboratories where large quantities of many different radioisotopes are used (and often wasted), and industrial uses like petroleum well testing.

Liquid scintillation counting has become an important biomedical tool and produces large volumes of "mixed" hazardous and radioactive waste: contaminated organic solvents like toluene that have a low specific radioactivity. The Resource Conservation and Recovery Act (RCRA) prohibits landfill disposal of the solvents in an LLW site, and there is considerable public opposition to incineration of any radioactive material, no matter how low the specific activity.

Naturally occurring radionuclides may also pose threats to public health. By far the largest fraction of background ionizing radiation comes from radon (Rn-222), a member of the uranium decay series found ubiquitously in rock. Chemically, Rn-222 is an inert gas, like helium, and does not become chemically bound. When uranium-bearing minerals or rock are crushed or machined, Rn-222 is released; there is even a steady release of Rn-222 from undisturbed rock outcrops. Buildings that are insulated to prevent excessive heat loss often have too little air circulation to keep the interior purged of Rn-222; thus Rn-222 concentrations in homes and commercial buildings with restricted circulation can become quite high. Rn-222 itself has a short half-life, but decays to much longer-lived metallic radionuclides.

Coal combustion, copper mining, and phosphate mining release isotopes of uranium and thorium into the environment. K-40, C-14, and H-3 are found in many foodstuffs. Although atmospheric nuclear testing was discontinued about 40 years ago, radioactive fallout from past tests continues to enter the terrestrial environment.

RADIOACTIVE WASTE MANAGEMENT

The objective of radioactive waste management is to prevent human exposure to radioactive materials and therefore to minimize the possibility that radioactive materials will enter drinking water or the food chain or be inhaled (Figure 16–6). To date, isolation of the waste has been the only possible option for achieving this objective.

When the fissile uranium in fuel rods has been used to a point where the fission rate is too slow for efficient power generation, the rods are ejected into a water pool, where they remain until the short-lived radionuclides have decayed and until the rods are thermally cool enough to handle with ordinary machinery. This takes about six months, but the lack of any other storage facility for most spent fuel in the United States has resulted in onsite pool storage for as long as ten years. In addition, the United States is bound by nonproliferation agreements to accept and store spent fuel that it has supplied to nuclear reactors in

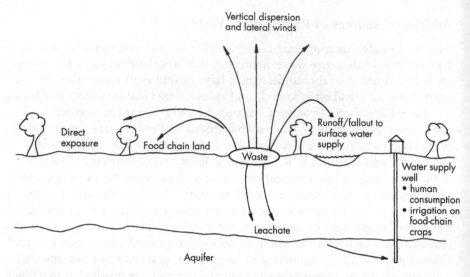

FIGURE 16–6. Potential movement of radioactive materials from waste storage and "disposal" areas to the accessible environment

other countries. At present, foreign spent fuel is stored at commercial reactors that have available pool storage space. Cooled spent fuel is also stored at plant sites in heavily shielded air-cooled concrete storage containers.

In the United States, a site at Yucca Mountain, Nevada, about 100 miles northwest of Las Vegas, is being investigated for permanent disposal of HLW—both commercial spent nuclear fuel and high-level defense reprocessing waste. However, this site has been under investigation since 1983, and there is as yet no determination of its suitability. The slow pace of this investigation is due in part to the difficulties inherent in assessing repository integrity for 10,000 years as required by regulation, in part to changes in the regulatory standard the proposed HLW repository must meet (the standard was changed in 1985, 1993, and 1997), and in part to opposition to the siting by the State of Nevada. The investigation is funded by a 1 mil/kwh tax on nuclear utilities, and the utilities that are financially supporting the project have become impatient with its lack of progress and have sought alternative solutions such as above-ground monitored storage. The Nuclear Regulatory Commission (NRC) has approved several dry-surface storage facilities at nuclear power plants.

Low-Level Radioactive Waste

Unlike HLW, commercial LLW has been the responsibility of the private sector since 1960. It is disposed of in appropriate containers in shallow burial trenches. Three sites are open today, at Hanford, Washington, Barnwell, South Carolina, and Beatty, Nevada. The last two sites are nearing capacity and are due to be closed. California, North Carolina, and New York are planning sites.

The Low Level Radioactive Waste Policy Act of 1980 (and its subsequent amendments), Part 61 of volume 10 of the U.S. Code of Federal Regulations, regulates shallow land disposal of LLW, provides the method of selecting LLW disposal sites, and regulates the environmental safety of LLW handling.

Techniques now being applied to LLW include impervious packaging, compaction, incineration, and stabilization in an asphalt or cement matrix. Compaction can achieve volume reduction of about 8:1, and incineration can reduce volumes by factors of 30 or more. Incineration appears to be an excellent treatment option for mixed radioactive and organic chemical wastes like liquid scintillators. Incineration emissions require careful monitoring and decontamination factors of 10,000 or more. Air emission controls are discussed further in Chapter 20.

In addition to the operating disposal facilities at Hanford, Barnwell, and Beatty, the states of California, Texas, and New York have located sites for LLW shallow land disposal, and the California site has been approved by the state and NRC. Investigation of the California site began in 1985, and as of this writing the third environmental impact study of the site is near completion. If and when the facility opens, it will be the first under the 1980 Low-Level Waste Policy Act. The remaining hurdles in the way of its opening are primarily the mistrust of some citizens' groups and the political considerations that appear to embroil all radioactive waste policy.

TRANSURANIC WASTE

With the end of the Cold War, the United States faced the task of dismantling its nuclear weapons production facilities and sufficiently cleaning the sites of those facilities so they could be used for other purposes. At the present time, neither plutonium nor weapons are being produced, and several sites in the weapons complex are dedicated entirely to clean-up.

Plutonium production and fabrication into weapons produced waste that was not nearly as highly radioactive as HLW, but it contained fairly large quantities of very long-lived alpha-emitting radionuclides, primarily Pu-238, Pu-239, Pu-240, and Am-241. This waste is in the form of contaminated machinery used in fabrication; dewatered and solidified sludges from plutonium purification; and contaminated laboratory equipment, tools, and material that had been cleaned up from spills. Since 1970, this material has been stored in drums above ground at various Department of Energy facilities. Because the drums have a surface dose of 200 mrem or less and can be handled directly, this waste is designated as contact-handled transuranic waste.

Since 1978, a bedded salt formation in southeastern New Mexico has been under investigation for a disposal facility for this material. The salt is expected to close around the waste and essentially encapsulate it; only intrusion by drilling could interfere with total encapsulation. In 1992, EPA was granted authority to permit and regulate the facility (the Waste Isolation Pilot Plant, WIPP) and has promulgated a regulation that sets release limits for it that allow for conditions under which there would be inadvertent drilling into the repository (drilling for oil or gas prospecting, for example). The WIPP was scheduled to begin receiving waste in 1998.

WASTE FORM MODIFICATION

Several methods for encapsulating radioactive waste in an impervious material are under investigation. The materials being studied are borosilicate glass, ceramic, asphalt, grout, portland cement, and a bakelite-type matrix called Envirostone. Encapsulation is especially appropriate for heterogeneous low-level waste, for which predicting mobility over long periods is very difficult. Glass and ceramic matrices are being investigated for high-level waste which is a byproduct of plutonium production and are both radioactively and thermally "hot." A disposal cask is being designed for commercial spent nuclear fuel.

CONCLUSION

Disposal of radioactive waste appears to be far more costly than was ever envisioned. Although public opposition has played a role, the cost is due primarily to two factors: the increased knowledge of geochemical processes that can mobilize radioactive materials stored underground and the complex models needed to predict repository behavior for 10,000 years or more. Burial in the ground, even in the deepest-mined repository, does not ensure adequate environmental protection for municipal solid waste, chemically hazardous waste, or radioactive waste. Satisfactory engineering of both a repository and a waste form is difficult and expensive, and reutilization options are becoming increasingly attractive. Reprocessing radioactive waste materials for use is no longer out of the question.

Reprocessing of commercial spent fuel has not been entertained as a serious option in the United States since the program was canceled in the mid-1970s. However, as the expense of storage and disposal increase, and as a viable disposal option seems further and further in the future, interest in reprocessing and formation of mixed (uranium and plutonium) oxides fuel is reviving. Mixed oxide fuel also provides an option for reducing the U.S. plutonium stockpile, whose stewardship is proving resource intensive and expensive.

Some radionuclides lend themselves quite readily to separation from waste: Radiocesium and radiostrontium are purified at Hanford, Washington, and used throughout the United States and the world as x-ray and heat sources. Many radiopharmaceuticals and radiochemicals are purified from by-product materials. Research into separation methods may prove as cost effective as research into encapsulation and disposal methods. Fuel combustion for electricity contributes to the concentration of CO_2 in the earth's atmosphere, while nuclear electric generation does not. Thorough and objective comparisons of electric generating methods should guide decisions about power generation. Generating electricity, no matter how it is done, causes irrevocable environmental damage on a scale directly proportional to the power produced.

PROBLEMS

16.1   Show that after 10 half-lives, about 0.1% of the initial radioactive material is left and that after 20 half-lives, 10^{-4}% is left.

16.2   Fe-55 has a half-life of 2.4 years. Calculate the disintegration constant. If a reactor core vessel contains 16,000 curies of Fe-55, how many curies will be left after 100 years? How many becquerels will be left?

16.3   I-131 is used in the diagnosis and treatment of thyroid disease, and a typical diagnostic amount is 1.9×10^6 Bq injected intravenously. If the *physiological* half-life of iodine is 15 days and a patient injected with I-131 produces about 3.5 liters of urine a day, what will be the specific activity (in Bq/L) of the patient's urine on the tenth day after injection? What will be the specific activity in Ci/L?

16.4   The dose in Problem 16.3 is adjusted not to exceed 0.5 Gy in 24 hours. Calculate the somatic risk in LCF from this dose.

LIST OF SYMBOLS

α	alpha radiation
β	beta radiation
γ	gamma radiation
amu	atomic mass unit
Bq	becquerel
esu	electrostatic unit of charge
eV	electron volt = 1.60×10^{-19} joule
Gy	gray: unit of absorbed energy; 1 joule/kg
HLW	high-level waste
K_b	fraction of atoms that disintegrate per second = $0.693/t_{1/2}$
LET	linear energy transfer
LLW	low-level waste
M	mass of a radionuclide in grams
MeV	million electron volts
N^0	Avogadro's number, 6.02×10^{23} atoms/gram atomic weight
Q	number of curies or becquerels
QF	quality factors
rad	unit of absorbed energy: 1 erg/gram
RBE	relative biological effectiveness
rem	roentgen equivalent man
SI	International System of Units
Sv	sievert; unit of dose equivalent
$t_{1/2}$	radiological half-life of a radionuclide
TLD	thermoluninescent detectors
TRU	transuranic material or transuranic waste
W	atomic weight of the radionuclide, in grams/gram = atom
WIPP	Waste Isolation Pilot Plant

Chapter 17

Solid, Hazardous, and Radioactive Waste Law and Regulations

Laws controlling environmental pollution are discussed in this text in terms of their evolution from the courtroom and Congressional committees to administrative agencies. The gaps in common law are filled by statutory laws adopted by Congress and state legislatures and implemented by administrative agencies such as EPA and state departments of environmental quality and natural resources. For several reasons, this evolutionary process was particularly rapid in the area of solid waste law.

For decades, and in fact centuries, solid waste was disposed of on land no one really cared about. Municipal refuse was historically taken to a landfill in the middle of a woodland, and industrial waste, which was often hazardous, was generally "piled out back" on land owned by the industry itself. In both cases, environmental protection and public health were not perceived as issues. Solid wastes were definitely out of sight and conveniently out of mind.

Only relatively recently has public interest in solid waste disposal sites reached the same level of concern as that for water and air pollution. Public interest was triggered in part by the rate at which land was being filled with waste and by the increasing cost of land for disposal. Some local courtrooms and zoning commissions have dealt with siting disposal facilities, but most decisions simply resulted in the city or industry hauling the waste a little farther away from the complaining public. At these remote locations, solid waste, unlike the smokestacks that emitted pollutants into the atmosphere and the pipes discharging wastewater into the rivers, was not visible. Pollution of the land and underground aquifers from solid waste disposal facilities was and is much more subtle.

Federal and state environmental statutes did not initially consider these subtle effects. The highly visible problems of air and water pollution were addressed in a series of clean air acts and water pollution control laws. Finally, as researchers dug deeper into environmental and public health concerns, they realized that even obscure landfills and holding lagoons have the potential to pollute the land significantly and to adversely impact public health. It became evident that subsurface and surface water supplies, and even local air quality, can be threatened by solid waste.

This chapter addresses solid waste law in three major sections: nonhazardous, chemically hazardous, and radioactive waste. The division reflects the regulatory philosophy for dealing with these three very distinct problems. Because of the general lack of common law in this area, we move directly to statutory controls of solid waste.

NONHAZARDOUS SOLID WASTE

The most significant solid waste disposal regulations were developed under the Resource Conservation and Recovery Act (RCRA) of 1976. This federal statute, which amended the elementary Solid Waste Disposal Act of 1965, reflected the concerns of the public in general and Congress in particular about (1) protecting public health and the environment from solid waste disposal, (2) closing the loopholes in existing surface water and air quality laws, (3) ensuring adequate land disposal of residues from air pollution technologies and sludge from wastewater treatment processes, and, most important, (4) promoting resource conservation and recovery.

The EPA implemented RCRA in a manner that reflected these concerns. Disposal sites were catalogued as landfills, lagoons, or landspreading operations, and the adverse effects of improper disposal were grouped into seven categories:

- *Floodplains* were historically prime locations for industrial disposal facilities because many industries elected to locate along rivers for water supply, power generation, or transportation of process inputs or production outputs. When the rivers flooded, the disposed wastes were washed downstream, with immediate adverse impact on water quality.
- *Endangered and threatened species* may be affected by habitat destruction as the disposal site is being developed and operated, or harmed by toxic or hazardous materials leaking from the site. Animals may be poisoned as they wander onto unfenced sites.
- *Surface water quality* may also be affected by certain disposal practices. Without proper control of runoff and leachate, rainwater has the potential to transport pollutants from the disposal site to nearby lakes and streams.
- *Groundwater quality* is of great concern because about half of the nation's population depends on groundwater for water supply. Soluble and partially soluble substances are leached from the waste site by rainwater; the leachate can contaminate underground aquifers.
- *Food chain crops* may be adversely affected by landspreading solid waste. In particular, leafy vegetables like lettuce and animal feed crops like alfalfa often bioconcentrate heavy metals and other trace chemicals.
- *Air quality* may be degraded by pollutants emitted from waste decomposition and may cause serious pollution problems downwind from a disposal

site. Methane generated at a site can cause explosions in downstream pipes. Uncovered waste in landfills can sustain spontaneous combustion, and landfill fires ("dump fires") can degrade air quality.
* *Health and safety* of onsite workers and the nearby public may be compromised by dump fires as well as explosions of gas generated at the site. Uncovered waste attracts flocks of birds that pose a hazard to aircraft.

The EPA guidelines spell out the operational and performance requirements to mitigate or eliminate these eight types of impact from solid waste disposal.

Ideally, operational standards, technologies, designs, and operating methods are sufficiently specific to protect public health and the environment. Any or all of a long list of operational considerations could go into a plan to build and operate a disposal facility: type of waste to be handled, facility location, facility design, operating parameters, and monitoring and testing procedures. The advantage of operational standards is that the best practical technologies can be used for solid waste disposal, and the state agency charged with environmental protection can determine compliance with a specified operating standard. The major drawback is that compliance is generally not measured by monitoring actual effects on the environment but rather as an either/or situation in which the facility either does or does not meet the required operational requirement.

On the other hand, performance standards are developed to provide given levels of protection to land, air, and water quality around the disposal site. Determining compliance with performance standards is not easy because the actual monitoring and testing of groundwater, surface water, and land and air quality are costly, complex undertakings. Because solid wastes are heterogeneous and because site-specific considerations are important in their adequate disposal, the federal regulatory effort allows state and local discretion in protecting environmental quality and public health. Therefore, EPA developed both operational and performance standards to minimize the effect of each of the eight potential impacts of solid waste disposal.

Floodplains. Floodplain protection focuses on limiting disposal facilities in such areas unless the local area has been protected against washout by the base flood. A base flood (sometimes called a "100-year flood") is one that has a 1 percent or greater chance of occurring in any year, or a flood equaled or exceeded only once in every 100 years on average.

Endangered and Threatened Species. Solid waste disposal facilities must be constructed and operated so that they will not contribute to the taking of endangered and threatened species. *Taking* means the harming, pursuing, hunting, wounding, killing, trapping, or capturing of species so listed by the U.S. Department of the Interior. Operators of a solid waste disposal facility may be required to modify operation to comply with these rules if nearby areas are designated critical habitats.

Surface Water. Solid waste disposal leads to the pollution of surface waters whenever rainwater percolates through the refuse and runoff occurs, or whenever spills occur during solid waste shipments. Point source discharges from disposal sites are regulated by the NPDES permit system discussed in Chapter 11. Nonpoint source, or diffused water, movement is regulated by the requirements implementing an area-wide or statewide water quality management plan approved by EPA. Generally, these plans regulate the degradation of surface water quality by facility design, operation, and maintenance. Artificial and natural runoff barriers such as liners, levees, and dikes are often required by states. If runoff waters are collected, the site becomes, by definition, a point source of pollution, and the facility must have an NPDES permit if this new point source discharges into surface watercourses.

Groundwater. To prevent contamination of groundwater supplies, the owners and operators of a solid waste disposal facility must comply with one or more of five design and operational requirements: (1) use of natural hydrogeologic conditions like underlying confining strata to block flow to aquifers, (2) collection and proper disposal of leachate by installing natural or synthetic liners, (3) reduction of rainwater infiltration into the solid waste with correct cover material, (4) diversion of contaminated groundwater or leachate from groundwater supplies, and (5) groundwater monitoring and testing procedures.

Land Used for Food Chain Crops. Federal guidelines attempt to regulate the movement of heavy metals and synthetic organics, principally polychlorinated biphenyl (PCB), from solid waste into soil used to produce human food chain crops. Typically, the concern here is sludge from municipal wastewater treatment facilities. To control heavy metals, the regulations give the facility operator two options: the controlled application approach and the dedicated facility approach.

In the controlled application approach, the operator must conform with an annual application rate of cadmium from solid waste. Cadmium is controlled basically as a function of time and the type of crop to be grown. Thus a cadmium addition of 2.0 kg/ha (1.8 lb/acre) is allowed for all food chain crops other than tobacco, leafy vegetables, and root crops grown for human consumption. For these accumulator crops an application rate of 0.5 kg/ha (0.45 lb/acre) is in force. The operator cannot exceed a cumulative limit on the amount of cadmium, which is a function of the cation exchange capacity (CEC) of the soil and the background soil pH. At low pH and low CEC, a maximum cumulative application rate of 5.0 kg/ha (4.50 lb/acre) is allowed; at high CEC and soils with high or near neutral pH levels, a 20 kg/ha (18 lb/acre) amount is allowed. The theory is that in soils of high CEC and high pH, the cadmium remains bound in the soil and is not taken up by the plants. All operators electing this option must ensure that the pH of the solid waste and soil mixture is 6.5 or greater at the time of application.

The second option is the dedicated facility, which relies on output control, or crop management, as opposed to input control, or limiting the amounts of cadmium that may be applied to the soil. It is designed specifically for a facility with the resources and the capabilities to closely manage and monitor the performance of its operations. The requirements include (1) growing only animal feed crops, (2) the pH of the solid waste and the soil mixture is 6.5 or greater at the time of application, (3) a facility operating plan that demonstrates how the animal feed will be distributed to preclude direct ingestion by humans, and (4) deed notifications for future property owners that the property has received solid wastes at high cadmium application rates and that food chain crops should not be grown because of possible health hazard.

Solid waste with PCB concentrations greater than or equal to 10 mg/kg of dry weight must be incorporated in the soil when applied to land used to produce animal feed, unless the PCB concentration is found to be less than 1.5 mg/kg fat basis in the produced milk. By requiring soil incorporation, the solid waste regulations attempt to ensure that the Food and Drug Administration levels for PCB in milk and animal feed will not be exceeded.

Air Quality. A solid waste disposal facility must comply with the Clean Air Act and the relevant air quality implementation plans developed by state and local air emission control boards. Open burning of residential, commercial, institutional, and industrial solid waste is generally prohibited, although several wastes—diseased trees, land-clearing debris, debris from emergency clean-up operations, and wastes from silviculture and agriculture operations—are excluded. Open burning is defined as the combustion of solid waste without (1) control of combustion in the air to maintain adequate temperatures for efficient combustion, (2) containment of the combustion reaction in an enclosed device to provide adequate residence time, or (3) control of the emissions from the combustion process.

Health. Federal rules require that the operation of a solid waste disposal facility protect the public health from disease vectors, that is, any routes by which disease is transmitted to humans (e.g., birds, rodents, flies, and mosquitoes). This protection must be achieved by minimizing the availability of food for the vectors. At landfills, an effective means to control vectors, especially rodents, is the application of cover material at the end of each day of operation. Other techniques include poisons, repellents, and natural controls such as predators. Treating sewage sludge with pathogen reduction processes serves to control the spread of disease from landscaping activities.

Safety. Fires at solid waste disposal facilities are a constant threat to public safety. In the past, death, injury, and vast property damage resulted from fires in open dumps. The ban on open burning reduces the chance of accidental fires. In terms of aircraft safety, facilities should not be located between airports and bird feeding, roosting, or watering sites. Flocks of birds continually fly around the working faces of even properly operated sanitary landfills, and they pose a severe hazard to nearby aircraft flying landing or takeoff patterns.

HAZARDOUS WASTE

Common law, in particular, is not well developed in the hazardous waste area. The issues discussed in Chapter 15 are newly recognized by society, and little case law has had a chance to develop. Since the public health impacts of solid waste disposal in general and hazardous waste disposal in particular were so poorly understood, few plaintiffs ever bothered to take a defendant into the common law courtroom to seek payment for damages or injunctions against such activities. For that reason hazardous waste law is discussed here as statutory law, specifically in terms of the compensation for victims of improper hazardous waste disposal and efforts to regulate hazardous waste generation, transport, and disposal. Historically, federal statutory law was generally lacking in guidelines for compensating victims of improper hazardous waste disposal. Thus a complex, repetitious, confusing list of federal statutes was a victim's only recourse.

The federal Clean Water Act covers only wastes that are discharged to navigable waterways. Surface water and ocean waters within 200 miles of the coast are alone considered, and a revolving fund is set up by the act and administered by the Coast Guard. Fines and charges are deposited into the fund to compensate victims of discharges, but the fund is available only if the discharger of the waste is clearly identifiable. It is used most often for compensating states for cleanup of spills from large tanker ships.

The Outer Continental Shelf (OCS) Lands Act sets up two funds to help pay for hazardous waste cleanup and to compensate victims. OCS leaseholders are required to report spills and leaks from petroleum-producing sites, and the Offshore Oil Spill Pollution Fund exists to finance cleanup costs and compensate injured parties for loss of the use of their property, loss of the use of natural resources, loss of profits, and a state or local government's loss of tax revenue. The U.S. Department of Transportation (DOT) is responsible for the administration of this fund, which is provided by a tax on oil produced at the OSC sites. If the operator of the site cooperates with DOT once a spill occurs, his or her liability is limited.

The Fisherman's Contingency Fund also exists under the OCS Lands Act to repay fishermen for loss of profit and equipment owing to oil and gas exploration, development, and production. The Secretary of the Department of Commerce is responsible for its administration. Assessments are collected from holders of permits and pipeline easements by the Department of the Interior. If a fisherman cannot pinpoint the site responsible for the discharge of hazardous waste, his or her claim against the fund may still be acceptable if the boat was operated within the vicinity of OCS activity and if he or she filed a claim within five days of the injury. That two agencies administer the fund makes the implementation especially confusing to the victims of hazardous waste incidents.

The Price-Anderson Act was passed to provide compensation to victims of discharge from nuclear facilities. Radioactive material/toxic spills and emissions are covered, as are explosions. The licensees of a nuclear facility are required by NRC to obtain insurance protection. If damages from an accident are greater than this insurance coverage, the federal government indemnifies the licensee up to $500 million. In no case does financial liability exceed $560 million. If an accident occurs such that liability exceeds that amount, federal disaster relief is called upon.

Under the Deepwater Ports Act, the Coast Guard is responsible for the removal of oil from deepwater ports. DOT administers a liability fund that helps pay for the cleanup and compensates injured parties. Financed by a 2-cent-per-barrel tax on oil loaded or unloaded at such a facility, the fund takes effect once the required insurance coverage is exceeded. The deepwater port itself must hold insurance for $50 million for claims against its waste discharges, and vessels using the port must hold insurance for $20 million for claims against their waste spills. Once these limits are exceeded, the fund established by the Deepwater Ports Act takes over. Again, the usefulness of the fund is dependent on how well two agencies work together and how well insurance claims are administered. At best administration is confusing to injured parties. After the 1990 oil spill in Alaska, the Exxon Corporation was required to provide additional compensation for loss of fisheries.

Statutes at the state level have generally paralleled these federal efforts. New Jersey has a Spill Compensation and Control Act for hazardous wastes listed by EPA. The fund covers cleanup costs, loss of income, loss of tax revenues, and the restoration of damaged property and natural resources, and is financed by a tax of 1-cent-per-barrel of hazardous substance. It pays injured parties if they file a claim within six years of the hazardous waste discharge and within one year of the discovery of the damage.

The New York Environmental Protection and Spill Compensation Fund is similar to the New Jersey fund, but differs in two respects. In New York, only petroleum discharges are covered and the generator may not blame a third party for a spill or discharge incident. Thus if a trucker or handler accidentally spills hazardous waste, the generator of the waste may still be held liable for damages.

Other states have made limited efforts to compensate victims of hazardous waste incidents. Florida's Coastal Protection Trust Fund of $35 million compensates victims for spills, leaks, and dumping of waste. However, coverage is limited to injury resulting from oil, pesticides, ammonia, and chlorine, which provides a loophole for many types of hazardous waste discussed in Chapter 15. Other states that have compensation funds generally spell out limited compensationable injuries and provide limited funds—often less than $100,000.

The value of these federal and state statutes has been questioned. Even when taken as a group, they do not provide a complete compensation strategy for dealing with hazardous waste. Few funds address personal injuries, and abandoned disposal sites are not considered. Moreover, huge administrative problems arise when several agencies attempt to respond to a hazardous waste spill. Questions linger regarding which fund applies, which agency has jurisdiction, and what injuries may receive compensation.

In 1980, the federal Comprehensive Environmental Response, Compensation and Liability Act (CERCLA), commonly called "Superfund," was enacted. Originally, it was to serve two purposes: to provide money for the cleanup of abandoned hazardous waste disposal sites and to establish liability so dischargers could be required to pay for injuries and damages. Three considerations went into the act's development:

- The type of incidents to be covered by Superfund include accidental spills and abandoned sites as well as onsite toxic pollutants in harbors and rivers. Fires and explosions caused by hazardous materials also have become a focus as the Superfund has evolved.
- The type of damages to be compensated include environmental cleanup costs; economic losses associated with property use, income, and tax revenues; and personal injury in the form of medical costs for acute injuries, chronic illness, death, and general pain and suffering.
- The sources of payment for the compensation include federal appropriations to the fund, industry contributions for sales tax, and income tax surcharges. Federal and state cost sharing is also possible, as is the establishment of fees for disposal of hazardous waste at permitted disposal facilities.

A fund of between $1 and $4 billion has been established, financed 87 percent by a tax on the chemical industry and 13 percent by general revenues of the federal government. Small payments are permitted out of the fund, but only for out-of-pocket medical costs and partial payment for diagnostic services. The Superfund Act does not set strict liability for spills and abandoned hazardous wastes. The Superfund Amendments and Reauthorization Act (SARA) of 1986 reauthorized and strengthened CERCLA.

Compensation for damages is one concern; regulations that control generators, transporters, and disposers of hazardous waste are another. RCRA, discussed in this chapter as it relates to solid waste, also is the principal statute that deals with hazardous waste through the 1984 Hazardous and Solid Waste Act (HSWA), which amended RCRA to cover generators of hazardous waste. RCRA requires EPA to establish a comprehensive regulatory program of hazardous waste control. This 1984 act is a good example of the types of reporting and recordkeeping mandated in federal environmental laws. The Clean Water Act and the Clean Air Act place requirements on water and air polluters similar to those for generators of hazardous waste.

It is worth noting that a large number of facilities and processes that emit hazardous materials are exempted from RCRA control by EPA regulation (40 CFR 261.4). They include

- Spent sulfuric acid from sulfuric acid manufacture.
- Wastes from production of coal, coke, and coal tar fuel.
- Ash from MSW incineration.
- Any solid waste from fossil fuel combustion.

- Any waste associated with "exploration, development, or production of crude oil, natural gas, or geothermal energy."
- Chromium waste from production or use of trivalent chromium, including chrome shavings and wastewater treatment sludges from tanning.
- Solid waste from extraction, beneficiation, and processing of ores (up to the point of smelting).
- Slag from primary copper, zinc, and lead smelting; bauxite refining, phosphorus, and phosphoric acid production.

Many constituents of these wastes are carcinogenic or otherwise hazardous and may be regulated under RCRA as part of some other waste. Moreover, most of these effluents are controlled under clean air or clean water legislation, or both. Nevertheless, these exemptions indicate the arbitrary nature of some RCRA regulations. The possibility that RCRA may be incompatible with other legislation, including the Atomic Energy Act (AEA), was envisioned in RCRA itself, and EPA is directed to integrate all provisions of RCRA for purposes of administration and enforcement.

In addition to CERCLA and RCRA control of hazardous waste, the Toxic Substances Control Act (TOSCA) regulates hazardous substances before they become hazardous waste. Under one of its regulatory programs, the Premanufacturing Notification (PMN) System, all manufacturers must notify EPA before marketing any substance not included in EPA's 1979 inventory of toxic substances. In this notification, the manufacturer must analyze the predicted effect of the substance on workers, on the environment, and on consumers. This analysis must be based on test data and all relevant literature.

RADIOACTIVE WASTE

The first legislation passed to protect public safety and health from damage done by exposure to ionizing radiation was the Atomic Energy Act of 1954, which created the seven-member Atomic Energy Commission (AEC). The AEC had jurisdiction over all aspects of radioactivity and radioactive materials, both commercial and defense-related, and reported to the House–Senate Joint Committee on Atomic Energy. In 1974, the Energy Reorganization Act abolished the House–Senate Joint Committee and separated the regulatory function of the AEC from the research and defense-related functions. The AEC was divided into two agencies—the five-member Nuclear Regulatory Commission (NRC) and the Energy Research and Development Administration (ERDA). Initially, the NRC was given regulatory authority over commercial nuclear power and the nondefense use of radioactive materials; ERDA was given all other AEC functions, including the purification, fabrication, and maintenance of plutonium and fissile uranium. In 1977, ERDA was combined with the Federal Energy Administration (formerly the White House Federal Energy Office) to form the Department of Energy (DOE). The Secretary of Energy has cabinet rank.

The first legislation dealing specifically with radioactive waste was the Low Level Radioactive Waste Policy Act of 1980 that introduced the notion of regional LLW sites and interstate compacts for LLW management. Recent amendments to this law have provided some deadlines and guidelines for accepting out-of-region wastes. This law provides NRC with the authority to issue regulations for LLW.

The Nuclear Waste Policy Act of 1982 is the first law that addressed spent nuclear fuel from commercial nuclear plants. The act required DOE to construct a repository for high-level radioactive waste (HLW), required EPA to set standards for the repository, and required NRC to license the repository to the EPA standards. It also required DOE to take title to commercial spent fuel by 1998, funded involvement in repository decisions by states and Native American tribes, and set up the Nuclear Waste Trust Fund to fund the repository program. The Act was amended in 1987 to concentrate repository investigations at Yucca Mountain, Nevada, and again in 1992 to charge EPA with revising the standard for the repository.

The WIPP Land Withdrawal Act of 1992 authorized the transfer of land for the Waste Isolation Pilot Plant (WIPP) from the Department of Interior to DOE and authorized EPA to certify the site. As discussed in Chapter 16, the WIPP is a repository for transuranic waste from weapons fabrication. The Land Withdrawal Act was amended in 1996 to remove the requirement for a no-migration variance under RCRA from the WIPP.

CONCLUSION

Solid waste statutory law developed rapidly once the public, along with scientists and engineers, realized the real and potential impacts of improper disposal. Air, land, and water quality from such disposal became a key concern to local health officials and federal and state regulators. Hazardous and nonhazardous solid wastes are now controlled under a complex system of federal and state statutes that impose operating requirements on facility operators.

How does EPA implement RCRA? Most states have been given the responsibility of controlling hazardous and nonhazardous waste within their boundaries, and the next few years will show how they respond. The "joint and several liability" provision of Superfund, which makes any and all waste disposers at a site liable irrespective of the amount of waste for which they are responsible, has already stalled or slowed some implementation. In the end, what will be the public health, environmental quality, and economic impacts of RCRA and the Superfund? The laws discussed in this chapter and throughout this book may be found on the World Wide Web.

PROBLEMS

17.1   Local land use ordinances often play a key role in limiting the number of sites available for a solid or hazardous waste disposal facility. Discuss the types of these zoning restrictions that apply in your hometown.

17.2   Assume you work for a firm that contracts to clean the inside and outside of factories and office buildings in the state capital. Your boss thinks she could make more money by expanding her business to include the handling and transportation of hazardous wastes generated by her current clients. Outline the types of data you must collect to advise her to expand or not expand.

17.3   Laws that govern refuse collection often begin with controls on the generator—that is, rules that each household must follow if city trucks are to pick up its solid waste. If you were asked by a town council to develop a set of such "household rules," what controls would you include? Emphasize public health concerns, minimizing the cost of collection, and even resource recovery considerations.

17.4   The oceans have long been viewed as bottomless pits into which the solid wastes of the world may be dumped. Engineers, scientists, and politicians are divided on the issue. Should ocean disposal be banned? Develop arguments pro and con.

17.5   RCRA allows exemption of radioactive waste. However, virtually all radioactive waste contains some RCRA-controlled substances. Should HLW, LLW, and TRU waste disposal facilities have to comply with RCRA also? Give arguments pro and con.

LIST OF SYMBOLS

AEA	Atomic Energy Act
AEC	Atomic Energy Commission
CEC	cation exchange capacity
CERCLA	Comprehensive Environmental Response, Compensation and Liability Act ("Superfund")
DOE	Department of Energy
DOT	U.S. Department of Transportation
ERDA	Energy Research and Development Administration
EPA	U.S. Environmental Protection Agency
HLW	high-level radioactive waste
HSWA	Hazardous and Solid Waste Act
MSW	municipal solid waste
NRC	Nuclear Regulatory Commission
OCS	Outer Continental Shelf
PMN	Premanufacturing Notification System
RCRA	Resource Conservation and Recovery Act
SARA	Superfund Amendments and Reauthorization Act
TOSCA	Toxic Substances Control Act
WIPP	Waste Isolation Pilot Plant

Chapter 18

Air Pollution

Be it known to all within the sound of my voice,
whoever shall be found guilty of burning coal
shall suffer the loss of his head.

—King Edward II

Air pollution is not a new problem. Indeed, King Edward II of England (1307–1327) tried to abate what his wife Eleanor of Aquitaine called Britain's "unendurable smoke" by prohibiting the burning of coal while Parliament was in session. Richard II (1377–1399) levied a tax on smoke production, and Henry V (1413–1422) formed the first of a number of commissions to study the problem. However, none of these commissions actually reduced the level of air pollution in England. Under Charles II (1630–1685), John Evelyn published a pamphlet in 1661, "Fumifugium, or the Inconvenience of Aer and Smoak of London Dissipated, together with some Remedies Humbly Proposed," which suggested, among other "remedies," moving industry to the outskirts of London and establishing greenbelts around the city. None of his proposals were implemented. In 1845 a commission suggested that locomotives "consume their own smoke," and two years later this requirement was extended to chimneys. In 1853 it was decreed that offending chimneys be torn down (this decree bears an uncanny resemblance to modern city ordinances that limit the use of wood stoves).

In spite of the rhetoric, there was little air pollution abatement in England, or anywhere else in the industrialized world, until after World War II. Action was finally prompted in part by two major air pollution episodes, in which human deaths were directly attributed to high levels of pollutants.

The first of these occurred in Donora, a small steel town of 14,000 in western Pennsylvania. Donora, on a bend of the Monongahela River, had three heavy industries in 1948: a steel mill, a wire mill, and a zinc plating plant. During the last week of October 1948, a heavy smog settled in the area and a weather inversion prevented the movement of pollutants out of the valley (see Chapter 19 for a definition of inversion). By midweek, the smog had become so intense that streamers of carbon appeared to hang motionless in the air and visibility was so poor that even natives could not find their way around town. By Friday, hospitals and doctors' offices were flooded with calls for medical help, yet no alarm was sounded, and a Friday parade and football game were well attended (although the visiting coach complained that Donora had arranged the smog so that his players couldn't see the ball).

The first death occurred at 2 A.M. that Friday, and more followed. By midnight seventeen persons were dead, and four more died before the smog abated. By this time the emergency had been recognized and medical help was rushed to the scene.

Although the Donora episode helped to focus attention on air pollution in the United States, England took action only four years later after a similar disaster. The "Killer Smog" of 1952 occurred in London under conditions similar to the Donora smog episode. Bitter cold and a dense fog at ground level trapped the smoke from coal burners, causing a "pea soup" more severe than the usual London smog that lasted over a week. Visibility during daylight hours was cut to a few yards. Sulfur dioxide concentrations increased to nearly seven times their normal levels, and carbon monoxide concentration was about twice as high as usual. Figure 18–1 shows the relationship between sulfur dioxide and deaths during the killer smog. It is unlikely, however, that sulfur dioxide alone caused the fatalities. Other air pollutants acting synergistically, and factors like the severe cold, probably affected the death rate as well.

By the mid-1950s the air in New York City seemed exceedingly dirty, particularly in the summers as electrically powered air conditioning became ubiq-

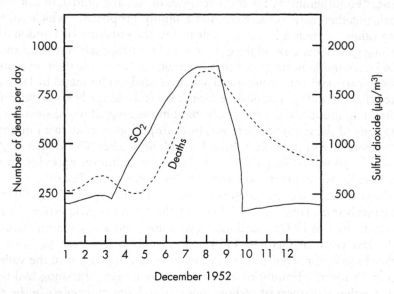

FIGURE 18–1. Deaths and SO_2 concentration during the London air pollution episode of 1952. [From Wilkins, E. T., *Journal of the Royal Sanitary Institute*, no. 74, 1(1954).]

uitous. New York burned sulfur-containing oil to generate electricity, and the sulfurous smog (much like that in London) became increasingly troublesome. At the same time, the city of Los Angeles noticed an increasing haze that decreased visibility, tasted acrid, and seemed quite different from New York air pollution. In 1972, Arie Haagen-Smit produced this "Los Angeles smog" in his laboratory at the California Institute of Technology by irradiating Mylar bags of automobile exhaust (from his own car) with ultraviolet radiation and ordinary sunlight. Two kinds of dirty air had now been identified: London-type smog, mainly sulfurous smoke, and Los Angeles-type smog, consisting of photochemical by-products of car exhaust gases.

Clean air is difficult to define. The clean air found in nature has more components than scientifically defined pure air (Table 18–1). Particulate matter and dust of all sorts may become suspended in the air by natural processes. For example, the blue haze that gives the Blue Ridge Mountains of Appalachia their name is a natural emission from the mountain forests. Air pollutants may be more appropriately defined as substances that exist in the air in sufficient concentration to cause an unwanted effect. Undesirable concentrations of air contaminants usually result from some human activity and are called *anthropogenic pollutants*. Occasional natural phenomena, like volcanic eruptions and lightning-caused forest fires, can also pollute the air to an unwanted extent. Air pollutants may be in the form of gases or suspended particulate matter (liquid or solid particles).

TABLE 18–1. Composition of Normal Dry Tropospheric Air

Element	Concentration (ppm)
Nitrogen	780,900
Oxygen	209,400
Argon	9300
Carbon dioxide	315
Neon	18
Helium	5.2
Methane	1.0–1.2
Krypton	1.0
Nitrous oxide	0.5
Hydrogen	0.5
Xenon	0.008
Nitrogen dioxide	0.02
Ozone	0.01–0.04

TYPES AND SOURCES OF GASEOUS AIR POLLUTANTS

Table 18–2 lists some common gaseous air pollutants. In the United States, these five are recognized as being pollutants of national concern:

- Sulfur dioxide
- Oxides of nitrogen (NO and NO_2)
- Carbon monoxide
- Hydrocarbons
- Ozone and other photochemical oxidants
- Carbon dioxide
- Chlorofluorocarbons

Sulfur Oxides

In the United States and the developed nations of the world, the largest single anthropogenic source of sulfur dioxide is the combustion of sulfur-containing fossil fuel, for both electric power generation and process heat. Sulfur dioxide generation is a function of the amount of sulfur in the fuel, which amount can be significant, as demonstrated in Example 18.1.

Example 18.1

It is estimated that during the 1952 London smog episode 25,000 metric tons of coal, with an average sulfur content of 4%, was burned. The mixing depth (the height of the inversion layer or cap over the city that prevented the escape of pollutants) was about 150 m over an area of about 1200 km^2. What was the approximate SO_2 concentration after the coal was burned?

The amount of sulfur in the coal was

$$(25,000 \text{ metric tons})(10^6 \text{ g/metric ton})(0.04 \text{ g S/g coal}) = 10^9 \text{ g S} \qquad (18.1)$$

Each mole of sulfur yields 1 mole of SO_2 when burned completely. Sulfur has an atomic weight of 32 g/g-atom; oxygen, of 16 g/g-atom. Thirty-two grams of S thus produces 64 grams of SO_2, so the weight of SO_2 produced by the burning coal is

$$(10^9 \text{ g S})(64 \text{ g } SO_2/32 \text{ g S}) = 2 \times 10^9 \text{ g } SO_2 \qquad (18.2)$$

and the SO_2 concentration is

$$\frac{(2 \times 10^9 \text{ g } SO_2)(10^6 \, \mu g/g)}{(150 \text{ m})(1200 \text{ km}^2)(10^6 \text{ m}^2/km^2)} = 11,000 \, \mu g/m^3 \qquad (18.3)$$

However, the measured peak concentration of SO_2 during the London episode was less than 2000 $\mu g/m^3$. Where did all the SO_2 go? This question is addressed in Chapter 19.

TABLE 18–2. Gaseous Air Pollutants

Name	Formula	Properties of Importance	Significance as Air Pollutant
Sulfur dioxide	SO_2	Colorless gas, intense acrid odor, forms H_2SO_3 in water	Damage to vegetation, building materials, respiratory system
Sulfur trioxide	SO_3	Soluble in water to form H_2SO_4	Highly corrosive
Hydrogen sulfide	H_2S	Rotten egg odor at low concentrations, odorless at high concentrations	Extremely toxic
Nitrous oxide	N_2O	Colorless; used as aerosol carrier gas	Relatively inert; not a combustion product
Nitric oxide	NO	Colorless; sometimes used as anaesthetic	Produced during combustion and high-temperature oxidation; oxidizes in air to NO_2
Nitrogen dioxide	NO_2	Brown or orange gas	Component of photochemical smog formation; toxic at high concentration
Carbon monoxide	CO	Colorless and odorless	Product of incomplete combustion; toxic at high concentration
Carbon dioxide	CO_2	Colorless and odorless	Product of complete combustion of organic compounds; implicated in global climate change
Ozone	O_3	Very reactive	Damage to vegetation and materials; produced in photochemical smog
Hydrocarbons	C_xH_y	Many different compounds	Emitted from automobile crankcase and exhaust
Hydrogen fluoride	HF	Colorless, acrid, very reactive	Product of aluminum smelting; causes reactive fluorosis in cattle; toxic

One effect of sulfur oxide pollution is the formation of acid rain, which results when sulfur dioxide (as well as other gases such as nitrogen dioxide) reacts with water and atmospheric oxygen to produce sulfuric acid—acid rain's main constituent. Uncontaminated rain has a pH of about 5.6, but acid rain can be as acidic as pH 2. Many lakes in Scandinavia and North America have become so acidic that they can no longer support fish life. Studies of Norwegian lakes show that lakes with pH 5.3 or higher contain fish, but 70% of the lakes with pH less than 4.5 contain none. Low pH not only affects fish directly, but contributes to the increased solubility of metals (from sediment to the water), potentially toxic to fish, in lakes and soil, thus magnifying the acid rain problem.

Storms that travel over Europe and Great Britain have been tracked and shown to deposit large amounts of acid precipitation on Norway. Figure 18–2 shows three such storms and the corresponding pH values of the rainwater. Recognition of the acid rain phenomenon calls into question the use of very tall stacks to dilute air pollutants and thus control local air pollution (Chapter 21).

Fish and aquatic plants in many New England lakes have been decimated by acid precipitation. The pH in these lakes has reached a sufficiently low level that trout and native aquatic plants have been replaced by mats of acid-tolerant algae.

In any given community, an industrial facility may be the principal source of SO_2, since many industrial processes emit it in significant quantities. Some important industrial emitters, in addition to fossil fuel combustion for power generation, are

- *Nonferrous smelters.* With the exception of iron and aluminum, metal ores are sulfur compounds. When the ore is reduced to the pure metal, its sulfur is ultimately oxidized to SO_2. Copper ore, for example, is CuS or Cu_2S. Since the atomic weight of Cu is 64 g/mol, every pound of copper refined from CuS yields a pound of SO_2, which goes into the air unless it is trapped first.
- *Oil refining.* Sulfur and hydrogen sulfide are constituents of crude oil, and H_2S is released as a gas during catalytic cracking. Since H_2S is considerably more toxic than SO_2, it is flared to SO_2 before release to the ambient air.
- *Pulp and paper manufacture.* The sulfite process for wood pulping uses hot H_2SO_3 and thus emits SO_2 into the air. The kraft pulping process produces H_2S, which is then burned (flared) to produce SO_2.

Two natural (nonanthropogenic) sources of SO_2 are volcanic eruptions and sulfur-containing geothermal sources like geysers and hot springs.

Oxides of Nitrogen

Nitric oxide is formed by the combustion of nitrogen-containing compounds (including fossil fuels) and by thermal fixation of atmospheric nitrogen. The equilibrium constant for the reaction is

$$N_2 + O_2 \Leftrightarrow 2NO \tag{18.4}$$

Thus all high-temperature processes produce NO, which is then oxidized further to NO_2 in the ambient air. In the United States, about half of the atmospheric nitrogen oxides are produced by stationary sources. A major global effect of nitrogen oxide emissions is their contribution to the formation of acid rain, discussed later.

FIGURE 18–2
Movement of three storms
in Europe bringing acid
precipitation to Kristiansand,
Norway

	pH of rain water at sampling site
"Normal"	5.8
Storm of 25 Aug.	4.2
Storm of 26 Aug.	4.0
Storm of 27 Aug	4.1

Carbon Monoxide

CO is a product of the incomplete combustion of carbon-containing compounds. Stationary combustion sources produce CO, which is oxidized to CO_2 while dispersing in the air from the stationary source. Stationary sources of CO are significant only quite near the source.

Most of the CO in the ambient air comes from vehicle exhaust. Internal combustion engines do not burn fuel completely to CO_2 and water; some unburned fuel will always be exhausted, with CO as a component. CO in vehicle exhaust can be reduced by using partially oxidized fuels like alcohol and by a variety of afterburner devices. It tends to accumulate in areas of concentrated vehicle traffic, in parking garages, and under building overhangs.

Hydrocarbons

Vehicles are also a major source of atmospheric hydrocarbons. Figure 18–3 shows some of the points of hydrocarbon emission in an automobile engine. Stationary sources of hydrocarbons include petrochemical manufacture, oil refining, incomplete incineration, paint manufacture and use, and dry cleaning.

Ozone (Photochemical Oxidant)

Photochemically formed organic oxidants, classified as ozone, are a secondary air pollutant. That is, ozone is not emitted directly into the air, but is the result of chemical reactions in the ambient air. The components of automobile exhaust are particularly important in the formation of atmospheric ozone and are the primary contributors to Los Angeles smog. Table 18–3 lists some of the key reactions in the formation of photochemical smog in simplified form.

The reaction sequence illustrates how nitrogen oxides formed in the combustion of gasoline and other fuel, and emitted to the atmosphere, are acted upon by sunlight to yield ozone (O_3). Since ozone is not emitted directly but formed from other pollutants in the air, it is considered a secondary rather than a primary pollutant like CO.

Ozone in turn reacts with hydrocarbons in the air to form a variety of compounds, including aldehydes, organic acids, and epoxy compounds. The atmosphere may be viewed as a huge reaction vessel in which new compounds are being formed while others are being destroyed. Unfortunately, no method has been found to convey the polluting ozone formed in the *troposphere* (the layer of air closest to the earth's surface) to the *stratosphere* to replenish the ozone layer.

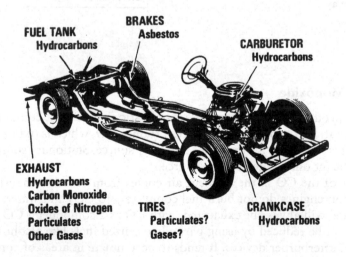

FUEL TANK
Hydrocarbons

BRAKES
Asbestos

CARBURETOR
Hydrocarbons

EXHAUST
Hydrocarbons
Carbon Monoxide
Oxides of Nitrogen
Particulates
Other Gases

TIRES
Particulates?
Gases?

CRANKCASE
Hydrocarbons

FIGURE 18–3. Locations of hydrocarbon emission in an automobile engine

TABLE 18–3. Production of Photochemical Smog

NO$_2$ (nitrogen dioxide)	+ Sunlight	→ NO (nitric oxide)	+ O (atomic oxygen)
O	+ O$_2$ (molecular oxygen)	→ O$_3$ (ozone)	
O$_3$	+ NO	→ NO$_2$	+ O$_2$
O	+ HC (hydrocarbons)	→ HCO· (radical)	
HCO·	+ O$_2$	→ HCO$_3$·	
HCO$_3$·	+ HC	→ Aldehydes and ketones	
HCO$_3$·	+ NO	→ HCO$_2$·	+ NO$_2$
HCO$_3$·	+ O$_2$	→ O$_3$	+ HCO$_2$·
HCO$_x$·	+ NO$_2$	→ Peroxyacetyl nitrates	

The formation of photochemical smog is a dynamic process. Figure 18–4 illustrates the diurnal variation of the atmospheric concentration of some of these compounds. As the morning rush hour begins, the NO levels increase, followed quickly by levels of NO$_2$. As the NO$_2$ reacts with sunlight, ozone and other oxidants are produced. The hydrocarbon level increases similarly at the beginning of the day and decreases in the evening.

Carbon Dioxide and Other "Greenhouse Gases"

Carbon dioxide, methane, and gas molecules that have a similar structure may influence the global climate by the following mechanism: Internal molecular vibration and rotation cause these molecules to absorb infrared radiation. When these gases form part of the atmosphere, they absorb some of the heat that the

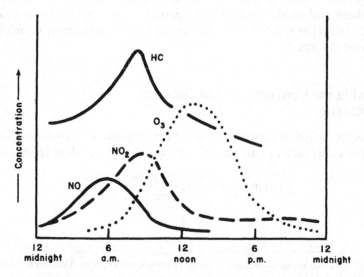

FIGURE 18–4. Formation of photochemical smog

earth normally radiates into space. Because they are heavier than oxygen and nitrogen, they are more concentrated close to the earth's surface and can be thought to form a "canopy" of gas, similar to the glass canopy of a greenhouse and with a similar result: Heat is trapped that would otherwise be lost. The effect of CO_2 on climate is discussed further in the section on climate change.

Chlorofluorocarbons

The loss of stratospheric ozone was first discovered in 1989, when measurement of the stratospheric ozone layer showed a large hole over Antarctica. Since then, the hole seems to reappear each spring, though it does not always seem to be the same size. The stratospheric ozone layer exists because of a dynamic equilibrium between ozone, molecular oxygen, and atomic oxygen:

$$O_3 \Leftrightarrow O_2 + O \tag{18.5}$$

Energy for the forward reaction is provided by the ultraviolet components of sunlight. The reverse reaction occurs because atomic oxygen (O) is highly reactive and recombines with molecular oxygen (O_2) on collision. Precisely because O is so reactive, it also combines with other molecules or free radicals with which it collides. When readily oxidizable molecules like NO and halogen free radicals are introduced into or near the stratosphere, they can scavenge atomic oxygen from the ozone layer which is then unavailable for recombination with molecular oxygen, and the reaction given by Equation 18.5 is driven to the right, depleting the concentration of ozone. The scavenging is apparently enhanced by ice crystals in the air over the poles. Studies have correlated the depletion of the ozone layer with worldwide production of chlorofluorocarbon (CFC) refrigerants.[1] These compounds are particularly good scavengers of atomic oxygen. The hydroxyl radical (OH) is also present in the atmosphere and oxidizes CFCs and related compounds. The mechanism of CFC action on the ozone layer is discussed further in the section on global atmospheric change.

Expressing the Concentration of Gaseous Air Pollutants

Gaseous air pollutant concentrations can be expressed in two ways: as micrograms per cubic meter of air ($\mu g/m^3$) and as parts per million (ppm), where

$$1 \text{ ppm} = \frac{1 \text{ volume of gaseous pollutant}}{10^6 \text{ total volumes}} \tag{18.6}$$

---

[1]CFCs are marketed in the United States under the trade name Freon. Dichlorodifluoromethane, for example, is Freon-12.

or 1 ppm = 0.0001% by volume. Conversion between $\mu g/m^3$ is by the ideal gas law

$$PV = nRT \qquad (18.7)$$

where   P = pressure of gas
         V = volume of gas
         n = number of moles
         R = gas constant
         T = absolute temperature of gas (in $^\circ K$ or $^\circ R$)

Example 18.2

The exhaust of an automobile is found to contain 2% by volume CO, at a temperature of 80°C. Express the CO concentration in the exhaust in $\mu g/m^3$.

    2% = 20,000 ppm = 20,000 L of CO/10^6 L of exhaust
    T = 273 + 80 = 353°K
    P = 1 atmosphere
    R = 0.082 L-atm/mole-°K
    Mol. weight of CO = 12 + 16 = 28 g/mole

From the ideal gas law,

$$PV = nRT = \frac{\text{Weight of CO}}{\text{Molecular weight of CO}} RT \qquad (18.7a)$$

Or, solving for the weight of CO,

$$\text{weight of CO} = \frac{(1 \text{ atm}) (20,000 \text{ L}) (28 \text{ g/mole})}{(0.082 \text{ L-atm-}^\circ\text{K})(353^\circ\text{K})} \qquad (18.7b)$$

$$= 1.93 \times 10^4 \text{g}$$

$$\frac{20,000 \text{ L of CO}}{10^6 \text{ L of exhaust}} = 1.93 \times 10^4 \text{ g/}10^6 \text{ L}$$
$$\qquad (18.7c)$$
$$= 19.3 \text{ g/m}^3$$

PARTICULATE MATTER

The form and size of airborne nongaseous pollutants are

1. *Dust,* consisting of solid particles that are (a) entrained by process gases directly from the material being handled or processed, like cement dust or grain from grain elevators; (b) direct offspring of a parent material undergoing a mechanical operation, like sawdust from woodworking; and (c) entrained materials used in a mechanical operation, like sand from

sandblasting. Dust particles are between 0.1 micron and 10 mm in diameter; they can be relatively large.

2. *Fume,* that is, solid particles formed by the condensation of vapors by sublimation, distillation, calcination, or other chemical reactions. Examples include zinc and lead oxides resulting from the oxidation and condensation of metal volatilized in a high-temperature process. Fume particles are from 0.03 to 0.3 μm in diameter.

3. *Mist,* liquid particles formed by the condensation of a vapor and sometimes by chemical reaction as, for example, the formation of sulfuric acid mist:

$$SO_3 \text{ (gas) } 22°C \rightarrow SO_3 \text{ (liquid)}$$
$$SO_3 \text{ (liquid) } + H_2O \rightarrow H_2SO_4$$

(18.8)

Sulfur trioxide gas becomes a liquid because its dew point is 22°C, and SO_3 particles are hygroscopic. Mists typically range from 0.5 to 3.0 μm in diameter.

4. *Smoke,* which consists of solid particles formed by incomplete combustion of carbonaceous materials. Although hydrocarbons, organic acids, sulfur dioxide, and oxides of nitrogen are also produced by combustion processes, only the solid particles resulting from incomplete combustion of carbonaceous materials are called smoke. Smoke particle diameters are between 0.05 μm and approximately 1 μm.

5. *Spray,* which is a liquid particle formed by the atomization of a parent liquid.

Virtually every industrial process is a potential source of dust, smoke, or aerosol emissions, including waste incineration, coal combustion, combustion of heavy (bunker-grade) oil, and smelting. Agricultural operations are a major source of dust, especially dry-land farming, as are demolition and construction. Traffic on roads, even when they are completely paved, is also a major source. In 1985, 40% of the total suspended particulate matter in the air of Seattle's downtown industrial center was identified as "crustal dust" raised by traffic on paved streets.

Fires are a major source of airborne particulate matter, as well as of hydrocarbon emissions, CO, and dioxin-related compounds. Forest fires are usually considered a natural (nonanthropogenic) source, but fires for land clearing, slash burns, agricultural burns, and trash fires contribute considerably. Since 1970, most communities in the United States have prohibited open burning of trash and dead leaves, and the Resource Conservation and Recovery Act of 1976 (RCRA) regulates the management of municipal waste landfills so that dump fires are a thing of the past.

Wood-burning stoves and fireplaces also produce smoke that contains partly burned hydrocarbons, aromatic compounds, tars, aldehydes, and dioxins, as well as smoke and ash. A growing number of cities in the United States now permit only low-polluting stoves ("certified" stoves) and prohibit the use of wood stoves and fireplaces altogether during unfavorable weather conditions.

The most important chemically identifiable particulate air pollutant is lead, usually lead oxide or lead chloride. Vehicle exhaust is the source of most airborne lead in the United States except in the vicinity of nonferrous smelters, although demolition of structures in which lead paint was used also puts lead dust in the air. The Bunker Hill lead smelter at Kellogg, Idaho, the largest in the United States, was shut down in 1980, eliminating a significant source of lead pollution in that city. Lead from vehicle exhaust is deposited for several hundred yards downwind of highways. At concentrations of freeway interchanges, in urban areas, deposited lead has been identified as a source of lead intoxication.

HAZARDOUS AIR POLLUTANTS

In addition to the air pollutants already mentioned, more than 300 chemical compounds have been identified as hazardous enough that controlling them as particulate matter is inadequate for human health protection. Specific control techniques are needed for these pollutants, which include asbestos, compounds of antimony, arsenic, beryllium, lead, mercury, manganese and nickel, benzene and most benzene derivatives (like toluene, phenols, and xylenes), halogen gases, alkyl and vinyl halides, organic nitrates, ethers, chloroform and its oxidation compounds, and a variety of pesticides and other organic compounds.

Hazardous substances are emitted into the air by a number of sources, chief among them the petrochemical industry. Some other specific sources of certain airborne hazardous materials are

- *Asbestos.* Sources are construction, demolition, remodeling of existing structures, replacement of pipes and furnaces, asbestos mining, refining and fabrication, and soil erosion.
- *Mercury.* Sources are chlor-alkali manufacture and battery manufacture, and solid waste incineration.
- *Hydrogen sulfide.* Sources are kraft paper manufacture, oil refining, and pipeline transportation.
- *Benzene.* Sources are petrochemical manufacture, industrial solvent use, and pharmaceutical manufacture.
- *Arsenic.* Sources are copper smelting and glass manufacture.
- *Fluorides.* Sources are primary aluminum smelting and phosphate fertilizer manufacture.

Radioactive substances, discussed in Chapter 16, are also considered hazardous air pollutants.

GLOBAL AND ATMOSPHERIC CLIMATE CHANGE

It has been known for some time that very large emissions of particulates into the atmosphere may be enough to induce climate change. Emission of volcanic ash and SO_2 by the 1990 eruption of Mt. Pinatubo in The Philippines appeared to have cooled the atmosphere by about 1°C by 1992.

Two aspects of global atmospheric change are generally considered together with air pollution: global climate change and destruction of the stratospheric ozone layer. Carbon dioxide has no other apparent adverse effects at low or moderate concentrations, but absorbs energy at the wavelength of 1.5 μm (heat energy) and may contribute to global temperature changes. All combustion of carbon-containing compounds produces CO_2. Fossil fuel and wood combustion have been singled out by some as a source of global effects because of the large amount of these substances burned in the world today.

Although global climate change resulting from CO_2 emissions has been a serious environmental concern for about a decade, there is still discussion about the causes of this change, or the extent to which it is happening. The "greenhouse gas" question seems to be answered. The average temperature at the earth's surface is estimated to have risen about 0.6°C during the past century. However, these temperatures are subject to seasonal fluctuations, latitudinal seasonal fluctuations, precession of the earth's orbit, and other periodic changes in climate. Measurements made in central England, Geneva, and Paris from about 1700 until the present indicate a general downward trend in surface temperature.[2] Measurements compiled by the Goddard Space Flight Center indicate fluctuation of ±0.4°C between 1950 and the present, closely correlated with natural phenomena like volcanic eruptions and the El Niño winds.[3] There is also a geological record of abrupt climate changes.[4]

There is evidence that atmospheric carbon dioxide has increased during the past century, and evidence for resulting global climate change is convincing. Climate is influenced by many factors, some of which exhibit periodicity over thousands of years. Paleontologic observations show periodic fluctuations of atmospheric CO_2, but they also show that climate has been strongly influenced by movement of the ice sheets, insolation, precession of the earth's orbit, volcanic eruptions, and so on.[5] Recently measured increases in atmospheric CO_2 may be the result of a high rate of fossil fuel combustion, or they may be part of a paleontological cycle. Global temperature change is difficult to determine in the stratosphere because temperatures can only be measured locally, and local temperatures are the result of local conditions and local microclimates. There is also evidence that observed nighttime warming during the last 50 years is correlated with changing cloud cover and is offset by cooler daytime temperatures.

Destruction of the stratospheric ozone layer, on the other hand, is much better understood. NO and the Cl and OH free radicals have been identified as

[2]Thompson, David J., "The Seasons, Global Temperature, and Precession," *Science* 268 (1995): 59.
[3]Kerr, Richard A., "1991: Warmth, Chill May Follow," *Science* 255 (1992): 281.
[4]Crowley, Thomas J., and North, Gerald R., "Abrupt Climate Change and Extinction Events in Earth's History," *Science* 240 (1988): 996.
[5]COHMAP, "Climatic Changes of the Last 18,000 Years: Observations and Model Simulations," *Science* 241 (1988): 1043–1052.

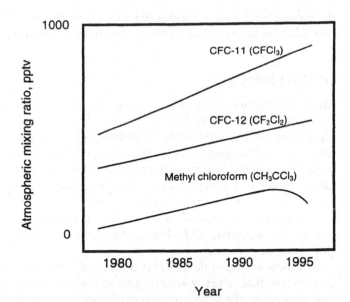

FIGURE 18–5
The atmospheric abundances of CFC-11, CFC-12, and methyl chloroform

scavengers that remove atomic oxygen from the ozone equilibrium. NO is emitted by the jet engines of high-flying aircraft; chlorine atoms are produced by the photochemical breakdown of CFC aerosols. Chlorine atoms can also react with ozone directly, and can ultimately be regenerated in a chain reaction:

$$Cl + O_3 \rightarrow ClO + O_2$$
$$ClO + O \rightarrow Cl + O_2$$
(18.9)

CF_2Cl_2 (dichlorodifluoromethane) is known commercially as Freon-12. During the 1960s and 1970s, it was an exceedingly popular aerosol propellant because, unlike carbon dioxide, it retains its propellant pressure and is chemically unreactive and nonflammable. By the mid-1970s, several investigators had observed thinning of the stratospheric ozone layer and correlated this thinning with Freon 12's worldwide production. Any Freon released into the air will undergo photochemical dissociation, but if it is contained, as in refrigeration systems, it does not dissociate unless it is released into the air. Between 1987 and 1992 the U.N. Environmental Programme formulated the Montreal Protocol for Protection of the Ozone Layer, which advises voluntary limitation on the use of CFCs and related compounds in the air.[6]

Recent studies of atmospheric halocarbons show that the ozone layer has the potential to recover. Moreover, atmospheric haloethane concentrations appear to have declined from a peak that occurred in 1991.[7] Figure 18–5 relates

---

[6]Prinn, R.G., et al., "Atmospheric Trends and Lifetime of CH_3CCl_3, and Global OH Concentrations," *Science* 269 (1995): 187.

[7]Ravishankara, A.R., and Albritton, D.L., "Methyl Chloroform and the Atmosphere," *Science* 269 (1995): 183.

the atmospheric concentrations of three halocarbons to the Montreal Protocol and other international agreements that limit halocarbon emissions.

HEALTH EFFECTS

Much of what we know about the health effects of air pollution comes from the study of acute air pollution episodes, like those in Donora, London, and New York, in which the illness appeared to be chemical irritation of the respiratory tract. The weather circumstances under which these episodes occurred were also similar: high pressure systems and inversion layers. An inversion (discussed in detail in Chapter 19) is a layer of warm air aloft that traps cold air beneath it close to the surface of the ground. The trapped cold air contains pollutants emitted near the surface that then cannot disperse upward. Inversions are usually accompanied by low surface winds, leading to reduced horizontal dispersion of pollutants.

In these episodes, the pollutants affected a specific segment of the public— those individuals already suffering from diseases of the cardiorespiratory system. Moreover, the adverse effects could not be blamed on any single pollutant. The health problems are thought to have been due to the synergistic action of particulate matter and sulfur dioxide.

A major catastrophic industrial air pollution episode was recorded in 1984 in Bhopal, India. A reactor vessel in which methyl isocyanate was being generated became overpressured and vented a cloud of the chemical to the air. More than 1500 deaths occurred within four miles of the source, and thousands of people suffered skin burns or temporary or permanent damage to their eyes, respiratory tracts, and nervous systems.

Until recently, scientists were able to evaluate the human health effects of air pollutants only from episodes like these. Laboratory studies using animals can yield some information, but the inter-species step from lab rat to human being involves more than the quantifiable parameters of body weight, metabolic rate, inhalation rate, and lifespan. For this reason, hospitals in large U.S. and European urban centers have undertaken long-term studies of the chronic effects of air pollution.

The Respiratory System

The major target of air pollutants is the respiratory system, pictured in Figure 18–6. Air and entrained pollutants enter the body through the throat and nasal cavities, and pass to the lungs through the trachea and bronchi. Entrained pollutant particles can be prevented from entering the lungs by the action of tiny hairs called cilia that sweep mucus out through the throat and nose. The bronchial cilia can be paralyzed by inhaled smoke, enhancing the synergistic effect between smoking and air pollution. Pollutant particles small enough to pass the bronchial cilia by (about 10 mμ or less in diameter) are

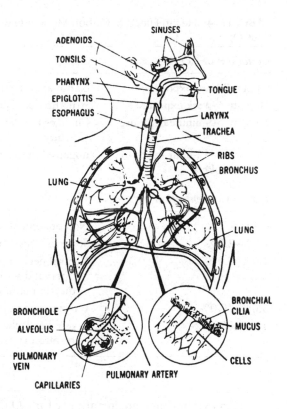

FIGURE 18–6
The respiratory system
[Courtesy the American Lung
Association.]

called respirable particles. In the lungs, the air moves through bronchial tubes
to the alveoli, small air sacs in which oxygen from the lungs is transferred to
the blood.

Carbon Monoxide

The effect of carbon monoxide inhalation on human health is directly proportional to the quantity of CO bound to hemoglobin. These effects are summarized
in Table 18–4.

Oxygen is transported in the blood as oxyhemoglobin (HbO_2), a semi-stable
compound in which O_2 is weakly bound to the Fe^{2+} in hemoglobin in red blood
cells. The O_2 is removed for cell respiration, and the regenerated hemoglobin is
available for more oxygen transport. CO reduces the oxygen-carrying capacity of
the blood by combining with hemoglobin and forming carboxyhemoglobin
(HbCO), which is stable. Hemoglobin that is tied up as HbCO cannot be regenerated and is not available for oxygen transport for the life of that particular red
blood cell. In this way CO effectively poisons the hemoglobin oxygen transport
system.

TABLE 18–4. Health Effects of Carbon Monoxide (CO) and Carboxyhemoglobin (HbCO)

Condition of CO Exposure	Effects of CO
9 ppm, 8-hour exposure	National ambient air quality standard
50 ppm, 6-week exposure	Structural changes in heart and brain of animals
50 ppm, 50-minute exposure	Changes in relative brightness threshold and visual acuity
50 ppm, 8–12-hour exposure, nonsmokers	Impaired performance on psychomotor tests

HbCO Concentration	Effects
< 1.0%	No apparent effect
1.0%–2.0%	Some evidence of effect on behavioral performance
2.0%–5.0%	Central nervous system effects; impairment of time-interval discrimination, visual acuity, brightness discrimination, and other psychomotor functions
5.0%–10.0%	Cardiac and pulmonary functional changes
10.0%–80.0%	Severe headaches, fatigue, drowsiness, coma, respiratory failure, death

Hemoglobin has a greater affinity for CO than for molecular oxygen. A person breathing a mixture of CO and oxygen will carry equilibrium concentrations of HbCO and HbO_2 given by

$$\frac{[HbCO]}{[HbO_2]} = \frac{M \times p(CO)}{p(O_2)} \tag{18.10}$$

where $p(CO)$ = partial pressure of CO
$p(O_2)$ = partial pressure of O_2
M = a constant whose range in human blood is from 200 to 250

HbCO levels as a function of exposure are shown in Figure 18–7.

Example 18.3
Estimate the saturation value of HbCO in the blood if the CO content of the air breathed is 100 ppm. Assume that the CO in the blood is at equilibrium.
Assume a value of 210 for M. Air is 21% O_2, so that Equation 18.2 becomes

$$[HbCO/HbO_2] = (210)(100)/210,000 = 0.1 \tag{18.11}$$

Thus $[HbCO/HbO_2]$ = 0.1, or 10%.

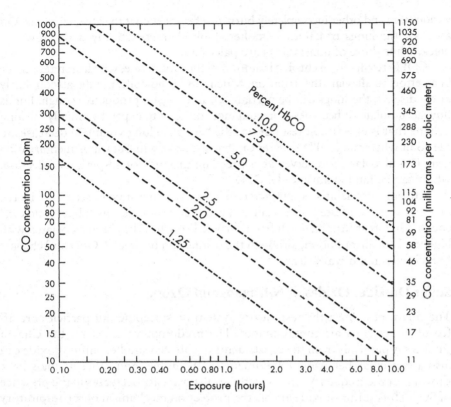

FIGURE 18–7. HbCO level as a function of exposure. [From National Air Pollution Control Administration, *Air Quality Criteria for Carbon Monoxide, AP-62,* Washington, DC (1970).]

Particulate Matter

The site and extent of deposition of particles in the respiratory tract are functions of certain physical factors, the most important of which is particle size. Alveolar deposition is particularly important since that region of the lungs is not provided with cilia to remove particulate matter. Particles deposited in the alveoli remain there and can become permanently embedded in the alveolar tissue.

A number of lung diseases are thought to have a causal association with particulate matter in the lung: lung cancer, "black lung" disease of miners, chronic obstructive pulmonary disease (COPD), of which emphysema is one form, and the diseases associated with specific particulate substances, like asbestosis. Even inert particles serve as irritants to lung tissue, and COPD is observed after chronic exposure of workers to chemically inert dust such as soil and road construction dust. Particles also carry adsorbed hazardous substances into the lungs. Tobacco smoke particles contain metal ash and when deeply inhaled serve as carriers for the carcinogenic incompletely burned organic compounds. Since cigarette smoke is known to paralyze the bronchial cilia, it travels into the lung alveoli, where the various tars and other incompletely burned substances coat the alveolar tissue. On autopsy, the lungs of habitual smokers look characteristically dark brown or black, while those of nonsmokers are paler pink.

COPD results from chronic irritation of the lung tissue, which secretes mucus in an effort to alleviate the irritation. Mucus and fluid collect in the alveoli. Early in the disease, the lungs can be partially cleared with a productive cough, but as fluid accumulates it becomes more and more difficult to expel. Eventually the lung volume that is effective in respiration (the "effective lung volume") decreases to the point where the COPD sufferer can hardly move without supplementary oxygen. The alveolar fluid serves as a breeding ground for respiratory infection, which can be fatal in advanced COPD.

Chronic bronchitis is characterized by excessive mucus secretion in the bronchi. It is manifested by a recurrent productive cough, much like "smokers' cough." In Great Britain, death from chronic bronchitis has been associated with high levels of air pollution, although the distinction between COPD and chronic bronchitis is not always clear.

Sulfur Dioxide, Oxides of Nitrogen, and Ozone

The cilia that protect the respiratory system by sweeping out particles are affected by gaseous air contaminants. The predominant effect on the cilia of smokers is paralysis from the constituents of tobacco smoke. Sulfur dioxide can also affect ciliar behavior. Experiments with rabbits and other animals have shown that the frequency with which the cilia oscillate decreases in the presence of SO_2. Thus sulfur dioxide affects the protection mechanism of the respiratory tract in addition to constricting the bronchi.

Nitrogen dioxide is a pulmonary irritant. Although little is known of the specific toxic mechanisms, high concentrations of NO_2 can produce pulmonary edema—an abnormally high accumulation of fluid in the lung tissue. NO_2 has never been found in such high concentrations in ambient air; as an air pollutant, it contributes synergistically to the effects of other pollutants. NO, the atmospheric precursor of NO_2, is not an irritant gas; in fact, it is often used as an anaesthetic commonly known as "laughing gas."

Ozone and photochemical oxidants are highly irritating, oxidizing gases. Concentrations of a few parts per million can produce pulmonary congestion, edema, and pulmonary hemorrhage. A one-hour exposure of human subjects to $2500 \, \mu g/m^3$ can decrease effective lung volume and decrease maximum breathing capacity. Symptoms of ozone and oxidant exposure are a dry throat, followed by headache, disorientation, and altered breathing patterns.

Hazardous Air Pollutants

Hazardous air pollutants encompass too wide a range of specific health effects to allow discussion in this text. Many of these air pollutants are known or potential carcinogens, and it is thus assumed that there is no threshold for an adverse effect. In virtually all cases, the effects cited have been observed only after occupational exposure to high concentrations of the pollutant, and not at the low concentrations in the ambient air. Examples of hazardous air pollutants and their specific health effects follow:

- *Arsenic and its inorganic compounds*, when inhaled, are associated with mesothelioma, a cancer of the pulmonary lining. Mesothelioma has been observed as a result of occupational exposure to airborne arsenic, but has not been observed among workers exposed to low concentrations or among the general public exposed to very low concentrations.
- *Mercury* in high concentrations causes central nervous system effects and ulceration of mucous membranes.
- *Asbestos* particles have a needle-like shape so that they lodge in the alveoli when inhaled and cause both asbestosis and malignant tumors. As with arsenic, these effects have been observed in workers exposed to high concentrations of asbestos but not among the general public.
- *Benzene* is a carcinogen.
- *Beryllium compounds* are severe oxidants.
- *Fluoride compounds* lodge in the bone and cause structural defects called fluorosis.

EFFECTS ON VEGETATION

Vegetation is injured by air pollutants in three ways: (1) necrosis (collapse of the leaf tissue), (2) chlorosis (bleaching or other color changes), and (3) alterations in growth. The types of injury caused by various pollutants differ markedly (Figure 18–8).

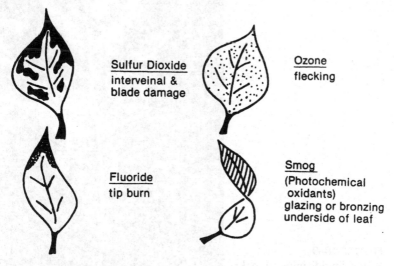

Sulfur Dioxide
interveinal &
blade damage

Ozone
flecking

Fluoride
tip burn

Smog
(Photochemical
 oxidants)
glazing or bronzing
underside of leaf

FIGURE 18–8. Typical air pollutant injury to vegetation

Sulfur dioxide produces marginal or interveinal blotches that are white to straw-colored on broad-leafed plants (Figure 18–9). Grasses injured by SO_2 show streaking (light tan to white) on either side of the midvein. Brown necrosis occurs on the tips of conifer needles with adjacent chlorotic areas. Alfalfa, barley, cotton, wheat, and apple are among the plants most sensitive to sulfur dioxide. Sensitive species are injured at concentrations of 780 μg/m³ for 8 hours.

In conifers and grasses, fluoride exposure produces an injury known as tip burn. In broad-leafed plants the fluoride injury is a necrosis at the periphery of the leaf. Among the plants most sensitive to fluorides are gladiolus, Chinese apricot, Italian prune, and pine.

At sufficient concentrations ozone produces tissue collapse and markings of the upper surface of the leaf known as stipple (pigmented red-brown) and flecking (bleached straw to white). One to two hours of exposure at air levels of 300 μg/m³ produce injury in sensitive species, including tomato, tobacco, bean, spinach, and potato.

Peroxyacyl nitrates (PAN) are present in photochemical smog and produce typical smog injury—a bronzing on the underside of vegetable leaves. In grasses, the collapsed tissue shows up as bands bleached tan to yellow. Smog exposure has also been shown to produce early maturity or senescence in plants. Even low

FIGURE 18–9
Sulfur dioxide injury to oak

PAN concentrations will injure sensitive species, among which are petunia, Romaine lettuce, pinto bean, and annual bluegrass. Acute devastation to vegetation has occurred around polluting industries like copper smelters.

EFFECTS ON ANIMALS

Air pollutants affect animals much the same as they affect people, and lethal or damaging doses to animals depend on the animal's size and respiratory rate (a canary will die from inhaling a toxic gas after a lower total intake than will kill a person). An additional hazard to animals is chronic poisoning from ingesting forage contaminated by particulate pollutants. Important in this connection are the heavy metals: arsenic, lead, and molybdenum, and compounds like organic emissions and fluorides.

EFFECTS ON MATERIALS

Perhaps the most familiar effect of air pollution on materials is the soiling of building surfaces, clothing, and other articles. Soiling results from the deposition of smoke on surfaces over time as surfaces become discolored or darkened. Cleaning of exterior building materials requires sandblasting that can damage or remove part of the building surface.

Acidic precipitation and pollutants like sulfur dioxide can accelerate the corrosion of metals. Ozone and PAN have resulted in rubber cracking, which can be used to measure ozone concentration (Chapter 20).

Fabrics are also affected by air pollutants. Cities like New York that exhibited high airborne SO_2 concentrations because high sulfur content oil was burned for power generation had air sufficiently acidic to cause nylon stockings to run. Fabrics and dyes also bleach and discolor under the influence of various pollutants.

Hydrogen sulfide, in the presence of moisture, reacts with lead dioxide in paint to form lead sulfide, producing a familiar brown to black discoloration.

EFFECTS ON VISIBILITY

Any air traveler in the United States is familiar with the permanent haze that exists over large urban areas. The visibility reduction within this haze results from the scattering of light off very fine particles 0.3 μm to 0.6 μm in diameter.

INDOOR AIR POLLUTION

Concentrations of pollutants with adverse health effects have these effects indoors as well as out of doors. Indeed, the limited volume of air inside a building, especially one with poor air circulation, causes pollutant concentrations to increase rapidly. One of the primary indoor air pollutants is "second-hand" to-bacco smoke: the CO, organic tars, metal oxide particles, and other con-stituents of cigarette, cigar, and pipe smoke that are either exhaled or not inhaled by the smoker. "Smoke-filled rooms" can quickly exceed the federal ambient air standard for total suspended particulate matter, which is why since 1988 the United States has strictly regulated smoking. All airplane flights within the continental United States are nonsmoking flights; states require places of public accommodation like restaurants to provide smoke-free areas for pa-trons; and federal government buildings and most state and local government buildings limit smoking to closed-off areas or prohibit it altogether.

Radon-222 is present naturally in rock and soil, and it can build up in un-ventilated or poorly ventilated structures. Adequate ventilation is the only method available for radon mitigation.

Recent concern about energy conservation has led to limits on the ventila-tion of buildings when energy is used for heating or cooling. As a result, CO, oxides of nitrogen, and the organic gases that all synthetic materials emit tend to build up indoors. In addition, inadequately cleaned building ventilation sys-tems release spores and bacteria into the building air, causing infectious diseases like Legionnaire's Disease. Wood stoves also increase indoor air concentrations of CO, ash particles, oxides of nitrogen, and some tars that result.from incom-pletely burned wood. Adequate ventilation appears to provide the most com-prehensive approach to keeping indoor air clean.

CONCLUSION

While the effects of air pollution on materials, vegetation, and animals can be measured, health effects on humans can only be estimated from epidemiologi-cal evidence. Most of the evidence comes from occupational exposure to much higher concentrations of pollutants than the general public is exposed to. More-over, the health effects of smoking and other lifestyle characteristics and expo-sures confound the observations of air pollutant effects. Ethical considerations preclude deliberate exposure of human subjects to concentrations of pollutants that might produce adverse effects, so evidence from sources other than epi-demiology is virtually impossible to obtain. All of the evidence we have suggests that air pollutants threaten human health and well-being to an extent that con-trol of these pollutants is necessary.

PROBLEMS

18.1   If a 1974 (pre-modern emission standards) car is driven an average of 1000 miles a month, how many grams of CO and HC are emitted during the year? The average emission standards for that model year are 3.4 g/mile for HC and 39 g/mile for CO.

18.2   A 2.5% level of CO in hemoglobin (HbCO) has been shown to cause impairment in time-interval discrimination. The level of CO on crowded city streets sometimes reaches 100 ppm CO. An approximate relationship between CO and HbCO at equilibrium after prolonged exposure is

$$HbCO\ (\%) = 0.5 + 16 \times ppm(CO) \qquad (18.12)$$

What blood level of HbCO will a bicyclist sustain while traveling for two hours in heavy traffic? A police officer directing traffic at a heavily traveled intersection for eight hours? What assumptions did you make in performing these estimates?

18.3   Washington, DC, offices and industries have introduced "flex-time" in an effort to mitigate traffic jams and air pollution. Flex-time allows the 8-hour workday to begin as early as 7 A.M. and last as late as 7 P.M. Assuming that the average automobile commute in the DC area lasts an hour, draw graphs like that in Figure 18–5 for the normal working day (8 A.M. to 5 P.M. with an hour for lunch) and for a flex-time scheme in which 1/4 of the workers who commute by automobile start at 7 A.M.; 1/4, at 8 A.M., 1/4, at 9 A.M., and 1/4, at 10 A.M. Assume one hour for lunch.

18.4   The state of Florida has set an annual average ambient SO_2 standard of 0.003 ppm. At ordinary ambient Florida temperatures (70°F), what is this in $\mu g/m^3$?

18.5   The state of Michigan set ambient H_2S standards at 2 ppb. Assuming that the H_2S is diluted by a factor of 10,000 after emission from the stack, what is the allowed H_2S emission from the stack of a kraft pulp mill, expressed in $\mu g/m^3$?

Chapter 19

Meteorology and Air Pollution

The earth's atmosphere is about 100 miles deep, which should be enough to dilute all of the garbage thrown into it. However, 95% of this air mass is within 12 miles of the earth's surface. Called the *troposphere*, this layer is where we have our weather and air pollution problems.

Weather patterns determine how air contaminants are dispersed and move through the troposphere; thus they determine the concentration of a particular pollutant that is breathed or the amount that is deposited on vegetation. An air pollution problem involves three parts: its source, its movement or dispersion, and its recipient (Figure 19–1). While the source and effects of pollutants are discussed in the previous chapter, this chapter concerns itself with the transport mechanism: how the pollutants travel through the atmosphere.

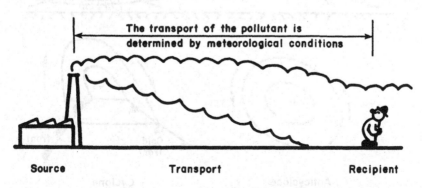

FIGURE 19–1. Meteorology of air pollutants

BASIC METEOROLOGY

Pollutants circulate the same way the air in the troposphere circulates. Air movement is caused by solar radiation and the irregular shape of the earth and its surface, which causes unequal absorption of heat by the earth's surface and atmosphere. This differential heating and unequal absorption create a dynamic system.

The dynamic thermal system of the earth's atmosphere also yields differences in barometric pressure. We associate low-pressure systems with both hot and cold weather fronts. Air movement around low-pressure fronts in the Northern Hemisphere is counterclockwise and vertical winds are upward, where condensation and precipitation take place. High-pressure systems bring sunny and calm weather—stable atmospheric conditions—with winds (in the Northern Hemisphere) spiraling clockwise and downward. Low-pressure and high-pressure systems, commonly called *cyclones* and anticyclones, are illustrated in Figure 19–2. Anticyclones are weather patterns of high stability, in which dispersion of pollutants is poor, and they are often precursors to air pollution episodes.

HORIZONTAL DISPERSION OF POLLUTANTS

The earth receives light energy at high frequency from the sun and converts it to heat energy at low frequency, which is then radiated back into space. Heat is transferred from earth's surface by radiation, conduction, and convection.

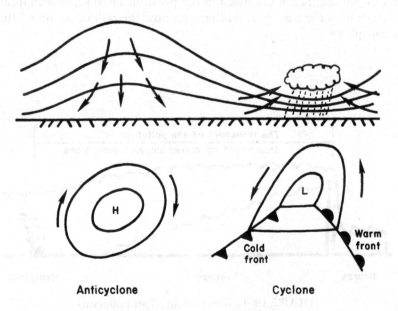

Anticyclone   Cyclone

FIGURE 19–2. Anticyclone and cyclone

Radiation is direct transfer of energy and has little effect on the atmosphere; conduction is the transfer of heat by physical contact (the atmosphere is a poor conductor since the air molecules are relatively far apart), whereas convection is transfer of heat by movement of warm air masses.

Solar radiation warms the earth and thus the air above it. This heating is most effective at the equator and least at the poles. The warmer, less dense air rises at the equator and cools, becomes more dense, and sinks at the poles. If the earth did not rotate, the surface wind pattern would be from the poles to the equator. However, the earth's rotation continually presents new surfaces to be warmed, so that a horizontal air pressure gradient exists along with the vertical pressure gradient. The resulting motion of the air creates a pattern of winds around the globe, as shown by Figure 19–3.

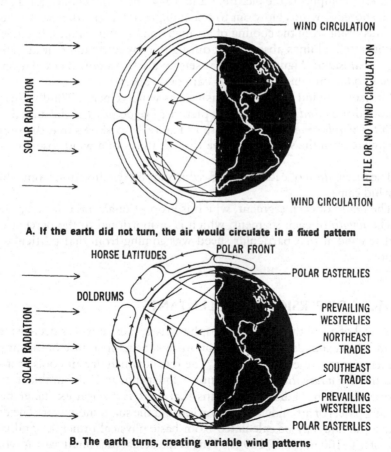

FIGURE 19–3. Global wind patterns [Courtesy the American Lung Association.]

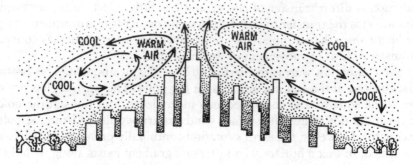

FIGURE 19–4. Heat island formed over a city

Seasonal and local temperature, pressure and cloud conditions, and local topography complicate the picture. Land masses heat and cool faster than water so that shoreline winds blow out to sea at night and inland during the day. Valley winds result from the cooling of air high on mountain slopes. In cities, brick and concrete buildings absorb heat during the day and radiate it at night, creating a *heat island* (Figure 19–4) that sets up a self-contained circulation called a *haze hood* from which pollutants cannot escape.

Horizontal wind motion is measured as wind velocity. Wind velocity data are plotted as a *wind rose*, a graphic picture of velocities and the direction *from which the wind came*. The wind rose in Figure 19–5 shows that the prevailing winds were *from* the southwest. The three features of a wind rose are

- The *orientation* of each segment, which shows the direction from which the wind came.
- The *width* of each segment, which is proportional to the wind speed.
- The *length* of each segment, which is proportional to the percent of time that wind at that particular speed was coming from that particular direction.

VERTICAL DISPERSION OF POLLUTANTS

As a parcel of air in the earth's atmosphere rises, it experiences decreasing pressure and thus expands. This expansion lowers the temperature of the air parcel, and therefore the air cools as it rises. The rate at which dry air cools as it rises is called the *dry adiabatic lapse rate* and is independent of the ambient air temperature. The term "adiabatic" means that there is no heat exchange between the rising parcel of air under consideration and the surrounding air. The dry adiabatic lapse rate may be calculated from basic physical principles and is equal to about 1°C/100 m (5.4°F/100 ft). This is the actual measured rate at which air cools as it rises in the troposphere, irrespective of the actual temperature of the

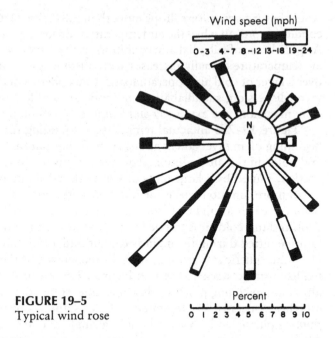

FIGURE 19–5
Typical wind rose

air. The dry adiabatic lapse rate is a function only of the elevation; move a parcel of air up 100 m and the dry air will always cool 1°C. Move a parcel of air down 100 m and it will warm 1°C, always (irrespective of the temperature of the surrounding air).

The ambient temperature in the atmosphere can be anything, of course. This *prevailing* or *ambient* temperature lapse rate (change of temperature with elevation above the earth) is measured with a thermometer. The relationships between the ambient and the dry adiabatic lapse rates essentially determine the stability of the air and the speed with which pollutants disperse. These relationships are shown in Figure 19–6.

When the ambient lapse rate and the dry adiabatic lapse rate are exactly the same, the atmosphere has *neutral* stability. *Superadiabatic* conditions prevail

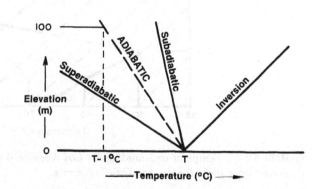

FIGURE 19–6
Ambient lapse rates and the
dry adiabatic lapse rate

when the air temperature drops more than 9.8°C/km (1°C/100 m); *subadiabatic* conditions prevail when the air temperature drops at a rate less than 1°C/100 m. A special case of subadiabatic conditions is the *temperature inversion*, when the air temperature actually increases with altitude and a layer of warm air exists over a layer of cold air. Superadiabatic atmospheric conditions are unstable and favor dispersion; subadiabatic conditions are stable and result in poor dispersion; inversions are extremely stable and trap pollutants, inhibiting dispersion.

Figure 19–7 is an actual temperature sounding for Los Angeles. Note the beginning of an inversion at about 1000 ft that puts an effective cap on the city and holds in the air pollution. This type of inversion is called a *subsidence inversion*, caused by a large mass of warm air subsiding over a city.

A more common type is the *radiation inversion*, caused by radiation of heat from the earth at night. As heat is radiated, the earth and the air closest to it cool, and this cold air is trapped under the warm air above it (Figure 19–8). Pollution emitted during the night is caught under the "inversion lid."

Atmospheric stability may often be recognized by the shapes of plumes emitted from smokestacks, as seen in Figure 19–9. Neutral stability conditions usually result in *coning* plumes, while unstable (superadiabatic) conditions result in a highly dispersive *looping* plume. Under stable (subadiabatic) conditions, the *fanning* plume tends to spread out in a single flat layer. One potentially serious condition is called *fumigation*, in which, owing to strong lapse rate, pollutants

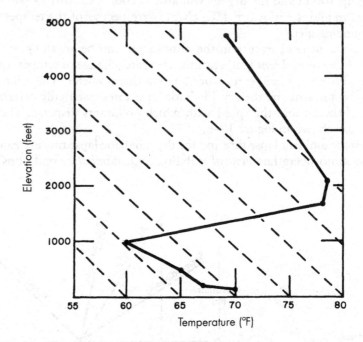

FIGURE 19–7. Temperature sounding for Los Angeles, 4 P.M., October 1962. The dotted lines show the dry adiabatic lapse rate.

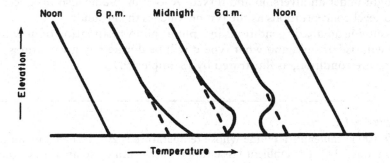

FIGURE 19–8. Typical ambient lapse rates during a sunny day and clear night

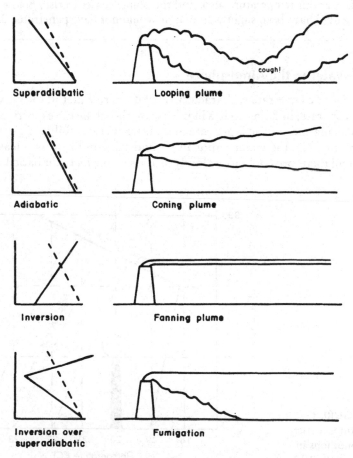

FIGURE 19–9. Plume shapes and atmospheric stability

are caught under an inversion and mixed. A looping plume also produces high ground-level concentrations as the plume touches the ground.

Assuming adiabatic conditions in a plume allows estimation of how far the plume will rise or sink, and what type it will be during any given atmospheric temperature condition, as illustrated by Example 19.1.

Example 19.1

A stack 100 m tall emits a plume whose temperature is 20°C. The temperature at the ground is 19°C. The ambient lapse rate is –4.5°C/km up to an altitude of 200 m. Above this the ambient lapse rate is +20°C/km. Assuming perfectly adiabatic conditions, how high will the plume rise and what type will it be?

Figure 19–10 shows the various lapse rates and temperatures. The plume is assumed to cool at the dry adiabatic lapse rate: 10°C/km. The ambient lapse rate below 200 m is subadiabatic; the surrounding air is cooler than the plume, so it rises and cools as it does so. At 225 m, the plume has cooled to 18.7°C, but the ambient air is at this temperature also, and the plume ceases to rise. Below 225 m, the plume would have been slightly coning. It would not have penetrated 225 m.

Effect of Water in the Atmosphere

The *dry* adiabatic lapse rate is characteristic of dry air. Water in the air will condense or evaporate; in doing so it will release or absorb heat, respectively, complicating calculations of the lapse rate and atmospheric stability. In general, as a parcel of air rises, the water vapor in it condenses and heat is released. The rising air will therefore cool more slowly as it rises; the *wet* adiabatic lapse rate

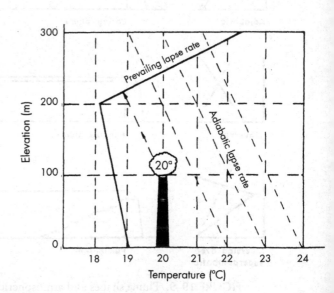

FIGURE 19–10
Atmospheric conditions in Example 19.1

will in general be less negative than the *dry* adiabatic lapse rate. The wet adiabatic lapse rate has been observed to vary between $-6.5°C/km$ and $-3.5°C/km$.

Water in the atmosphere affects air quality in other ways as well. Fogs form when moist air cools and the moisture condenses. Aerosols provide the condensation nuclei, so that fogs tend to occur more frequently in urban areas. Serious air pollution episodes are almost always accompanied by fog (remember that the roots of the word "smog" are "smoke" and "fog") because the tiny water droplets in fog participate in the conversion of SO_3 to H_2SO_4. Fog sits in valleys and stabilizes inversions by preventing the sun from warming the valley floor, thus often prolonging the episodes.

ATMOSPHERIC DISPERSION

Dispersion is the process by which contaminants move through the air and a plume spreads over a large area, thus reducing the concentration of the pollutants it contains. The plume spreads both horizontally and vertically. If it is gaseous, the motion of the molecules follows the laws of gaseous diffusion.

The most commonly used model for the dispersion of gaseous air pollutants is the Gaussian, developed by Pasquill, in which gases dispersed in the atmosphere are assumed to exhibit ideal gas behavior. Rigorous derivation of the model is beyond the scope of this text, but the principles on which the model is based are as follows:

- The predominant force in pollution transport is the wind; pollutants move predominantly downwind.
- The greatest concentration of pollutant molecules is along the plume centerline.
- Molecules diffuse spontaneously from regions of higher concentration to regions of lower concentration.
- The pollutant is emitted continuously, and the emission and dispersion process is steady state.

Figure 19–11 shows the fundamental features of the Gaussian dispersion model, with the geometric arrangement of source, wind, and plume. We can construct a Cartesian coordinate system with the emission source at the origin and the wind direction along the x-axis. Lateral and vertical dispersion are along the y- and z-axes, respectively. As the plume moves downwind, it spreads both laterally and vertically away from the plume centerline as the gas molecules move from higher to lower concentrations. Cross-sections of the pollutant concentration along both the y- and z-axes thus have the typical bell shape of Gaussian curves, as shown in the figure.

Since stack gases are generally emitted at temperatures higher than ambient, the buoyant plume will rise some distance before beginning to travel downwind. The sum of this vertical travel distance and the geometric stack height is

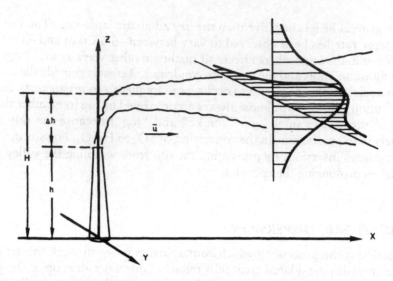

FIGURE 19–11. Gaussian dispersion model

H, the *effective* stack height. The source of the pollutant plume is, in effect, a source elevated above the ground at elevation

$$z = H \tag{19.1}$$

and the downwind concentration emanating from this elevated source may be written

$$C(x,y,z) = \frac{Q}{2\pi u \sigma_y \sigma_z} e^{-\frac{y^2}{2\sigma_y^2}} \left[e^{-\frac{(z+H)^2}{2\sigma_x^2}} + e^{-\frac{(z-H)^2}{2\sigma_z^2}} \right] \tag{19.2}$$

where $C(x, y, z)$ is the concentration at some point in space with coordinates x, y, z, and
  Q = emission rate of the pollution source, in g/sec
  u = average wind speed, in m/sec
  σ_y = standard deviation of the plume in the y direction, in meters
  σ_z = standard deviation of the plume in the z direction, in meters

The units of concentration are g/m^3. Pollution concentrations are usually measured at ground level; that is, for z = 0, Equation 19.2 reduces to

$$C(x,y,0) = \frac{Q}{\pi u \sigma_y \sigma_z} e^{\left[-\frac{y^2}{2\sigma_y^2} \right]} e^{\left[-\frac{H^2}{2\sigma_z^2} \right]} \tag{19.3}$$

This equation takes into account the reflection of gaseous pollutants from the surface of the ground.

We are usually interested in the greatest value of the ground-level concentration in any direction, and this is the concentration along the plume center-line, that is, for y = 0. In this case, Equation 19.3 reduces to

$$C(x, 0, 0) = \frac{Q}{\pi u \sigma_y \sigma_z} e^{-\frac{H^2}{2\sigma_z^2}} \qquad (19.4)$$

Finally, for a *source of emission* at ground level, H = 0, and the ground-level concentration of pollutant downwind along the plume centerline is given by

$$C(x, 0, 0) = \frac{Q}{\pi u \sigma_y \sigma_z} \qquad (19.5)$$

For a release above ground level the maximum downwind ground-level concentration occurs along the plume centerline when the following condition is satisfied:

$$\sigma_z = \frac{H}{\sqrt{2}} \qquad (19.6)$$

The standard deviations σ_y and σ_z are measures of the plume spread in the crosswind (lateral) and vertical directions. They depend on atmospheric stability and on distance from the source. Atmospheric stability is classified in categories A through F, called stability classes. Table 19–1 shows the relationship between stability class, wind speed, and sunshine conditions. Class A is the least stable; class F, the most stable. In terms of ambient lapse rates, classes A, B, and C are associated with superadiabatic conditions; class D, with neutral condition, and classes E and F, with subadiabatic conditions. A seventh class, G, has been proposed for conditions of extremely severe temperature inversion. Urban and suburban populated areas rarely achieve greater stability than class D because of the heat island effect; stability classes E and F are found in rural and unpopulated areas. Values for the lateral and vertical dispersion constants, σ_y and σ_z, are given in Figures 19–12 and 19–13. Use of the figures is illustrated in Examples 19.2 and 19.3.

TABLE 19–1. Atmospheric Stability Under Various Conditions

Surface Wind Speed (at 10 m) (m/sec)	Day[a] Incoming Solar Radiation (Sunshine)			Night[a]	
	Strong	Moderate	Slight	Thin Overcast or 1/2 Low Cloud	3/8 Cloud
< 2	A	A–B	B	—	—
2–3	A–B	B	C	E	F
3–5	B	B–C	C	D	E
5–6	C	C–D	D	D	—
> 6	C	D	D	D	D

[a]The neutral category, D, should be assumed for overcast conditions during day or night.

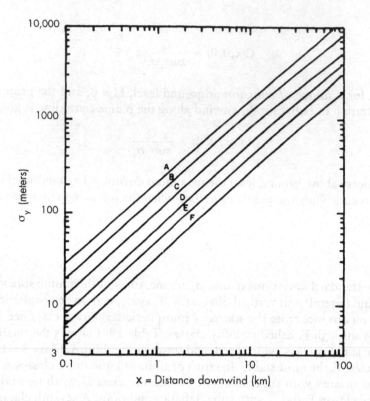

FIGURE 19–12. Standard deviation or dispersion coefficient, σ_y, in the crosswind direction as a function of downwind distance

---

Example 19.2

An oil pipeline leak results in emission of 100 g/hr of H_2S. On a very sunny summer day, with a wind speed of 3.0 m/sec, what is the concentration of H_2S 1.5 km directly downwind from the leak?

From Table 19–1, we may assume class B stability. Then, from Figure 19–12, at x = 1.5 km, σ_y is approximately 210 m, and, from Figure 19–13, σ_z is approximately 160 m, and

$$Q = 100 \text{ g/hr} = 0.0278 \text{ g/sec} \tag{19.7}$$

Applying Equation 19.5, we have

$$C(1500, 0, 0) = \frac{0.0278 \text{ g/sec}}{\pi(3.0 \text{ m/sec}) (210 \text{ m}) (160 \text{ m})} = 8.77 \times 10^{-9} \text{ g/m}^3 \tag{19.8}$$

or $0.088 \ \mu g/m^3$.

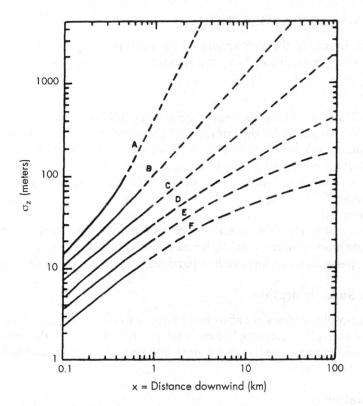

FIGURE 19–13. Standard deviation or dispersion coefficient, σ_z, in the vertical direction as a function of downwind distance

---

Example 19.3

A coal-burning electric generating plant emits 1.1 kg/min of SO_2 from a stack with an effective height of 60 m. On a thinly overcast evening, with a wind speed of 5.0 m/sec, what is the ground-level concentration of SO_2 500 m directly downwind from the stack?

From Table 19–1, we may assume class D stability. Then, from Figures 19–12 and 19–13, at x = 0.5 km, σ_y is approximately 35 m and σ_z is approximately 19 m, and

$$Q = 1.1 \text{ kg/min} = 18 \text{ g/sec} \qquad (19.9)$$

In this problem, the release is elevated, and H = 60 m.

Applying Equation 19.4, we have

$$C(0.5, 0, 0) = \frac{18 \text{ g/sec}}{\pi(5.0 \text{ m/sec}) (35 \text{ m}) (19 \text{ m})} e^{\left(-\frac{(60)^2}{2(19)^2}\right)} = 11.8 \times 10^{-6} \text{ g/m}^3 \qquad (19.10)$$

or 11.8 μg/m³.

CLEANSING THE ATMOSPHERE

Processes by which the atmosphere cleans itself include the effect of gravity, contact with the earth's surface, and removal by precipitation.

Gravity

Particles in the air, if they are larger than about a millimeter in diameter, are observed to settle out under the influence of gravity; the carbon particles from elevated diesel truck exhaust are a good example of such settling. However, most particles of air pollutants are small enough that their settling velocity is a function of atmospheric turbulence, viscosity, and friction, as well as of gravitational acceleration. This settling can be exceedingly slow. Particles smaller than 20 μm in diameter will seldom settle out by gravity alone. Gases are removed by gravitational settling only if they are adsorbed onto particles or if they condense into particulate matter. Sulfur trioxide, for example, condenses with water and other airborne particulates to form sulfate particles, a major component of acid rain.

Surface Sink Absorption

Many atmospheric gases are absorbed by the features of the earth's surface, including stone, soil, vegetation, bodies of water, and other materials. Soluble gases like SO_2 dissolve readily in surface waters, and such dissolution can result in measurable acidification.

Precipitation

Precipitation removes contaminants from the air by two methods. *Rainout* is an "in-cloud" process in which very small pollutant particles become nuclei for the formation of rain droplets that grow and eventually fall as precipitation. *Washout* is a "below-cloud" process in which rain falls through the pollutant particles and molecules, which are entrained by the impinging droplets or actually dissolve in them.

The relative importance of these removal mechanisms was illustrated by a study of SO_2 emissions in Great Britain, where the surface sink accounted for 60% of the SO_2, 15% was removed by precipitation, and 25% blew away from Great Britain, heading northwest toward Norway and Sweden.

CONCLUSION

Polluted air results from both emissions into the air and meteorological conditions that control the dispersion of those emissions. Pollutants are moved predominantly by wind, so very light wind results in poor dispersion. Other conditions conducive to poor dispersion are

- Little lateral wind movement across the prevailing wind direction
- Stable meteorological conditions, resulting in limited vertical air movement

- Large differences between day and night air temperatures, and the trapping of cold air in valleys, resulting in stable conditions
- Fog, which promotes the formation of secondary pollutants and hinders the sun from warming the ground and breaking inversions
- High-pressure areas resulting in downward vertical air movement and absence of rain for washing the atmosphere

Air pollution episodes can now be predicted, to some extent, on the basis of meteorological data. EPA and many state and local air pollution control agencies are implementing early warning systems and acting to curtail emissions and provide emergency services in the event of a predicted episode.

PROBLEMS

19.1   Consider the following temperature soundings: ground level, 21°C; 500 m, 20°C; 600 m, 19°C; 1000 m, 20°C. If we release a parcel of air at 600 m, will it rise or sink or remain at 600 m? If a 70-m stack releases a plume with a temperature of 30°C, what type of plume results? If the ambient temperature profile is adiabatic, how high does the plume rise?

19.2   Plot the ambient lapse rate given the following temperature sounding:

Elevation (m)	Temperature (°C)
0	20
50	15
100	10
150	15
200	20
250	15
300	20

What type of plume would you expect if the exit temperature of the plume at the stack were 15°C and the smokestack were 40 m tall? 120 m tall? 240 m tall?

19.3   A power plant burns 1000 tons of coal per day, 2% of which is sulfur. All of the sulfur is burned completely and emitted into the air from a stack with an effective height of 100 m. For a wind speed of 6 m/sec, calculate (a) the ground-level SO_2 concentration along the plume centerline 10 km downwind, (b) the maximum ground-level SO_2 concentration for class B stability, using a conservative value for the wind speed, and (c) the downwind distance of maximum ground-level SO_2 for class B stability.

19.4   The odor threshold of hydrogen sulfide is about 0.7 μg/m³. If 0.08 g of H_2S is emitted per second out of a stack with an effective height of 40 m during an overcast night with a wind speed of 3 m/sec, calculate the maximum ground-level concentration of H_2S and estimate how far downwind this concentration occurs. Estimate the area (in terms of x and y coordinates) where H_2S

can be detected by its odor. A programmable calculator or a computer spreadsheet may facilitate this calculation, or you can write a computational program.

19.5 The concentration of H_2S 200 m downwind from an abandoned oil well is $3.2/m^3$. What is the emission rate on a partially overcast afternoon when the wind speed is 2.5 m/sec at ground level?

LIST OF SYMBOLS

C   concentration of pollutant, in g/m^3
H   effective stack height, in m
Q   emission rate, in g/sec or kg/sec
T   temperature
u   average wind speed, in m/sec
σ_y standard deviation, y direction, in m
σ_z standard deviation, z direction, in m

Chapter 20

Measurement of Air Quality

Air quality measurements are designed to measure all types of air contaminants, with no attempt to differentiate between naturally occurring contaminants and those that result from human activity. Measurements of air quality fall into three classes:

- *Measurement of emissions.* This is called *stack sampling* when a stationary source is analyzed. Samples are drawn out through a hole or vent in the stack for on-the-spot analyses. Mobile sources like automobiles are tested by sampling exhaust emissions while the engine is running and working against a load.
- *Meteorological measurements.* The measurement of meteorological factors—wind speed, wind direction, lapse rates, and so forth—is necessary to determine how pollutants travel from source to recipient.
- *Measurement of ambient air quality.* Ambient air quality is measured by a variety of monitors, which are discussed in this chapter. Almost all evidence of the health effects of air pollution is based on correlation of these effects with measured ambient air quality.

Air quality monitoring instrumentation has been developed in three phases, or "generations." First-generation devices were developed when little or no precedent existed for measuring very small quantities of gases in the atmosphere, nor was much money available for their development. Accordingly, they are simple, inexpensive, and usually do not require power to operate. They are also inconvenient, slow, and of questionable accuracy.

Second-generation measurement equipment evolved when more accurate data and more rapid data collection were required. It uses power-driven pumps and other collection devices, and can sample a larger volume of air in a relatively short time. Gas measurement is usually by wet chemistry—that is, collected gas is either dissolved into or reacted with a collecting fluid.

Third-generation devices differ from their predecessors in that they provide continuous readout. The measurement of pollutant concentrations is almost instantly translated by a readout device, so that the pollution may be measured while it is happening. Examples of all three types of device are discussed in this chapter.

MEASUREMENT OF PARTICULATE MATTER

First-generation devices for measuring particulate matter measure how much dust settles to the ground. Such *dustfall* measurements are by far the simplest means of evaluating air quality, and can also indicate the direction of the pollution qualitatively. Dust is collected either in open buckets which can contain water to trap the dust, or on a sticky surface. The buckets are sampled at 30-day intervals; the sticky tapes are usually read every 7 days. Dustfall jars are dried and weighed to determine the amount of dust in the jar, which is usually reported as tons of dust settled per square mile in 30 days. The data can also be reported as grams per cm^2 per 30 days. Sometimes called "low-volume samplers," dustfall jars provide low-quality data. Only one data point can be obtained each month, and extraneous material can easily get into the jar (pigeons and other birds do not cooperate with these measurements).

The second-generation particle sampling device, the *high-volume,* or *hi-vol, sampler,* is a substantial improvement over the dustfall jar, and has become the workhorse of particle sampling (see Figure 20–1). It operates much like a vacuum cleaner by pumping air at a high rate through a filter. About 2000 m^3 (70,000 ft^3) can be pumped in 24 hours, so sampling time can be cut to between 6 hours and 24 hours. Analysis is gravimetric: The filter is weighed before and after the sampling period; the weight of particles collected is then the difference between these two weights.

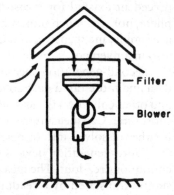

FIGURE 20–1. High-volume sampler

Air flow through the filter is measured with a flow meter, usually calibrated in cubic feet of air per minute. Because the filter collects dirt during its hour of operation, less air passes through it during the latter part of the test than in the beginning, and the air flow must therefore be measured at the start and the end of the test period and the values averaged.

The high-volume sampler operated in this way measures *total suspended particulate matter*, or TSP, as distinguished from other measurements of particulate matter.

Example 20.1
A clean filter is found to weigh 10.00 g. After 24 hours in a hi-vol, the filter plus the dust weighs 10.10 g. The air flow at the start and the end of the test was 60 and 40 ft^3/min, respectively. What is the concentration of particulate matter?

Weight of particulates (dust) = $(10.10 - 10.00)$g$\times10^6$ µg/g = 0.1×10^6 µg

Average air flow = $(60 + 40)/2 = 50$ ft^3/min

Total air through the filter = 50 ft^3/min $\times$ 60 min/hr $\times$ 24 hr/day $\times$ 1 day

$\qquad$ = 72,000 ft^3 $\times$ 0.0283 m^3/ft^3 = 2038 m^3

Total suspended particulate matter = $(0.1\times10^6$ µg$)/2038$ m^3 = 49 µg/m^3

Another measure of airborne dust widely used in environmental health assessments is that of fine particles, including particles less than 10 µm in diameter, identified by EPA as PM_{10} *particles*, and for which the ambient standard is more stringent than for TSP (see Chapter 22). Fine particles also include *respirable particles* less than 1.0 µm in diameter—small enough to penetrate the lung. They may be measured with a series of stacked filters, of which the first removes particles larger than those for which measurement is desired.

No third-generation gravimetric particle-measuring devices have yet been developed because of the inherent difficulty of making continuous gravimetric measurements, particularly of very small quantities.

The *nephelometer* (Figure 20–2) is an indirect measuring device for particles and can make continuous measurements of real-time data. A nephelometer measures the intensity of light scattered by fine particles in the air, and the scattered light intensity is proportional to the concentration of smoke or very fine particulate matter in the air. Fine particles interfere with visibility by scattering light; this scattering is what we know as haze. In a nephelometer, the scattered light intensity is measured at a 90° angle from the incident light. The instrument can be calibrated either in units of percent visibility decrease or in units of µg/m^3. However, it does not measure particulates but uses visibility as a surrogate for particulates, which can be quite erroneous, especially if visibility is attenuated by water droplets (fog).

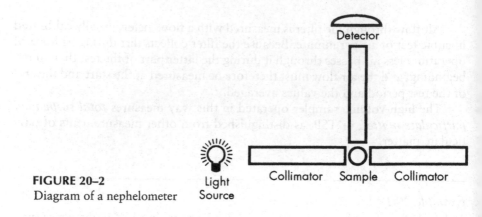

FIGURE 20–2
Diagram of a nephelometer

Light Source   Collimator   Sample   Collimator

MEASUREMENT OF GASES

First-generation methods for measuring the concentration of gaseous air pollutants are simple and use no power, yet they provide a quantitative measure of concentration. Two of the most ingenious devices were developed to measure ozone (O_3) and SO_2.

Ozone attacks and cracks rubber. To measure it, specially prepared and weighed strips of rubber are hung out of doors and the cracks formed in each strip are measured. The strips may be calibrated by exposure for certain periods of time to known ozone concentrations.

Sulfur dioxide may be measured by impregnating filter papers with chemicals that react with SO_2 and change color. For example, lead peroxide reacts to form dark lead sulfate by the reaction

$$PbO_2 + SO_2 \rightarrow PbSO_4 \qquad (20.1)$$

The extent of dark areas on the filter and the depth of color may be related to the SO_2 concentration. Like the ozone measurement, this type of SO_2 measurement takes several days of exposure to yield measurable results.

Second-generation devices are much faster, requiring hours rather than days of sampling, and they usually involve a *gas bubbler*, shown in Figure 20–3. The air sample is bubbled through a solution that reacts chemically with the particular gaseous pollutant being measured. The concentration is then measured with further wet chemical techniques. For example, SO_2 may be measured by bubbling air through hydrogen peroxide, so that the following reaction occurs:

$$SO_2 + H_2O_2 \rightarrow H_2SO_4 \qquad (20.2)$$

The amount of sulfuric acid formed can then be determined by titrating against a base of known concentration.

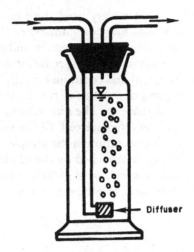

FIGURE 20–3
A typical bubbler used for measurement
of gaseous air pollutants

Diffuser

Figure 20–4 is a schematic diagram of the pararosaniline technique, which is a standard method for measuring atmospheric SO_2. In this method, air is bubbled through a solution of tetrachloromercurate (TCM). The SO_2 and TCM combine to form a complex that is then reacted with pararosaniline to form a colored solution. The intensity of the color is proportional to the SO_2 concentration, absorbs light at wavelength 560 mμ, and can be measured with a colorimeter or spectrophotometer. A similar colorimetric technique for measuring ammonia concentration is described in Chapter 4.

Most bubblers are not 100% efficient; not all of the gas bubbled through the liquid will be absorbed, and some will escape. The quantitative efficiency of a bubbler is established by testing and calibrating with known concentrations of various gases in air. Gas chromatography is a newer and very useful second-generation measurement method, particularly since trapping the pollutant in a bubbler is not necessary. Long sampling times are not needed, and the air sample can usually be introduced directly into the gas chromatograph.

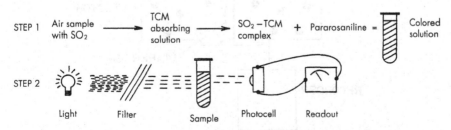

FIGURE 20–4. Schematic diagram of the pararosaniline method for measuring SO_2

One widely used third-generation device is *nondispersive infrared spectrophotometry (nondispersive IR)*, used for measurement of CO, such as in routine automobile inspection and maintenance. Like all asymmetric gas molecules, CO absorbs at the specific infrared frequencies that correspond to molecular vibrational and rotational energy levels. As shown in Figure 20–5, the air sample is pumped into one of two chambers in the detector. The other chamber contains a reference gas like nitrogen. Infrared lamps shine through both the sample cell and the reference cell. CO in the sample will absorb IR in direct proportion to its concentration in the sample. After passing through the two cells, the radiant energy is absorbed by the gas in the two detector cells, both of which contain CO. Absorption of radiant energy by the CO causes the gas in the detector cells to expand, but the detector under the reference cell receives more energy (since

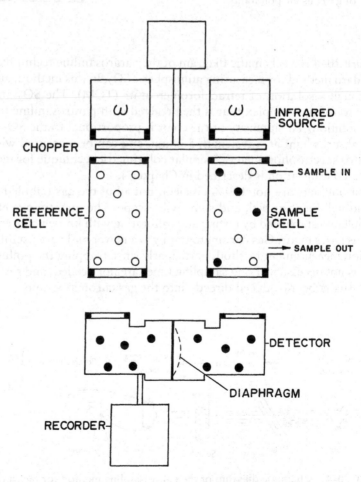

FIGURE 20–5. Nondispersive infrared spectrophotometer for CO measurement

the reference gas did not absorb) and expands, moving the diaphragm separating the two detector cells. This movement is detected electronically, and the continuously transmitted signal is read out on a recorder.

REFERENCE METHODS

Many methods are available for measuring air pollutant concentration, and new, more rapid and more accurate and precise methods are always being developed. EPA has designated a series of reference methods to which results from all other methods can be compared. Though the EPA standard reference methods may not always be the most sensitive, they have been standardized and have a history of independently duplicated results. Some change with each annual edition of the Code of Federal Regulations. Reference methods are especially important when there is a question of compliance with air quality standards and when analyzing for hazardous air pollutants.

GRAB SAMPLES

It is often necessary to obtain a sample of a gas for analysis in the laboratory, but obtaining a grab sample such as this presents some difficulties. Collection of grab samples of exhaust gases, as from automobiles, is relatively straightforward; care must be taken only that the container can withstand the temperature of the sample, so plastic or aluminum-coated bags are often used. The gas is pumped or exhausted with some positive pressure into the bag and allowed to escape through a small hole. By displacing several volumes of gas in this way, contamination problems can be avoided.

Evacuated containers can be used for grab samples. The air to be sampled is drawn into a previously evacuated container, usually attached to a vacuum system. Some contamination is always possible, since no container can be completely evacuated and the concentration of air pollutants is usually small. If gases being sampled are not soluble in water, allowing gas to displace water in a container is a useful sampling method. Unfortunately, most air pollutants measured in grab samples are water soluble.

STACK SAMPLES

An art worthy of individual attention, sampling of gas directly from a stack is necessary to evaluate compliance with emission standards and to determine the efficiency of air pollution control equipment. In moderate- to large-diameter industrial stacks, when the gas is exhausted at relatively high temperatures, the concentration of its constituents is not uniform across the gas stream, and the sampler must take care to ensure a representative sample. A thorough survey is made of the flow, temperature, and pollutant concentration both across the effluent stream and at various locations within the stack. As shown in Figure 20–6,

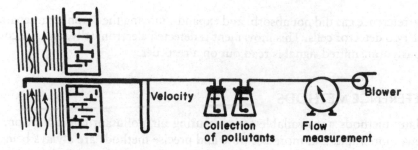

FIGURE 20–6. A stack sampling train

a train of instruments is often used for stack sampling, so that a number of measurements may be determined at each positioning of the intake nozzle.

SMOKE AND OPACITY

Visible smoke from a stack is often the only immediate evidence of a pollution violation that is external to an industrial source. Effluent gases cannot be sampled and analyzed without a complicated system that always involves notifying the emitter. However, since smoke issuing from a stack can be seen the opacity of a smoke plume is still the only enforcement method that may be used without the emitter's knowledge (and, often, cooperation). Therefore, many regulations are still written on the basis of visually estimated smoke density.

The density of black or gray smoke is measured on the Ringelmann scale of opacity. One end of the scale, Ringelmann 0, corresponds to complete transparency, or no visible plume. The other end, Ringelmann 5, corresponds to a completely opaque black plume through which no light at all is transmitted. A typical opacity standard is Ringelmann 1, a barely visible plume of about 20% opacity, with allowances for Ringelmann 2 (40% opacity) for very short periods of time. The Ringelmann test was at one time conducted by comparing the blackness of a card, like those shown in Figure 20–6, with the blackness of the observed plume. Modern practice consists of training enforcement agents to recognize Ringelmann opacities by repeated observation of smoke of predetermined opacity. White smoke is reported as "percent opacity" rather than by Ringelmann number.

Correspondence between the Ringelmann scale and smoke opacity is shown in Figure 20–7. Visual estimates can usually be made to within 5%, or 1/4 of a Ringelmann number. Opacity may also be measured continuously by installing a photometer in the stack breach and calibrating the emitted smoke by the Ringelmann or opacity scale.

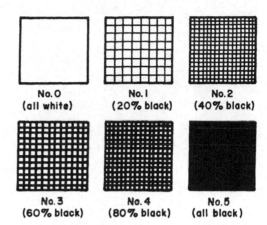

FIGURE 20–7
Ringelmann scale for measuring
smoke opacity

CONCLUSION

As with water pollution, the analytical tests of air quality can be only as good as the samples or sampling techniques used. Moreover, the prevailing analytical techniques leave much to be desired in both precision and accuracy. Most measurements of environmental quality, including these, are at best reasonable estimates.

PROBLEMS

20.1   An empty 6-inch-diameter dustfall jar weighs 1560 g. After sitting outside for the prescribed amount of time, the jar weighs 1570 g. Report the dustfall.

20.2   A hi-vol clean filter weighs 20.0 g. After exposure in the sampler for 24 hours, the filter (now dirty) weighs 20.5 g. The initial and final air flows through the hi-vol were 70 ft³/min and 50 ft³/min, respectively. What was the concentration of particulate matter in the air? Were any National Ambient Air Quality Standards exceeded? (See Chapter 22.)

20.3   A hi-vol sampler draws air at an average rate of 70 ft³/min. If the measured particle concentration is 200 μg/m³ and the hi-vol was operated for 12 hours, what was the weight of dust on the filter?

LIST OF SYMBOLS

EPA   U.S. Environmental Protection Agency
TSP   total suspended particulate matter

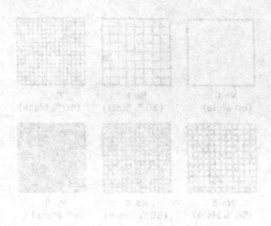

Chapter 21

Air Pollution Control

Limiting emissions into the air is both technically difficult and expensive. However, since rain is the only air-cleaning mechanism available—and not very efficient—good air quality depends on pollution prevention and on limiting what is emitted. The control of air emissions may be realized in a number of ways. Figure 21–1 shows five separate possibilities for control. Dispersion is discussed in Chapter 19, and the four remaining control points are discussed individually in this chapter.

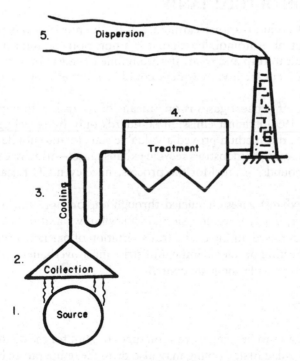

FIGURE 21–1. Points of possible air pollution control

SOURCE CORRECTION

Changing or eliminating a process that produces a polluting air emission is often easier than trying to trap the pollutant. A process or product may be necessary, but could be changed to control emissions. For example, automobile exhaust once caused high lead levels in urban air. Elimination of lead from gasoline, which was needed for proper catalytic converter operation, also reduced lead. Similarly, removal of sulfur from coal and oil before the fuel is burned has reduced the amount of SO_2 emitted. In these cases, the source of air pollution was corrected.

Processes may also be modified to reduce air pollution. Odors from municipal incinerators may be controlled by operating the incinerator at a high enough temperature to effect more complete oxidation of odor-producing organic compounds. The 1990 Clean Air Act mandates the use of oxygenated fuels in urban areas in order to limit CO emissions from automobiles.

Strictly speaking, such measures as process change, raw material substitution, and equipment modification to meet emission standards are known as *controls*. In contrast, *abatement* refers to all devices and methods for decreasing the quantity of pollutant reaching the atmosphere once it has been generated by the source. For simplicity, we refer to all of the procedures as controls.

COLLECTION OF POLLUTANTS

Collection of pollutants for treatment can be the most serious problem in air pollution control. Automobile exhaust is a notorious polluter mainly because it is so difficult to trap and treat. If automobile exhaust could be channeled to a central treatment facility, treatment could be more efficient in controlling individual cars.

Recycling of exhaust gases is one means of control. Although automobiles cannot meet 1990 exhaust emission standards only by recycling exhaust and blow-by gases, this method proved a valuable start to automobile emission controls. Many stationary industries recycle exhaust gases—usually CO and volatile organic compounds—as fuel for the process, since even CO releases heat when burned to CO_2.

Process exhaust gases channeled through one or more stacks are relatively easy to collect, but fugitive emissions from windows and doors, cracks in the walls, and dust raised during onsite transportation of partially processed materials pose a collection problem. Some industries must overhaul the entire plant air flow system to provide adequate control.

COOLING

The exhaust gases to be treated are sometimes too hot for the control equipment and must be cooled first. Cooling may also drop the temperature below the condensation point of some pollutants so that they can be collected as liquids. Dilution, quenching, and heat exchange, shown in Figure 21–2, are all acceptable

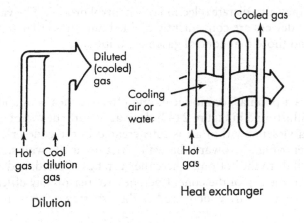

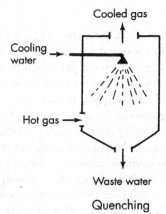

FIGURE 21–2. Cooling hot waste gases

cooling methods. Quenching has the added advantage of scrubbing out some gases and particulate matter, but may yield a dirty, hot liquid that itself requires disposal. Cooling coils are probably the most widely used cooling method and are especially appropriate where heat can be conserved.

TREATMENT

Selection of the correct treatment device requires matching its features with the characteristics of the pollutant. Pollutant particles vary in size over many orders of magnitude, from ideal gas molecules to macroscopic particles several mm in diameter. One device will not be effective and efficient for all pollutants, or even for all pollutants coming from the same stack. The chemical behavior of pollutants may also dictate selection of a control process. The various air pollution control devices are conveniently divided into those that control particulate matter and those that control gaseous pollutants.

Cyclones

The cyclone is a popular, economical, and effective means of controlling particulates. As illustrated in Figure 21–3, dirty air enters the cyclone off-center at the bottom; a violent swirl of air is thus created in the cone and particles are accelerated centrifugally toward the wall. Friction at the wall slows the particles and they slide to the bottom, where they can be collected, and clean air exits at the center of the top of the cone. Cyclones are reasonably efficient for large particle collection and are widely used as the first stage of dust removal.

Fabric Filters

Fabric filters used for controlling particulate matter (Figure 21–4) operate like a vacuum cleaner. Dirty gas is blown or sucked through a fabric filter bag, which collects the dust. The dust is removed periodically when the bag is shaken. Fabric filters can be very efficient collectors for even sub-micron-sized particles and are widely used in industrial applications.

The basic mechanism of dust removal in fabric filters is thought to be similar to that of sand filters in water quality management, as discussed in Chapter 6. Dust particles adhere to the fabric because of surface force that results in entrapment. They are brought into contact with the fabric by impingement or Brownian diffusion. The removal mechanism cannot be simple sieving, since fabric filters commonly have an air-space-to-fiber ratio of 1 to 1.

Very hazardous or toxic particulate matter of a diameter less than 1 μm sometimes must be controlled to better than 99.9%. A single stage of High Efficiency Particle Attenuation (HEPA) micropore or glass frit filters, through which the precleaned gas is forced or sucked by vacuum, can achieve this level of control, and four to six HEPA filter stages in series can achieve 99.9999% control. HEPA filters are commonly used to control emission of radioactive particles, for example.

Wet Collectors

The spray tower, or scrubber, pictured in Figure 21–5 can effectively remove larger particles. More efficient scrubbers promote the contact between air and water by violent action in a narrow throat section into which the water is introduced. Generally, the more violent the encounter, the smaller the gas bubbles or water droplets, hence the more effective the scrubbing. A *venturi scrubber*, as shown in Figure 21–6, is a frequently used high-energy wet collector. Gas flow is constricted through a venturi throat section and water is introduced as high-pressure streams perpendicular to the gas flow. The venturi scrubber is essentially 100% efficient in removing particles greater than 5 μm in diameter.

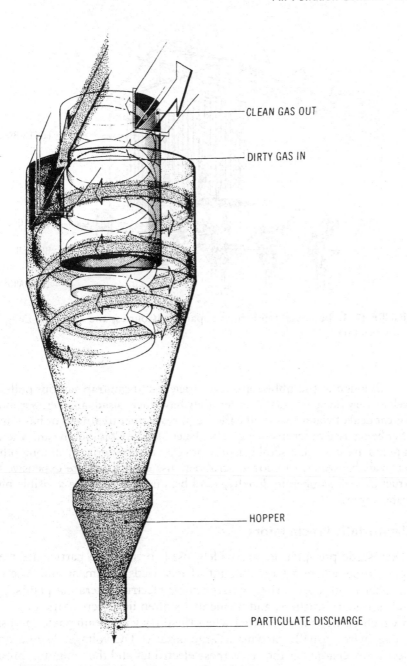

CLEAN GAS OUT

DIRTY GAS IN

HOPPER

PARTICULATE DISCHARGE

FIGURE 21–3. Cyclone [Courtesy the American Lung Association.]

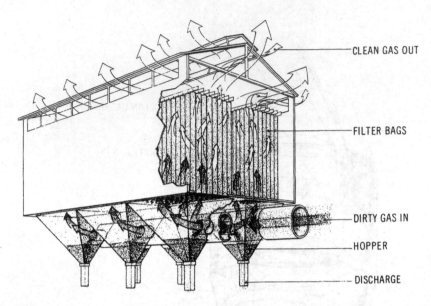

FIGURE 21–4. Industrial fabric filter apparatus [Courtesy the American Lung Association.]

Although wet scrubbers are very efficient and can trap gaseous pollutants as well as very fine particulate matter, they have their disadvantages. Scrubbers use a great deal of water that itself either requires further treatment or has limited use after being used to scrub dirty gas. In places like the Colorado Basin, where water supplies are limited, a scrubber may have a very low priority among other uses for available water. Moreover, scrubbers use energy and are expensive to construct as well as operate. Finally, scrubbers usually produce a visible plume of water vapor.

Electrostatic Precipitators

Electrostatic precipitators are widely used to trap fine particulate matter in applications where a large amount of gas needs treatment and where a wet scrubber is not appropriate. Coal-burning electric generating plants, primary and secondary smelters, and incinerators often use electrostatic precipitators, in which particles are removed when the dirty gas stream passes across high-voltage wires, usually carrying a large negative DC voltage. The particles are electrically charged as they pass these electrodes and then migrate through the electrostatic field to a grounded collection electrode. The collection electrode can be either a cylindrical pipe surrounding the high-voltage charging wire or a flat plate like that shown in Figure 21–7. In either case, it must be periodically rapped with small hammer-heads to loosen the collected particles from its surface.

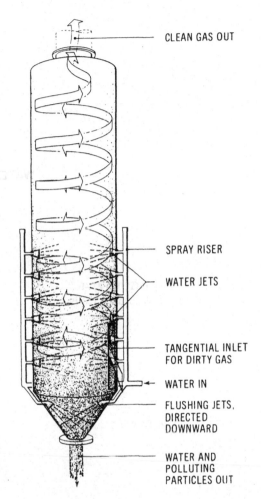

CLEAN GAS OUT

SPRAY RISER

WATER JETS

TANGENTIAL INLET
FOR DIRTY GAS

WATER IN

FLUSHING JETS,
DIRECTED
DOWNWARD

WATER AND
POLLUTING
PARTICLES OUT

FIGURE 21–5
Spray tower [Courtesy the American
Lung Association.]

As the dust layer builds up on the collecting electrode, the collection efficiency may decrease, particularly if the electrode is the inside of a cylindrical pipe. Moreover, some dust has a highly resistive surface and does not discharge against the collection electrode but sticks to it. Heated or water-flushed electrodes may solve this difficulty. Electrostatic precipitators are efficient collectors of very fine particles. However, since the amount of dust collected is directly proportional to the current drawn, the electrical energy used by an electrostatic precipitator can be substantial, with resulting high operating cost.

Comparison of Particulate Control Devices

Figure 21–8 shows the approximate collection efficiencies, as functions of particle size, for the devices discussed. Costs of collection also vary widely.

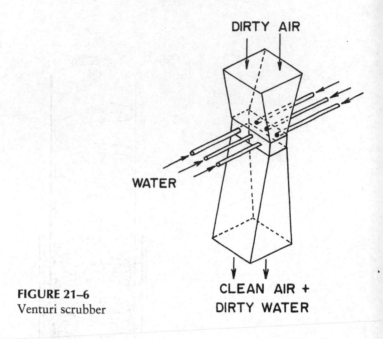

FIGURE 21–6
Venturi scrubber

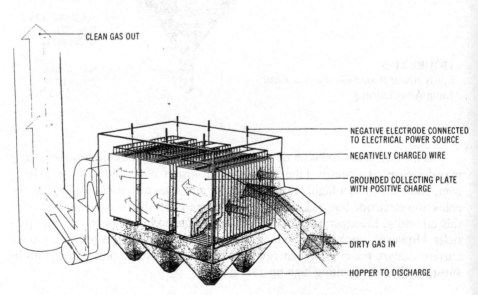

FIGURE 21–7. Flat-plate electrostatic precipitator [Courtesy the American Lung Association.]

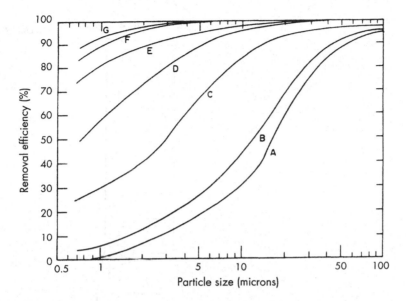

FIGURE 21–8. Comparison of removal efficiencies: (A) baffled settling chamber, (B) cyclone "off the shelf," (C) carefully designed cyclone, (D) electrostatic precipitator, (E) spray tower, (F) venturi scrubber, (G) bag filter

CONTROL OF GASEOUS POLLUTANTS

Gaseous pollutants may be removed from the effluent stream by trapping, by chemical change, or by a change in the process that produces them.

The *wet scrubbers* discussed can remove pollutants by dissolving them in the scrubber solution. SO_2 and NO_2 in power plant off-gases are often controlled in this way. *Packed scrubbers*—spray towers packed with glass platelets or glass frit—carry out such solution processes more efficiently than ordinary wet scrubbers; an example is the removal of fluoride from aluminum smelter exhaust gases. *Adsorption*, or chemisorption, is the removal of organic compounds with an adsorbent like activated charcoal.

Incineration (Figure 21–9), or *flaring*, is used when an organic pollutant can be oxidized to CO_2 and water, or in oxidizing H_2S to SO_2. *Catalytic combustion* is a variant of incineration in which the reaction is facilitated energetically and carried out at a lower temperature by surface catalysis. It is discussed further in connection with mobile source control. Figure 21–10 compares incineration and catalytic combustion.

Control of Sulfur Dioxide

Sulfur dioxide (SO_2) is ubiquitous and a serious pollution hazard. Its largest single source in the United States, and probably in the industrialized world, is

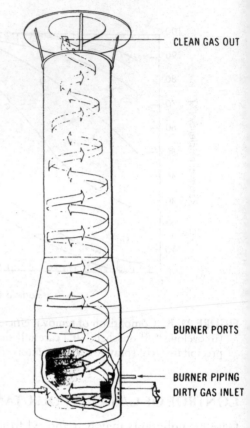

CLEAN GAS OUT

BURNER PORTS

BURNER PIPING
DIRTY GAS INLET

FIGURE 21–9
Incinerator for controlling gaseous
pollutants [Courtesy the American
Lung Association.]

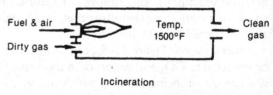

FIGURE 21–10
Comparison of incineration
and catalytic combustion

generation of electricity by burning oil or coal, both of which contain sulfur. Increasingly strict standards for control have prompted the development of a number of options and techniques for reducing SO_2 emissions. Among these options are

- *Change to low-sulfur fuel.* Natural gas is exceedingly low in sulfur, while oil burned for industrial heat and electric power generation contains between 0.5% and 3% sulfur; coal, between 0.3% and 4%. Low-sulfur fuel, however, is expensive and the supply is uncertain.
- *Desulfurization.* Sulfur may be removed from heavy industrial oil by a number of chemical methods similar to those used to lower the H_2S content of crude oil. In coal, sulfur may be either inorganically bound, as pyrite (FeS_2), or organically bound. Pyrite can be removed by pulverizing the coal and washing with a detergent solution. Organically bound sulfur can be removed by washing with concentrated acid. Preferred methods are coal gasification, which produces pipeline-quality gas, or solvent extraction, which produces low-sulfur liquid fuel.
- *Tall stacks.* Although a tall stack does reduce ground-level concentration of gaseous pollutants (see Equation 19.2), it is not an appropriate pollution control measure.
- *Flue gas desulfurization.* The off-gases from combustion or other SO_2-producing processes are called flue gases. SO_2 may be cleaned from flue gas by chemical processes, some of which are discussed later in this chapter. The reaction of SO_2 to form sulfuric acid or other sulfates is the most frequently used flue-gas desulfurization method. It has two limitations: The flue gas must be cleaned of particulate matter before entering an acid-producing plant, and acid formation is energetically favorable only for a fairly concentrated gas stream (about 3%—30,000 ppm—SO_2). The reactions for acid formation are

$$SO_2 + \tfrac{1}{2}O_2 \rightarrow SO_3 \qquad (21.1)$$

$$SO_3 + H_2O \rightarrow H_2SO_4 \qquad (21.2)$$

A double-contact acid plant can produce industrial-grade 98% sulfuric acid. The nonferrous smelting and refining industry has made the most use of this control method. Analogous reactions can be carried out to produce $(NH_4)_2SO_4$, a fertilizer, and gypsum, $CaSO_4$. The SO_2 in flue gas from fossil fuel combustion is too dilute to permit trapping as acid or commercial fertilizer or gypsum. Coal combustion off-gases are also too dirty.

A typical method is single-alkali scrubbing, for which the reactions are

$$SO_2 + Na_2SO_3 + H_2O \rightarrow 2NaHSO_3 \qquad (21.3)$$

$$2NaHSO_3 \rightarrow Na_2SO_3 + H_2O + SO_2 \qquad (21.4)$$

The concentrated SO_2 that is recovered from this process can be used industrially in pulp and paper manufacture and in sulfuric acid manufacture. Figure 21–11 is a diagram for single alkali scrubbing.

SO_2 may also be removed from flue gas by dissolution in an aqueous solution of sodium citrate by the reaction

$$SO_2(g) + H_2O(l) \rightarrow HSO_3^- + H^+ \tag{21.5}$$

Citrate does not itself enter into the reaction but buffers the solution at about pH 4.5. The citrate buffer is readily regenerated. Removal efficiency of the citrate process is between 80% and 99%—much better than the 75% achievable by alkali scrubbing.

Control of Nitrogen Oxides

Wet scrubbers absorb NO_2 as well as SO_2, but are usually not installed primarily for NO_2 control. An effective method for controlling nitrogen oxides, often used on fossil fuel burning power plants, is off-stoichiometric burning. This method controls NO_2 formation by limiting the amount of air (or oxygen) in the combustion process to just a bit more than is needed to burn the hydrocarbon fuel in question. For example, the reaction for burning natural gas,

$$CH_4 + O_2 \rightarrow 2O_2 + CO_2 \tag{21.6}$$

competes favorably with the high-temperature combination of nitrogen in the air with oxygen in the air to form NO (eventually oxidized to NO_2):

$$N_2 + O_2 \rightarrow 2NO \tag{21.7}$$

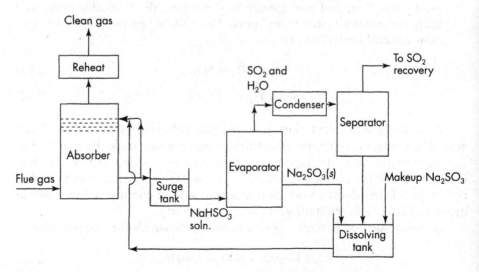

FIGURE 21–11. Simplified diagram for single alkali scrubbing of flue gas with regeneration

The stoichiometric ratio of oxygen needed in natural gas combustion is

$$32 \text{ g of } O_2 : 16 \text{ g of } CH_4 \tag{21.8}$$

A slight excess of oxygen in the combustion air will cause virtually all of the oxygen to combine with fuel rather than with nitrogen. In practice, off-stoichiometric combustion is achieved by adjusting the air flow to the combustion chamber until any visible plume disappears.

Control of Volatile Organic Compounds and Odors

Volatile organic compounds and odors are controlled by thorough oxidation—either incineration or catalytic combustion—since they are only slightly soluble in aqueous scrubbing media.

CONTROL OF MOVING SOURCES

Mobile sources pose special pollution control problems, and one, the automobile, has received particular attention. Pollution control for other mobile sources, such as light-duty trucks, heavy trucks, and diesel engine vehicles, requires controls similar to those for automobile emissions. The important pollution control points in an automobile are shown in Figure 21–12 and are

- Evaporation of hydrocarbons (HC) from the fuel tank
- Evaporation of HC from the carburetor
- Emission of unburned gasoline and partly oxidized HC from the crankcase
- CO, HC, and NO/NO$_2$ from the exhaust

Evaporative losses from the gas tank and carburetor often occur when the engine has been turned off and hot gasoline in the carburetor evaporates. These vapors may be trapped in an activated-carbon canister and can be purged periodically with air and then burned in the engine, as shown schematically in Figure 21–13. The crankcase vent can be closed off from the atmosphere and the blowby gases recycled into the intake manifold. The positive crankcase ventilation (PCV) valve is a small check valve that prevents buildup of pressure in the crankcase.

The exhaust accounts for about 60% of the emitted hydrocarbons and almost all of the NO, CO, and lead. Thus it poses the most difficult control problem of mobile sources. Exhaust emissions depend on the engine operation, as shown in Table 21–1. During acceleration, the combustion is efficient, CO and HC are low, and high compression produces a lot of NO/NO$_2$. On the other hand, deceleration results in low NO/NO$_2$ and high HC because of the presence of unburned fuel in the exhaust. This variation in emissions has prompted EPA to institute a standard acceleration-deceleration cycle for measuring emissions. Testing proceeds from a cold start through acceleration, cruising at constant speeds (on a dynamometer in order to load the engine), deceleration, and a hot start.

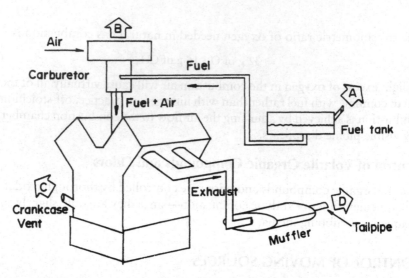

FIGURE 21–12. Diagram of the internal combustion engine showing four major emission points

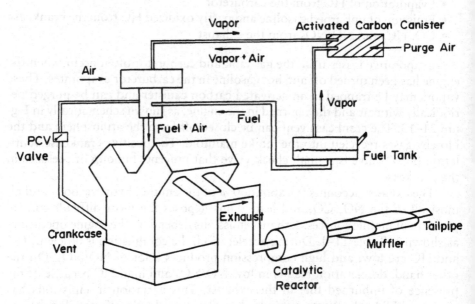

FIGURE 21–13. Internal combustion engine, showing methods of controlling emissions

TABLE 21–1. Effect of Engine Operation on Exhaust Emissions, Shown as Fraction of Emissions at Idle

	CO	HC	NO_x
Idling	1.0	1.0	1.0
Accelerating	0.6	0.4	100
Cruising	0.6	0.3	66
Decelerating	0.6	11.4	1.0

Emission control techniques include engine tune-ups, engine modifications, exhaust gas recirculation, and catalytic reactors. A well-tuned engine is the first line of defense for emission control.

A wide range of acceptable engine modifications is possible. Injection of water can reduce emission of NO, and fuel injection (bypassing or eliminating the carburetor) can reduce CO and HC. However, fuel injection is not compatible with water injection since water may clog the fuel injectors. The stratified charge engine operates on a very lean air/fuel mixture, thus reducing CO and HC, but does not increase NO appreciably. The two compartments of the engine (the "stratification") accomplish this result: The first receives and ignites the air/fuel mixture; the second provides a broad flame for an efficient burn. Better than 90% CO reduction can be achieved by this engine.

Recirculating the exhaust gas through the engine can achieve about 60% reduction of CO and hydrocarbons. The only major modification to an ordinary engine required by exhaust gas recirculation (EGR), in addition to the necessary fittings, is a system for cooling the exhaust gas before recirculation to avoid heat deformation of the piston surfaces. Exhaust gas recirculation, although it increased the rate of engine wear, was a popular and acceptable emission control method until 1980, but present-day emission standards require 90% CO control, which cannot be realized by this method.

New cars sold in the United States since 1983 have required the use of a catalytic reactor ("catalytic converter") to meet exhaust emission standards, and the device is now standard equipment on new cars. The modern three-stage catalytic converter performs two functions: *oxidation* of CO and hydrocarbons to CO_2 and water, and *reduction* of NO to N_2. A platinum-rhodium catalyst is used, and NO reduction is accomplished in the first stage by burning a fuel-rich mixture, thereby depleting the oxygen at the catalyst. Air is introduced in the second stage, and CO and hydrocarbons are oxidized at a lower temperature. Catalytic converters are rendered inoperable by inorganic lead compounds, so that cars using catalytic converters require the use of unleaded gasoline.

Diesel engines produce the same three major pollutants as gasoline engines, although in somewhat different proportions. In addition, diesel-powered heavy-duty vehicles produce annoying black soot—essentially unburned carbon. Control of diesel exhaust was not required in the United States until passage of the 1990 Clean Air Act (nor is it required anywhere else in the world), and therefore little research on diesel exhaust emission control has reached the stage of operational devices.

An emission-free *internal* combustion engine is something of a contradiction in terms. Drastic lowering of emissions to produce a virtually pollution-free engine might be attained with an *external* combustion engine that can achieve better than 99% control of all three major exhaust pollutants. However, although work began in 1968 on such an engine, a commercial model has yet to be built.

Natural gas is used in some cities to fuel fleets of cars (like those owned by the local utility) and some buses, but the limited supply of natural gas serves a number of competing uses. A complete changeover to natural gas would require a different refueling system from that used for gasoline.

Electric cars are clean, but can store only limited power and have limited range. Generation of the electricity to power such cars also generates pollution, and the world's supply of battery materials would be strained to provide for a changeover.

The 1990 Clean Air Act requires that cities in violation of the National Ambient Air Quality Standards sell *oxygenated fuel* during the winter months. Oxygenated fuel is gasoline containing 10% ethanol (CH_3CH_2OH), and its use results in somewhat more efficient conversion of CO to CO_2.

CONTROL OF GLOBAL CLIMATE CHANGE

The two types of compounds involved in global climate change are those that produce free halogen atoms by photochemical reaction, and thus deplete the stratospheric ozone layer, and those that absorb energy in the near infrared spectral region, that may ultimately produce global temperature change. The first group comprises mostly chlorofluorocarbons (CFCs). Control of chlorofluorocarbon emission involves control of leaks, as from refrigeration systems, and eliminating use of the substances, as suggested by the Montreal Protocol. Chlorofluorocarbon aerosol propellants may be useful and convenient, but they are no longer used for aerosol deodorant, cleaners, paint, hairspray, and so on. Roll-on deodorant, wipe-on cleaners, brushed-on or rolled-on paint, atomized liquids, and hair mousse have been found to do the job without affecting the ozone layer.

CONCLUSION

Most of this chapter describes air pollution control by "bolt-on" devices, but these are usually the most expensive methods. A general pollution control truism is that the least expensive and most effective control point is always the farthest up the process line. Most effective control is achieved at the beginning of the process or, better yet, by finding a less polluting alternative to the process: mass transit as a substitute for cars, for example, and energy conservation instead of infinite expansion of generating capacity. Not only are such considerations good engineering and good economics, but they provide a sensible and enlightened analysis of the impact of modern lifestyles on the environment.

PROBLEMS

21.1   Taking into account cost, ease of operation, and ultimate disposal of residuals, what type of control device do you suggest for the following emissions?

a. Dust particles with diameters between 5 and 10 μm
b. Gas containing 20% SO_2 and 80% N_2
c. Gas containing 90% HC and 10% O_2
d. Gas containing 80% N_2 and 20% O_2

21.2   An industrial emission has the following characteristics: 80% HC, 15% O_2, and 5% CO. What type of air pollution control equipment do you recommend? Why?

21.3   A whiskey distillery has hired you as a consultant to design air pollution control equipment for a new facility, to be built upwind from a residential area. What problems will you encounter and what will your control strategy be?

21.4   A copper smelter produces 500 tons of copper per day from ore that is essentially CuS_2. The sulfur dioxide produced in this process is trapped in a sulfuric acid plant that produces 98% by weight sulfuric acid, which has a specific gravity of 2.3. If 75% of the SO_2 produced is trapped by the acid plant, how many liters of 98% H_2SO_4 are produced each day?

the process of, better yet, by finding a less polluting alternative to the chemicals that cannot be substituted for use, for example, and energy conservation, increased of famous application of generation schemes. We could, instead of introducing wood engineering and good resource substitutes provide a simple and quantitated analysis of the impact of a certain lifestyle on the environment.

PROBLEMS

21.1 Taking into account costs, risk inherent, and ultimate disposal of pollutants, what types of control do you suggest for the following gas streams?

    a. Gas primarily with Hg vapor between 3 and 15 ppm

    b. Gas containing 20% CO_2 and 80% N_2

    c. Gas containing 99% HCl and 1% Cl_2

    d. Gas containing 50% N_2 and 50% CO

21.2 A plant for conversion has the following exhaust gas: 70% N_2, 10% HCl, 5% Cl_2, and 5% CO_2. What type of air pollution control equipment do you use, commend it by?

21.3 A thick grinding wheel has the force, 50 m/s, of manufacture to you in production control equipment to cause factory to be built now and build residential area. What problems will you encounter and what will you control measure be?

21.4 A copper smelter produces several tons of copper per day from ore that is usually Cu_2S. Sulfur dioxide evolved is then processed to form and a dilute sulfuric acid or diluted 93% concentrated sulfuric acid, which has approximately 2.3 p.p. each of the O_2 produced is pumped into the acid plant flux, at a flux of 93%, 14:50, are produced as follow?

Chapter 22

Air Pollution Law and Regulations

As with water pollution, a complex system of laws and regulations governs the use of air pollution abatement technologies. In this chapter, the evolution of this system is described from its roots in common law through the passage of federal statutory and administrative initiatives. Problems encountered by regulatory agencies and polluters are addressed, with particular emphasis on the impacts the system may or may not have on future economic development. Figure 22–1 offers a road map to be followed through this maze.

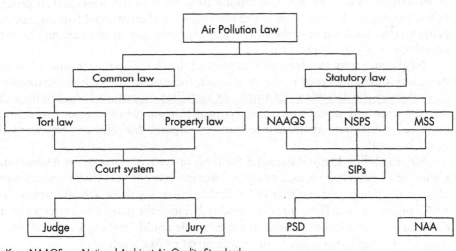

Key: NAAQS – National Ambient Air Quality Standards
NSPS – New (stationary) Source Performance Standards
MSS – Moving Source Standards
SIPs – State Implementation Plans
PSD – Prevention of Significant Deterioration
NAA – Nonattainment Areas

FIGURE 22–1. Diagram of the federal Clean Air Act

AIR QUALITY AND COMMON LAW

When relying on common law, an individual or group of individuals injured by a source of air pollution may cite general principles in two branches of that law: tort and property. These branches have developed over the years and may apply in particular cases. The harmed party, the plaintiff, can enter a courtroom and seek remedies from the defendant for damaged personal well-being or damaged property.

Tort Law

A tort is an injury incurred by one or more individuals. Careless accidents and exposure to harmful airborne chemicals are the types of wrong included under this branch of common law. A polluter can be held responsible for the damage to human health under three broad categories: tort liability, negligence, and strict liability.

Intentional liability requires proof that somebody did a wrong to another party *on purpose*. This proof is especially complicated in the case of damages from air pollution. The fact that a "wrong" actually occurred must first be established—a process that may rely on direct statistical evidence or strong inference, such as the results of laboratory tests on rats. Additionally, intent to do the "wrong" must be established, which involves producing evidence in the form of written documents or direct testimony from the accused individual or group of individuals. Such evidence is not easily obtained. If intentional liability can be proven to the satisfaction of the courts, actual as well as punitive damages can be awarded to the injured plaintiff.

Negligence may involve mere inattention by the air polluter who allowed the injury to occur. Proof in the courtroom focuses on the lack of reasonable care taken by the defendant. Examples in air pollution include failure to inspect the operation and maintenance of electrostatic precipitators or failure to design and size an adequate abatement technology. Again, damages can be awarded to the plaintiff.

Strict liability does not consider the *fault* or state of mind of the defendant. Under certain extreme cases, a court of common law has held that some acts are *abnormally dangerous* and that individuals conducting those acts are strictly liable if injury occurs. The court does tend to balance the danger of an act against the public utility associated with it. An example could be the emission of a radioactive or highly toxic gas from an industrial smoke stack.

Again, if personal damage is caused by air pollution from a known source, the damaged party may enter a court of common law and argue for monetary damages to be paid by the defendant or for an injunction to stop the polluter from polluting, or for both. Sufficient proof and precedent are often difficult if not impossible to muster, and in many cases tort law has been found inadequate in controlling air pollution and awarding damages.

Property Law

Property law, on the other hand, focuses on the theories of nuisance and property rights; nuisance is based on interference with the use or enjoyment of property, and property rights are based on actual invasion of the property. Property law is founded on ancient actions between landowners and involves such considerations as damage and trespassing. A plaintiff basing a case on property law takes chances, rolls the dice, and hopes the court will rule favorably as it balances social utility against the property rights of the individual.

Nuisance is the most widely used common law action concerning the environment. Public nuisance involves unreasonable interference with a right, such as the "right to clean air," common to the general public. A public official must bring the case to court and represent the public that is harmed by the air pollution. Private nuisance, on the other hand, is based on unreasonable interference with the use and enjoyment of private property. The key to a nuisance action is how the courts define "unreasonable." Based on precedents and the arguments of the parties involved, the common law court balances the equities, hardships, and injuries in the particular case, and rules in favor of the plaintiff or the defendant.

Trespass is closely related to nuisance. The major difference is that some physical invasion, no matter how minor, is technically a trespass. Recall that nuisance theory demands an unreasonable interference with land and the outcome of a particular case depends on how a court defines "unreasonable." Trespass is relatively uncomplicated. Examples include physical presence on the property, vibrations from nearby surface or subsurface strata, and possibly gases and microscopic particles flowing from an individual smoke stack.

In conclusion, common law has generally proven inadequate in dealing with problems of air pollution. The strict burdens of proof required in the courtroom often result in decisions that favor the defendant and lead to smoke stacks that continue to pollute the atmosphere. Additionally, the technicality and complexity of individual cases often limit the ability of a court to act; complicated tests and hard-to-find experts often leave a court and a plaintiff with their hands tied. Furthermore, the plaintiff has to have suffered material or bodily harm from the air pollution to have standing in the court (i.e., the right to be heard by a judge).

One key aspect of these common law principles is their degree of variation. Each state has its own body of common law, and individuals relying on the court system are generally confined to using the common laws of the applicable state. To deal with the shortcomings inherent in application of common law to air pollution cases, Congress adopted the federal Clean Air Act.

STATUTORY LAW

Federal statutory law controlling air pollution began with the 1963 and 1967 Clean Air acts. Although these laws provided broad goals and research money, they did not apply air pollution controls throughout the entire United States but only in particularly dirty communities. In 1970 the Clean Air Act was amended to cover the entire United States, and EPA was created to promulgate clean air regulations and enforce the act. Then, in 1977, provisions were added to the act to protect very clean areas (protection against significant deterioration, PSD), to enforce against areas that were not in compliance, and to extend the compliance dates for automobile emission standards. In 1990, amendments focused on toxic air pollutants and control of all vehicular emissions. The 1970 amendments, however, are the basis for existing federal clean air legislation.

National Ambient Air Quality Standards

EPA is empowered to determine allowable ambient concentrations of air contaminants that are pollutants throughout the United States. These are the National Ambient Air Quality Standards (NAAQS), which have been the focus of the nationwide strategy to protect air quality. The primary NAAQS are intended to protect human health; the secondary NAAQS, to "protect welfare." The latter levels are actually determined as those needed to protect vegetation. These standards are listed in Table 22–1.

NAAQS are set on the basis of extensive collections of information and data on the effects of these air pollutants and human health, ecosystems, vegetation, and materials. These collections are called *criteria documents* by EPA, and the NAAQS pollutants that exist are sometimes referred to as *criteria pollutants*. Data indicate that all criteria pollutants have some threshold below which there is no damage. Under the Clean Air Act, most enforcement power is delegated to the states by EPA, however, the states must show that they can clean up the air to the levels of the NAAQS. This showing is made in a State Implementation Plan (SIP), a document that contains all of the state's regulations governing air pollution control, including local municipal regulations. The SIP must be approved by EPA, but once approved it has the force of federal law.

Regulation of Emissions

Under Section 111 of the Clean Air Act, EPA has the authority to set *emission standards* (called performance standards in the act) only for new or markedly modified sources of the criteria pollutants. The states may set performance standards for existing sources and have the authority to enforce EPA's new source performance standards (NSPS). EPA has also delegated to certain states the authority to develop their own NSPS. A priority list of industries for which NSPS are to be set has been in place since 1971; new technology can also motivate NSPS revisions. The list of industries and NSPS is too long for this chapter, but Table 22–2 gives some examples.

Under Section 112 of the Clean Air Act, EPA has the authority to set *national emission standards* for hazardous air pollutants (NESHAPS) for all sources of those pollutants. The 1990 Clean Air Act Amendments required that EPA develop a list of substances to be regulated. Substances can be added to

TABLE 22–1. Selected National Ambient Air Quality Standards

Pollutant	Primary (ppm)	Primary ($\mu g/m^3$)	Secondary (ppm)	Secondary ($\mu g/m^3$)
Particulate matter ($\mu g/m^3$) measured as particles <10 microns in diameter (PM_{10})				
Annual geometric mean	—	75	—	50
Max 24-hr concentration	—	260	—	150
Sulfur dioxides				
Annual arithmetic mean	0.03	80	0.02	60
Max 24-hr concentration	0.14	365	0.1	260
Max 3-hr concentration	—		0.5	1300
Carbon monoxide				
Max 8-hr concentration	9	10,000	same as primary	
Max 1-hr concentration	35	40,000		
Ozone				
Max 1-hr concentration	0.12	235	same as primary	
Hydrocarbons				
Max 3-hr concentration	0.24	160	same as primary	
Nitrogen oxides				
Annual arithmetic mean	0.053	100	same as primary	
Lead				
Avg of 3 months		1.5	same as primary	

TABLE 22–2. Some Typical NSPS

Facility	Minimum Size	NSPS Particulate Matter	NSPS SO_2	NSPS NO_x
Coal-burning generating plants	250 MBtu/hr[a]	0.10 lb/MBtu[a] and 20% plume opacity	1.2 lb/MBtu[a]	0.7 lb/MBtu[a]
Oil-burning generating plants	250 MBtu/hr[a]		0.8 lb/MBtu[a]	0.3 lb/MBtu[a]
H_2SO_4 plants			0.5 lb/ton of acid produced	
MSW incinerators	250 tons/day	34 mg/dscf[b]		180 ppm

[a]"MBtu" means million Btu heat input.
[b]"dscf" means dry standard cubic foot, corrected to 7% oxygen.
From 40 CFR 60 (1991).

and removed from the list as a result of ongoing research. Currently the list contains hundreds of pollutants, including

Acetaldehyde	Hydrazine	Styrene
Aniline	Hydrogen fluoride	Toxaphene
Asbestos	Lindane	Vinyl chloride
Benzene	Methanol	Cadmium compounds
Ethylene glycol	Phenol	Mercury compounds

Categories of industrial facilities that emit these listed pollutants must use removal technologies if the facility emits 10 tons per year of any single hazardous substance or 25 tons per year of any combination. Categories of industrial sources, taken from the list of hundreds, include

- Industrial external combustion boilers
- Printing and publishing
- Waste oil combustion
- Gasoline marketing
- Glass manufacturing

Each "major source," as defined by the 10 ton/yr or 25 ton/yr limit, must achieve a maximum achievable control technology (MACT) for emissions. On a case-by-case basis, a State Implementation Plan could require a facility to install control equipment, change an industrial or commercial process, substitute materials in the production process, change work practices, and train and certify operators and workers. The MACT requirements will impact air pollution sources throughout the 1990s as the new Clean Air Act is implemented.

Prevention of Significant Deterioration

In 1973, the Sierra Club successfully sued EPA for failing to protect the cleanliness of the air in parts of the United States where the air was cleaner than the NAAQS. In response, Congress included prevention of significant deterioration (PSD) in the 1977 and 1990 Clean Air Act amendments.

For PSD purposes, the United States is divided into class I and class II areas, with the possibility of class III designation for some areas. Class I includes the so-called "mandatory class I" areas—all national wilderness areas larger than 5000 acres and all national parks and monuments larger than 6000 acres—and any area that a state or Native American tribe wishes to designate class I. The rest of the United States is class II, except that a state or tribe may petition EPA for redesignation to class III.

To date, the only pollutants covered by PSD are sulfur dioxide and particulate matter. PSD limits the increases in these, as indicated in Table 22–3. In addition, visibility is protected in class I areas.

TABLE 22–3. Maximum Allowed Increases Under PSD

	Particulate Matter		SO_2		
Class	Annual Mean	24-hr Max	Annual Mean	24-hr Max	3-hr Max
I	5	10	2	5	25
II	19	37	20	91	512
III	37	75	40	182	700

Note: Values are in $\mu g/m^3$.

An industry wishing to build a new facility must show, by dispersion modeling with a year's worth of weather data, that it will not exceed the allowed increment. On making such a showing, the industry receives a PSD permit from EPA. The PSD permitting system has had considerable impact in new facility siting. In addition, any new project in a nonattainment area must not only be the most stringent control technology, but it must also guarantee that "offsets" exist for any emissions from its facility. That is, even if the new facility uses MACT, it may still emit 15 tons of particles per year. Some other facility must reduce its emissions of particles by 15 tons/year before the new one can begin operation.

Where the ambient concentrations are already close to the NAAQS and the PSD limits would allow them to exceed the NAAQS, the PSD limits are clearly moot.

Nonattainment

A region in which the NAAQS for one or more criteria pollutants are exceeded more than once a year is called a nonattainment area for that pollutant. If there is nonattainment of the lead, CO, or ozone standard, a traffic reduction plan and an inspection and maintenance program for exhaust emission control are required. Failure to comply results in the state's loss of federal highway construction funds.

If nonattainment results from stationary source emission, an offset program must be initiated. Such a program requires a rollback in emissions from existing stationary sources such that total emissions after the new source operates will be less than before. New sources in nonattainment areas must attain the lowest achievable emission rate (LAER) without necessarily considering the cost of such emission control.

The offset program allows industrial growth in nonattainment areas, but offers a particular challenge to the air pollution control engineer. The offset represents emission reductions that would otherwise not be required. Actions that could generate offsets include tighter controls on existing operations at the same site, a binding agreement with another facility to reduce emissions, and the purchase of another facility to reduce emissions by installation of control equipment or by closing the facility down.

MOVING SOURCES

The emissions from cars, trucks, and buses are regulated under the 1990 Clean Air Act. Phase I of the car standards beginning in the year 1994 for CO, nonmethane hydrocarbons, and NO_x are

- 3.4 g/mile for CO
- 0.4 g/mile for NO_x
- 0.25 g/mile for nonmethane hydrocarbons

Phase II standards, being considered for implementation in the year 2003, are

- 1.7 g/mile for CO
- 0.2 g/mile for NO_x
- 0.125 g/mile for nonmethane hydrocarbons

Standards allowing slightly higher emissions for these pollutants, as well as a standard of 0.08 grams of particulate matter per mile, will be set for light-duty trucks. Hazardous air pollutants emitted from vehicles, particularly benzene and formaldehyde, will also be regulated.

TROPOSPHERIC OZONE

The 1990 Clean Air Act requires each state to establish categories of ozone nonattainment areas. Reasonably available control technologies (RACT) are to be required for all air polluters in such areas depending on the degree of nonattainment. Requirements for RACT are summarized in Table 22–4. Each state must classify areas relative to nonattainment for ozone concentration.

ACID RAIN

Fossil fuel electric generating facilities have been the major contributors to the acid rain problem; the 1990 Clean Air Act acknowledges the importance of these sources of sulfur dioxide and nitrogen oxides, and its regulations require SO_2

TABLE 22–4. Size of Pollution Sources That Must Use RACT

Level of Nonattainment	Size of Pollutant Source
Extreme	10 tons/year
Severe	25 tons/year
Serious	50 tons/year
Moderate	100 tons/year

emissions to be reduced by 9 million tons from the 1980 nationwide emission level, and NO_2 emissions to be reduced by 2 million tons from the 1980 levels.

Reductions are to be achieved at 111 key fossil fuel facilities in 21 states encompassing the Ohio River valley, New England and the northeastern seaboard, the Appalachian states, and the North Central states. Section 404 of the act issues regulated facilities "allowances" for sulfur dioxide emissions, which may be freely used by the facility, transferred by sale to other polluters, or held for future use. Allowances for the 111 facilities are based on historical operation data for the calendar years 1985 through 1987. An annual auction of these transferable discharge permits will be held, with anyone, including national and local environmental organizations, able to bid.

PROBLEMS OF IMPLEMENTATION

Both regulatory agencies and the sources of pollution have encountered difficulties in the implementation of air pollution abatement measures. Because most major stationary sources of air pollution have installed abatement equipment in response to federal pollution control requirements, regulatory agencies must develop programs that promote continuing compliance. However, the track records of the agencies indicate that current programs are not always effective.

Flaws in design and construction of control equipment contribute to significant noncompliance. Problems include the use of improper materials in constructing controls, undersizing of controls, inadequate instrumentation of the control equipment, and inaccessibility of control components for proper operation and maintenance. Design flaws such as these cast doubt on how effectively the air pollution control agencies evaluate the permit applications for these sources. Additionally, many permit reviewers working in government lack the necessary practical experience to fully evaluate proposed controls and tend to rely too heavily on the inadequate technical manuals available to them.

A source permit, as it is typically designed within the regulatory framework, could ensure that an emitter will install necessary control equipment. It could also require the emitter to keep records that facilitate proper operation and maintenance. However, agencies do not always use the permit program to accomplish these objectives. For example, in some states, a majority of the sources fail to keep any operating records at all that would enable an independent assessment of past compliance. In other states, although many sources keep good operating records, they are not required to do so by the control agency.

An agency's ability to detect violations of emission requirements may be critical to the success of its regulatory effort. However, many agencies rely on surveillance by sight and smell as opposed to stack testing or monitoring. Air pollution control agencies, for example, often issue notices of violation for such problems as odor, dust, and excessive visible emissions, even though it is recognized that this approach neglects other, perhaps more detrimental pollutants.

The emitters of air pollution also face problems in complying with regulations. Although most large stationary sources of pollution have installed control equipment, severe emission problems may be caused by design or upsets with the control equipment or in the production process. Poor design of the control equipment is probably the primary cause of excess emissions in the greatest number of sources, and process upsets and routine component failure appear to be a close second. Other causes include improper maintenance, lack of spare parts, improper construction materials, and lack of instrumentation. Additionally, wide variations in the frequency and duration of excess emissions are generally experienced by different industrial categories.

CONCLUSION

Air pollution law is a complex web of common and statutory law. Although common law has offered and continues to offer checks and balances between polluters and economic development, shortcomings do exist. Federal statutory law has attempted to fill the voids and, to a certain extent, has been successful in cleaning the air. Engineers must be aware of the requirements placed on industry by this system of laws. Particular attention must be paid to the siting of new plants in different sections of the nation.

PROBLEMS

22.1 Acid rain is a mounting problem, particularly in the northeastern states. Discuss how this problem can be controlled under the system of common law and compare this approach with the remedy under federal statutory law. Which would be more successful and why?

22.2 Pollution from tobacco smoke can significantly degrade the quality of certain air masses, and many ordinances exist that limit where one can smoke. Develop sample town ordinances to improve the quality of air in shopping malls and over the spectators at both indoor and outdoor sporting events. Rely on both structural and nonstructural alternatives—solutions that mechanically, chemically, or electrically clean the air—and those that prevent all or part of the pollution in the first place.

22.3 Would you favor an international law that permits open burning (combustion without emission controls) of hazardous waste on the high seas? Would you favor such a law if emissions were controlled? Discuss your answer in detail; would you require burning permits that specify time of day or distance from shore, or banning certain wastes from incineration?

22.4 Emissions from a nuclear power generating facility pose unique problems for local health officials. Radioactive materials are included in the list of hazardous air pollutants. What precautions would you take if you were charged with protecting the air quality of nearby residents? How would your set of rules consider direction of prevailing winds, age of residents, and siting of new schools? Suggest a model NESHAPS standard for nuclear power plant emissions.

22.5   Assume you live in a small town that has two stationary sources of air pollution: a laundry/dry cleaning establishment and a regional hospital. Automobiles and front porches in the town have a habit of turning black literally overnight. What air laws apply in this situation, and, presuming that they are not being enforced, how do you determine if, in fact, ambient or emissions standards, or both, are being violated and, if either are, advise the residents to respond to see that the laws are enforced?

22.6   Emission standards for automobiles are set in units of grams per mile. Why not ppm or grams per passenger-mile? What could be the advantages and disadvantages of these units of measurement? Which would you recommend and why?

LIST OF SYMBOLS

EPA	U.S. Environmental Protection Agency
LAER	lowest achievable emission rate
MACT	maximum achievable control technology
MSS	Moving Source Standards
NAA	nonattainment areas
NAAQS	National Ambient Air Quality Standards
NESHAPS	National Emission Standards for Hazardous Air Pollutants
NSPS	New (stationary) Source Performance Standards
PSD	prevention of significant deterioration
RACT	reasonably available control technologies
SIP	State Implementation Plans

Chapter 23

Noise Pollution and Control

The ability to make and detect sound enables humans to communicate with each other as well as to receive useful information from the environment. Sound can provide warning, as from a fire alarm, information, as from a whistling tea kettle, and enjoyment, as from music. In addition to such useful and pleasurable sounds there is noise, often defined as unwanted or extraneous sound. We generally classify as noise the unwanted sound made by such products of our civilization as trucks, airplanes, industrial machinery, air conditioners, and similar sound producers.

Damage to the ear is a measurable problem associated with noise; annoyance, while not necessarily easily measured, may be of equal concern. In one Danish investigation of annoyance factors in the workplace, 38 percent of workers interviewed put noise at the top of the list.

Urban noise is not a modern phenomenon. Legend has it that Julius Caesar forbade the driving of chariots on Rome's cobblestone streets at night so he could sleep. Before the ubiquitous use of automobiles, the 1890 noise pollution levels in London were described by an anonymous contributor to the *Scientific American* as follows:

> The noise surged like a mighty heart-beat in the central districts of London's life. It was a thing beyond all imaginings. The streets of workaday London were uniformly paved in "granite" sets . . . and the hammering of a multitude of iron-shod hairy heels, the deafening side-drum tattoo of tyred wheels jarring from the apex of one set (of cobblestones) to the next, like sticks dragging along a fence; the creaking and groaning and chirping and rattling of vehicles, light and heavy, thus maltreated; the jangling of chain harness, augmented by the shrieking and bellowing called for from those of God's creatures who desired to impart information or proffer a request vocally-raised a din that is beyond conception. It was not any such paltry thing as noise.[1]

Noise may adversely affect humans physiologically as well as psychologically. It is an insidious pollutant; damage is usually long range and permanent.

---

[1]Taylor, R., *Noise*, New York: Penguin Books (1970).

THE CONCEPT OF SOUND

The average citizen has little if any concept of what sound is or what it can do. This is exemplified by the tale of a well-meaning industrial plant manager who decided to decrease the noise level in his factory by placing microphones in the plant and channeling the noise through loudspeakers to the outside.

Sound is the transfer of energy without transfer of mass. For example, a rock thrown at you would certainly get your attention, but this would require the transfer of the rock's mass. Alternatively, your attention may be gained if you were poked with a stick, in which case the stick would not be lost, but energy would be transferred from the poker to the target. In the same way, sound travels through a medium such as air without a transfer of mass. Just as the stick has to move back and forth, so must air molecules oscillate in waves to transfer energy.

The small displacement of air molecules that creates pressure waves in the atmosphere is illustrated in Figure 23–1. As the piston is forced to the right in the tube, the air molecules next to it are reluctant to move and instead pile up on the face of the piston (Newton's First Law). These compressed molecules now act as a spring and release the pressure by jumping forward, creating a wave of compressed air molecules that moves through the tube. The potential energy has been converted to kinetic energy. These pressure waves move down the tube at a velocity of 344 m/sec (at 20°C). If the piston oscillates at a *frequency* of, say, 10 cycles/sec, there will be a series of pressure waves in the tube, each 34.4 m apart. This relationship is expressed as

$$\lambda = \frac{c}{v} \tag{23.1}$$

where λ = wavelength, in m
c = velocity of the sound in a given medium, in m/sec
v = frequency, in cycles/sec

Sound travels at different speeds in different materials, depending on the physical properties of the material.

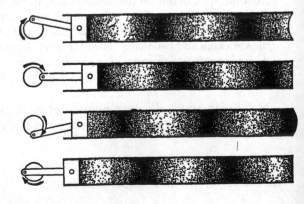

FIGURE 23–1
A piston creates pressure waves that are transmitted through air

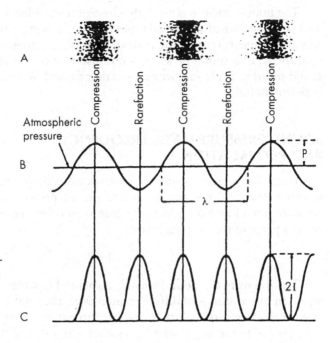

FIGURE 23–2
Sound waves. All parts
of the figure show the
spatial variation along
the wave at a particular
instant of time.
(A) regions of compres-
sion and rarefaction in
the air. (B) pressure
wave; P = pressure ampli-
tude and λ = wavelength.
(C) Intensity wave; the
average intensity is
denoted I.

Example 23.1
In cast iron, sound waves travel at about 3440 m/sec. What is the wavelength of a
sound from a train if it rumbles at 50 cycles/sec and one listens to it placing an
ear on the track?

$$\lambda = \frac{c}{v} = \frac{3440}{50} \approx 69 \text{ m/cycle} \qquad (23.1a)$$

In acoustics, the frequency as cycles per second is denoted by the term
hertz,[2] and written Hz. The range of frequency audible to the human ear is between 20 and 20,000 Hz. The middle A on the piano (concert pitch) is 440 Hz.
Frequency is one of the two basic parameters that describe sound. *Amplitude*,
how loud the sound is, is the other.

If the amplitude of a pressure wave of a sound with only one frequency is
plotted against time, the wave is seen to produce a sinusoidal trace, as shown
in Figure 23–2. All other sounds are made up of a number of suitable sinusoidal
waves, as demonstrated originally by Fourier. Although both nonrandom and
random combinations of sinusoidal waves can be pleasing to the ear, noise is
usually a random combination.

---

[2]In honor of the German physicist Heinrich Hertz (1857–1894).

The human ear is a remarkable instrument, able to detect sound pressures over seven orders of magnitude, but it is not a perfect receptor of acoustic energy. In the measurement and control of noise, it is therefore important not only to know what a sound pressure is but also to have some notion of how loud a sound *seems* to be. Before we address that topic, however, we must review some basics of sound.

SOUND PRESSURE LEVEL, FREQUENCY, AND PROPAGATION

Figure 23–2 represents a wave of pure sound: a single frequency. A sound wave is a compression wave, and the amplitude is a pressure amplitude, measured in pressure units like N/m^2. As is the case with other wave phenomena, intensity is the square of the amplitude, or

$$I = P^2 \tag{23.2}$$

The intensity of a sound wave is measured in watts, a unit of power. When a person hears sounds of different intensities, the total intensity heard is not the sum of the intensities. Rather, the human ear tends to become overloaded or saturated with too much sound. Another statement of this phenomenon is that human hearing sums up sound intensities logarithmically rather than linearly. A unit called the *bel* was invented to measure sound intensity. Sound intensity level (IL) in bels is defined in Equation 23.3:

$$IL_b = \log_{10}\left(\frac{I}{I_0}\right) \tag{23.3}$$

where   I = sound intensity, in watts
I_0 = intensity of the least audible sound, usually given as $I_0 = 10^{-12}$ watts

The bel is an inconveniently large unit. More convenient, and now in common usage, is the *decibel (dB)*. Sound intensity level in dB is defined as

$$IL = 10 \log_{10}\left(\frac{I}{I_0}\right) \tag{23.4}$$

Since intensity is the square of pressure (Equation 23.2), an analogous equation may be written for sound pressure level (SPL) in dB:

$$SPL \text{ (as dB)} = 20 \log_{10}\left(\frac{P}{P_{ref}}\right) \tag{23.5}$$

where   SPL (dB) = sound pressure level, in dB
P = pressure of sound wave, N/m^2
P_{ref} = some reference pressure, generally chosen as the threshold of hearing, $0.00002 \ N/m^2$

These relationships are also derivable from a slightly different point of view. In 1825, E. H. Weber found that people can perceive differences in small weights, but if a person is already holding a substantial weight, that same increment is not detectable. The same idea is true with sound. For example, a 2-N/m^2 increase from an initial sound pressure of 2 N/m^2 is readily perceived, whereas the same 2-N/m^2 difference is not noticed if it is added to a background of 200-N/m^2 sound pressure. Applying this principle, we can also define the decibel as

$$dB = 10 \log_{10} \frac{W}{W_{ref}} \tag{23.6}$$

where the power level, W, is divided by a constant, W_{ref}, which is a reference value, both measured in watts.

Since the pressure waves in air are half positive and half negative, adding these results in zero. Accordingly, sound pressures are measured as root-mean-square (rms) values, which are related to the energy in the wave. It may also be shown that the power associated with a sound wave is proportional to the square of the rms pressure. We can therefore write

$$dB = 10 \log_{10} \frac{W}{W_{ref}} = 10 \log_{10} \frac{P^2}{P^2_{ref}} = 20 \log_{10} \frac{P}{P_{ref}} \tag{23.7}$$

where P and P_{ref} are the pressure and reference pressure, respectively. If we define $P_{ref} = 0.00002$ N/m^2, the threshold of hearing, we can again express the *sound pressure level* (SPL), as in Equation 23.8:

$$SPL \ (dB) = 20 \log_{10} \left(\frac{P}{P_{ref}} \right) \tag{23.8}$$

Although it is common to see the sound pressure measured over a full range of frequencies, it is sometimes necessary to describe a noise by the amount of sound pressure present at a specific range of frequencies. Such a *frequency analysis*, shown in Figure 23–3, may be used for solving industrial problems or evaluating the danger of a certain sound to the human ear.

In addition to amplitude and frequency, sound has two more characteristics of importance. Both may be visualized by imagining the ripples created by dropping a pebble into a large, still pond. The ripples, analogous to sound pressure waves, are reflected outward from the source, and their magnitude is dissipated as they move farther away from the source. Similarly, sound levels decrease as the distance between the receptor and the source is increased.

In summary, the four important characteristics of sound waves are as follows:

- Sound pressure, the magnitude or amplitude of sound
- Pitch, determined by the frequency of the pressure fluctuations

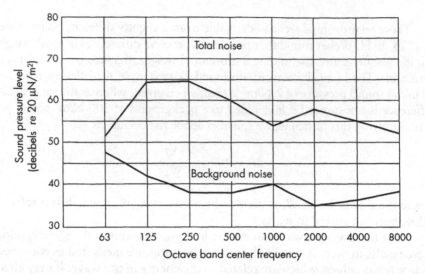

FIGURE 23–3. Typical analysis of a machine noise with background noise

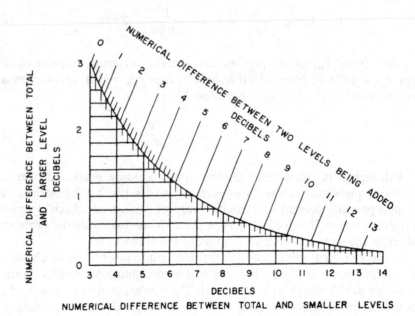

FIGURE 23–4. Chart for combining different sound pressure levels. For example: Combine 80 and 75 dB. The difference is 5 dB. The 5-dB line intersects the curved line at 1.2 dB; thus the total value is 81.2 dB. [Courtesy of General Radio.]

- Sound waves, which propagate away from the source
- Sound pressure, which decreases with increasing distance from the source

The mathematics of adding decibels is a bit complicated, but may be greatly simplified by the graph shown as Figure 23–4. As a rule of thumb, adding two equal sounds increases the SPL by 3 dB, and if one sound is more than 10 dB louder than a second, the contribution of the latter is negligible.

Background noise (or ambient noise) must also be subtracted from any measured noise. Using the above rule of thumb, if the SPL is more than 10 dB greater than the ambient level, the contribution may be ignored. The covering of a sound with a louder one is known as *masking*. Speech can be masked by industrial noise, for example, as shown in Table 23–1. These data show that an 80-dB SPL in a factory will effectively prevent conversation. Telephone conversations are similarly affected, with a 65-dB background making communication difficult and 80 dB making it impossible. In some cases, it has been found advisable to use *white noise*, a broad frequency hum from a fan, for example, to mask other more annoying noises.

TABLE 23–1. Sound Pressure Levels for Speech Masking

| Distance (ft) | Speech Interference Level (dB) | |
	Normal	Shouting
3	60	78
6	54	72
12	48	66

Example 23.2

A jet engine has a sound pressure level of 80 dB, as heard from a distance of 50 feet. A ground crew member is standing 50 feet from a four-engine jet. What SPL reaches her ear when the first engine is turned on? The second, so that two engines are running? The third? Then all four?

When the first engine is turned on, the SPL is 80 dB, provided there is no other comparable noise in the vicinity. To determine, from Figure 23–3, what the SPL is when the second engine is turned on, we note that the difference between the two engine intensity levels is

$$80 - 80 = 0$$

From the chart, a numerical difference of 0 between the two levels being added gives a difference of 3 between the total and the larger of the two. The total SPL is thus

$$80 + 3 = 83 \text{ dB}$$

When the third engine is turned on, the difference between the two levels is

$$83 - 80 = 3 \text{ dB}$$

yielding a difference from the total of 1.8, for a total IL of

$$83 + 1.8 = 84.8 \text{ dB}$$

When all four engines are turned on, the difference between the sounds is

$$84.8 - 80 = 4.8 \text{ dB}$$

yielding a difference from the total of 1.2, for a total IL of 86 dB.

These characteristics of sound (amplitude, frequency, and dissipation) are physical properties which have nothing to do with the human ear. We know that the ear is an amazingly sensitive receptor, but is it equally sensitive at all frequencies? Can we hear low and high sounds equally well? The answers to these questions lead us to the concept of *sound level*.

SOUND LEVEL

Suppose you are put into a very quiet room and subjected to a pure tone at 1000 Hz at a 40-dB SPL. If this sound is turned off and a pure sound at 100 Hz is piped in and adjusted in loudness until you judge it to be equally loud to the 40-dB, 1000-Hz tone you heard a moment ago, you will, surprisingly enough, judge the 100-Hz tone to be equally loud when it is at about 55 dB. In other words, more energy must be generated at the lower frequency in order to hear a tone at about the same perceived loudness, indicating that the human ear is less efficient at lower frequencies.

If such experiments are conducted for many sounds and with many people, the average equal loudness contours can be drawn (Figure 23–5). These contours are in terms of *phons*, which correspond to the sound pressure level in decibels of the 1000-Hz reference tone. Using Figure 23–5, a person subjected to a 65-dB SPL tone at 50 Hz will judge this to be as loud as a 40-dB, 1000-Hz tone. Hence, the 50-Hz, 65-dB sound has a loudness level of 40 phons. Such measurements are commonly called *sound levels* and are not based on physical phenomena only but have an adjustment factor, a "fudge factor," that corresponds to the inefficiency of the human ear.

Sound level (SL) is measured with a sound level meter consisting of a microphone, an amplifier, a frequency-weighing circuit (filters), and an output scale, shown in Figure 23–6. The weighing network filters out specific frequencies to make the response more characteristic of human hearing. Through use, three scales have become internationally standardized (Figure 23–7).

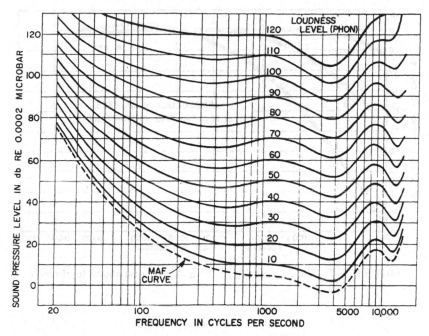

FIGURE 23–5. Equal loudness contours [Courtesy of General Radio.]

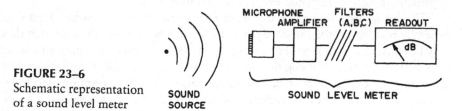

FIGURE 23–6
Schematic representation
of a sound level meter

Note that the A scale in Figure 23–7 corresponds closely to an inverted 40-phon contour in Figure 23–5. Similarly, the response reflected by the B scale approximates an inverted 70-phon contour. The C scale shows an essentially flat response, giving equal weight to all frequencies, and approximates the response of the ear to very loud sounds.

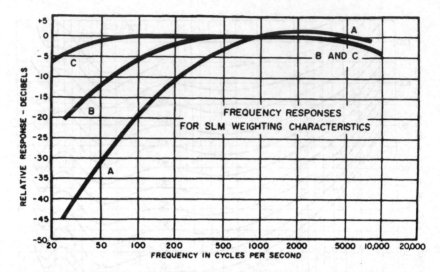

FIGURE 23–7. The A, B, and C filtering curves for a sound level meter

The results of noise measurement with the standard sound level meter are also expressed in terms of decibels but with the scale designated. If on the A scale the meter reads 45 dB, the measurement is reported as 45 dB(A).

Most noise ordinances and regulations are in terms of dB(A) because this scale is a good approximation of human response for sounds that are not very loud. For very loud noise the C scale is a better approximation. However, because the use of multiple scales complicates matters, scales other than A are seldom used. In addition to the A, B, and C scales, a new D scale has been introduced to approximate human response exclusively to aircraft noise.

This brings up another complication in the measurement of noise. We do not generally respond in the same way to different types of noise, even though they may be equal in sound level. For example, a symphony orchestra playing middle C at 120 dB(A) and a jet engine running at an equal 120 dB(A) will draw very different reactions from people. Numerous parameters have been devised to respond to this discrepancy in reactions, all supposedly "the best" means of quantitatively measuring human response to noise. Among these are

- Traffic Noise Index (TNI)
- Sones
- Perceived Noise Level (PNdB)
- Noise and Number Index (NNI)
- Effective Perceived Noise Level (EPNdB)
- Speech Interference Level (SIL)

Measuring methods have proliferated because the physical phenomenon, a pressure wave, is related to a physiological human response (hearing) and then to a psychological response (pleasure or irritation). Along the way, the science becomes progressively more subjective.

Measuring some noises, particularly *community noise* such as traffic and loud parties, is complicated further. Although we have thus far treated noise as a constant in intensity and frequency with time, this is obviously not true for *transient noises* such as trucks moving past a sound level meter or a loud party near the measurement location. An unusual noise that may occur during a period of general quiet is an *intermittent noise* and has a still different effect on people.

MEASURING TRANSIENT NOISE

Transient noise is measured with a sound level meter, but the results must be reported in statistical terms. The common parameter is the percent of time a sound level is exceeded, denoted by the letter L with a subscript. For example, $L_{10} = 70$ dB(A) means that 10% of the time the noise is louder than 70 dB as measured on the A scale. Transient noise data are gathered by reading the SL at regular intervals. These numbers are then ranked and plotted, and the L values are read off the graph.

Example 23.3

Suppose the traffic noise data in Table 23–2 are gathered at 10-second intervals. These numbers are then ranked as indicated in the table and plotted as in Figure 23–5. Note that since 10 readings are taken, the lowest reading (Rank No. 1) corresponds to an SL that is equaled or exceeded 90% of the time. Hence, 70 dB(A) is plotted vs. 90% in Figure 23–8. Similarly, 71 dB(A) is exceeded 80% of the time.

Alternatively, we can calculate the frequency as $[m/(n + 1)]*100$, where m is the rank and n is the number of observations. The frequency can be plotted as either $(m/n)*100$ or $(m/(n + 1)*100$. The error decreases as the number of data points increases.

Plotting the results, as in Figure 23–8, yields $L_{10} = 80$dB(A). That is, only 10% of the time was the SL equal to or greater then 80dB(A). Similarly $L_{50} = 75$ dB(A).

One widely used parameter for gauging the perceived level of noise from transient sources is the noise pollution level (NPL), which takes into account the irritation caused by impulse noises. The NPL is defined as

$$\text{NPL in dB(A)} = L_{50} + (L_{10} - L_{90}) + \frac{(L_{10} - L_{90})^2}{60} \qquad (23.9)$$

TABLE 23–2. Sample Traffic Noise Data and Calculations

Time (sec)	dB(A)	Rank	dB(A)	% of Time Equal to or Exceeded
10	71	1	70	90
20	75	2	71	70
30	70	3	74	70
40	78	4	74	60
50	80	5	75	50
60	84	6	75	40
70	76	7	76	30
80	74	8	78	20
90	75	9	80	10
100	74	10	84	0

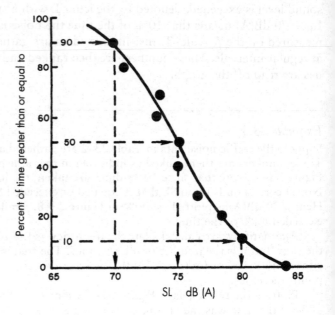

FIGURE 23–8
Results from a survey of transient noise. The data (from Table 23–2) are plotted as percent of time the SL is exceeded vs. the SL, in dB(A).

With reference to Figure 23–8, L_{10}, L_{50}, and L_{90} are 80 dB(A), 75 dB(A), and 70 dB(A), respectively. These may then be substituted into the equation, and the NPL may be calculated. It is always advisable to take as many data as possible; the 10 readings in the above example are seldom sufficient for a thorough analysis.

THE ACOUSTIC ENVIRONMENT

The types of sound around us vary from a Beethoven concerto to the roar of a jet plane. Noise, the subject of this chapter, is generally considered to be unwanted sound, or that incidental to our civilization that we would rather not have to endure. The intensities of some typical environmental noises are shown in Figure 23–9. Most local jurisdictions have ordinances against "loud and unnecessary noise." The problem is that most of them are difficult to enforce.

In the industrial environment, the federal Occupational Safety and Health Administration (OSHA) sets limits for noise in the workplace. Table 23–3 lists these limits. There is some disagreement as to the level of noise that should be allowed for an 8-hour working day. Some researchers and health agencies insist that 85 dB(A) should be the limit. This is not, as might seem, a minor quibble, since the jump from 85 to 90 dB is actually an increase of about four times the sound pressure, remembering the logarithmic nature of the decibel scale.

HEALTH EFFECTS OF NOISE

The human ear is an incredible instrument. Imagine having to design and construct a scale for weighing equally accurately a flea and an elephant—the range of performance to which we are accustomed from our ears. Only recently have we become aware of the devastating psychological effects of noise, although its effect on our ability to hear has been known for a long time.

The human auditory system is shown in Figure 23–10. Sound pressure waves caused by vibrations set the eardrum (*tympanic membrane*) in motion. This activates the three bones in the middle ear—the *hammer, anvil*, and *stirrup*—which physically amplify the motion received from the eardrum and transmit it to the inner ear. This fluid-filled cavity contains the *cochlea*, a snail-like structure in which the physical motion is transmitted to tiny hair cells. Much like seaweed swaying in the current, these hair cells deflect and certain cells are responsive only to certain frequencies. The mechanical motion of the hair cells is transformed to bioelectrical signals and transmitted to the brain by the auditory nerves. Acute damage may occur to the eardrum, but only with very loud, sudden noises. More serious is the chronic damage to the tiny hair cells in the inner ear. Prolonged exposure to noise of a certain frequency pattern may cause temporary hearing loss, which disappears in a few hours or days, or permanent loss. The former is called *temporary threshold shift*, and the latter is known as *permanent threshold shift*. Literally, the threshold of hearing changes, so some sounds cannot be heard.

Temporary threshold shift generally is not damaging to the ear unless exposure to the sound is prolonged. People who work in noisy environments commonly find that they hear worse at the end of the day. Performers in rock bands are subjected to very loud noises (substantially above the allowable OSHA levels) and commonly are victims of temporary threshold shift. In one study, the results of which are shown in Figure 23–11, the players suffered as much as 15-dB temporary threshold shift after a concert.

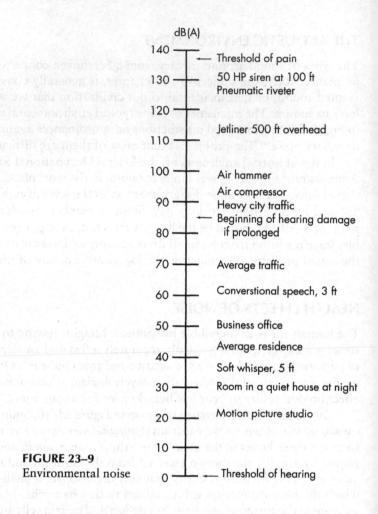

FIGURE 23–9
Environmental noise

dB(A)

140 — Threshold of pain
130 — 50 HP siren at 100 ft
        Pneumatic riveter
120
110 — Jetliner 500 ft overhead
100 — Air hammer
        Air compressor
90 — Heavy city traffic
     ← Beginning of hearing damage
80   if prolonged
70 — Average traffic
60 — Converstional speech, 3 ft
50 — Business office
     Average residence
40
     Soft whisper, 5 ft
30 — Room in a quiet house at night
20 — Motion picture studio
10
0 — ← Threshold of hearing

TABLE 23–3. OSHA Maximum Permissible
Industrial Noise Levels

Sound Level dB(A)	Maximum Duration During Any Working Day (hr)
90	8
92	6
95	4
100	2
105	1
110	0.5
115	0.25

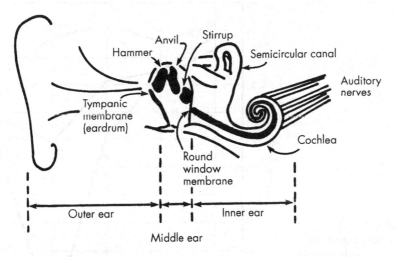

FIGURE 23–10. Cut-away drawing of the human ear

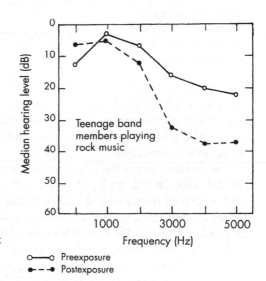

FIGURE 23–11
Temporary threshold shift for rock
band performers [Data from the
U.S. Public Health Service.]

Repeated noise over a long time leads to permanent threshold shift. This is especially true in industrial applications in which people are subjected to noises of a certain frequency. Figure 23–12 shows data from a study performed on workers at a textile mill. Note that the people who worked in the spinning and weaving parts of the mill, where noise levels were highest, suffered the most severe hearing loss, especially at around 4000 Hz, the frequency of noise emitted by the machines.

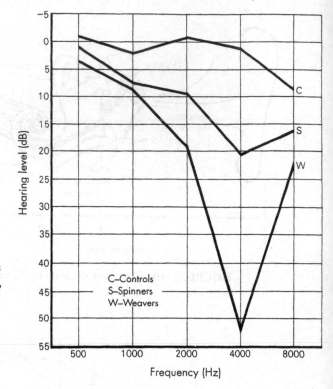

FIGURE 23–12
Permanent threshold
shift for textile workers
[From Burns, W., et al.,
"An Exploratory Study
of Hearing and Noise
Exposure in Textile
Workers," *Annual of
Occupational Hygiene*
7 (1958): 323.]

Hearing becomes less acute simply as an effect of aging. This loss of hearing, called *presbycusis*, is illustrated in Figure 23–13. Note that the greatest loss occurs at the higher frequencies. Speech frequency is about 1000 to 2000 Hz, so the loss is noticeable.

In addition to presbycusis, a serious loss of hearing can result from environmental noise. In one study, 11% of ninth graders, 13% of twelfth graders, and 35% of college freshmen had a greater than 15-dB hearing loss at 2000 Hz, which the study concluded had resulted from exposure to loud noises such as motorcycles and rock music. The researchers found that "the hearing of many of these students had already deteriorated to a level of the average 65-year-old person."[4]

Noise also affects other bodily functions, including those of the cardiovascular system. It alters the rhythm of the heartbeat, makes the blood thicken, dilates blood vessels, and makes focusing the eyes difficult. It is no wonder that excessive noise has been blamed for headaches and irritability. All of these reactions are those that ancestral cave dwellers also experienced. Noise meant danger, and senses and nerves were "up," ready to repel any threat. In the mod-

---

[4]Taylor, R., *Noise,* New York: Penguin Books (1970).

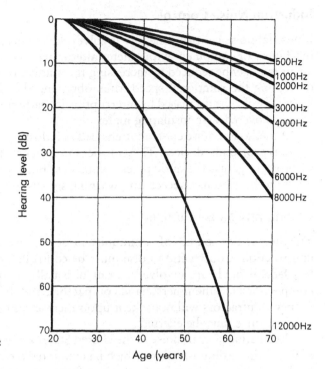

FIGURE 23–13

Hearing loss with age [From Hinchcliffe, R., "The Pattern of the Threshold of Perception of Hearing and Other Special Senses as a Function of Age," *Gerontologica* 2 (1958): 311.]

ern noise-filled world, we are always "up," and it is unknown how much if any of our physical ills are due to our response to noise. We do know that we cannot adapt to noise in the sense that our body functions no longer react a certain way to excessive noise. Thus people do not "get used to" noise in the physiological sense.

In addition to the noise-we-can-hear problem, it is appropriate to mention the potential problems of very high or very low frequency sound, out of our usual 20- to 20,000-Hz hearing range. The health effects of these, if any, remain to be documented.

NOISE CONTROL

The control of noise is possible at three different stages of its transmission:

1. Reducing the sound produced
2. Interrupting the path of the sound
3. Protecting the recipient

When we consider noise control in industry, in the community, or in the home, we should keep in mind that all problems have these three possible solutions.

Industrial Noise Control

Industrial noise control generally involves the replacement of noise-producing machinery or equipment with quieter alternatives. For example, the noise from an air fan may be reduced by increasing the number of blades or their pitch and decreasing the rotational speed, thus obtaining the same air flow. Industrial noise may also be decreased by interrupting its path; for example, a noisy motor may be covered with insulating material.

A method of noise control often used in industry is providing workers with hearing protection devices. These devices must have enough noise attenuation to protect against the anticipated exposures, but must not interfere with the ability to hear human speech and warning signals in the workplace.

Community Noise Control

The three major sources of community noise are aircraft, highway traffic, and construction. Construction noise must be controlled by local ordinances (unless federal funds are involved). Control usually involves the muffling of air compressors, jack hammers, hand compactors, and the like. Since mufflers cost money, contractors will not take it upon themselves to control noise, and outside pressures must be exerted.

Regulating aircraft noise in the United States is the responsibility of the Federal Aviation Administration, which has mounted a two-pronged attack on the problem. First, it has set limits on aircraft engine noise and does not allow aircraft exceeding these limits to use airports, forcing manufacturers to design engines for quiet operation as well as for thrust. Second, it has diverted flight paths away from populated areas and, whenever necessary, has had pilots use less than maximum power when the takeoff carries them over a noise-sensitive area. Often this approach is not enough to prevent significant damage or annoyance, and aircraft noise remains a real problem in urban areas.

Supersonic aircraft present a special problem. Not only are their engines noisy, but the sonic boom may produce considerable property damage. Damage from supersonic military flights over the United States has led to a ban on such flights by commercial supersonic aircraft.

The third major source of community noise is traffic. The car or truck exhaust system, tires, engine, gears, and transmission all contribute to a noise level, as does the very act of moving through the atmosphere. Elevated highways and bridges resonate with the traffic motion and amplify traffic noise. The worst offender on the highways is the heavy truck, which generates noise in all of these ways such that the total noise generated by vehicles may be correlated directly to the truck traffic volume. Figure 23–14 is a typical plot showing sound level as a function of traffic volume (measured in number of trucks per hour). Clearly, truck volume is of great importance. Note that this graph is plotted as "sound level exceeded 10% of the time." Peak sound levels could be a great deal higher.

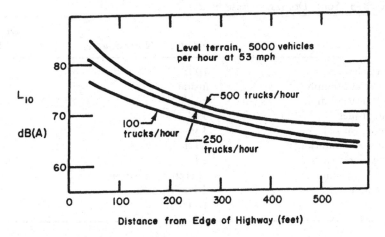

FIGURE 23-14. The effect of truck density and distance from a highway on noise

A number of alternatives are available for reducing highway noise. First, the source could be controlled by making quieter vehicles; second, highways could be routed away from populated areas; and third, noise could be baffled with walls or other types of barriers.

Vegetation, surprisingly, makes a very poor noise screen, unless the screen is 50 yards or more deep. The opposing lanes of the Baltimore-Washington Parkway are separated in many places by 100 to 200 yards of fairly dense vegetation, which provides an excellent noise and light screen. However, newer highways rarely have the luxury of so much right-of-way. The most effective solutions have been to lower the highway or to build physical wood or concrete barriers beside the road and thus screen the noise. All of these have limitations: Noise will bounce off the walls and create little or no noise shadow, and walls hinder highway ventilation, thus contributing to buildup of CO and other pollutants from car and truck exhaust. The Department of Transportation has established design noise levels for various land uses, as shown in Table 23-4.

TABLE 23-4. Design Noise Levels Set by the Federal Highway Administration

Land Category	Design Noise Level (L_{10})	Description of Land Use
A	60 dB(A) exterior	Activities requiring special qualities of serenity and quiet, like amphitheaters
B	70 dB(A) exterior	Residences, motels, hospitals, schools, parks, libraries
	55 dB(A) interior	Residences, motels, hospitals, schools, parks, libraries
C	75 dB(A) exterior	Developed land not included in categories A and B
D	No limit	Undeveloped land

TABLE 23–5. Some Domestic Noisemakers

Item	Distance from Noise Source	Sound Level dB(A)
Vacuum cleaner	10 ft	75
Quiet car at 50 mph	inside	65
Sports car at 50 mph	inside	80
Flushing toilet	5 ft	85
Garbage disposal	3 ft	80
Window air conditioner	10 ft	55
Ringing alarm clock	2 ft	80
Powered lawn mower	operator's position	105
Snowmobile	driver's position	120
Rock band	10 ft	115

Noise in the Home

Private dwellings are getting noisier because of internally produced sound as well as an external community sound. The gadgets in a modern American home read like a list of New Year's Eve noisemakers. Some examples of domestic noise are listed in Table 23–5. Similar products of different brands often will vary significantly in levels. Thus, when shopping for an appliance, it is just as important to ask how noisy it is as it is to ask how much it costs.

The Eraring Power Station near Dove Creek, Australia, is an interesting example of the complexity of environmental pollution and control: All aspects of a problem must be considered before a solution is possible. During calm early mornings, residents complained of a loud "rumbling noise" coming from the power station 1.5 km away. The noise was measured in Dove Creek, and for a single morning, with no changes in the operation of the plant, the sound level increased from 43 dB at 5:30 A.M. to 62 dB at 6:00 A.M. The problem: an atmospheric inversion "capturing" the noise and permitting it to reach Dove Creek's sleepy inhabitants. The problem was solved when the duct work within the power station was attached to reactive dissipative silencers—a series of chambers within the duct work system—which were tuned to the predominant frequencies of the noise source. After a lengthy research project, the problem in Dove Creek was solved.

CONCLUSION

While noise was considered just another annoyance in a polluted world, not much attention was given to it. We now have enough data to show that it is a definite health hazard and should be numbered among our more serious pollutants. With available technology, it is possible to lessen noise pollution. However, the solutions cost money, and private enterprise will not provide them until forced to by the government or the public.

PROBLEMS

23.1   If an office has a noise level of 70 dB and a new machine emitting 68 dB is added to the din, what is the combined sound pressure level?

23.2   Animals like dogs can hear sounds that have SPL *less than* zero. Show how this is possible.

23.3   Dogs can hear sounds at pressures close to $2{\times}10^{-6}$ N/m². What is this in decibels?

23.4   An air compressor emits pressure waves at 0.01 N/m². What is the SPL in dB?

23.5   Why was P_{ref} in Equation 23.9 chosen at 0.00002 N/m²? What if a mistake were made and P_{ref} should have been 0.00004 N/m²? This is a 100% error. How does this affect the numbers in Table 23–1?

23.6   Given the following data, calculate the L_{10} and the L_{50}.

Time (sec)	dB(A)	Time (sec)	dB(A)
10	70	60	65
20	50	70	60
30	65	80	55
40	60	90	70
50	55	100	50

23.7   Suppose your dormitory is 200 yards from a highway. What truck traffic is "allowable" in order to stay within the Federal Highway Administration guidelines?

23.8   In addition to the data listed in Example 23.3 (Table 23–2), the following SL measurements were taken:

Time (sec)	dB(A)	Time (sec)	dB(A)
110	80	160	95
120	82	170	98
130	78	180	82
140	87	190	88
150	92	200	75

Calculate the L_{50}, L_{10}, and NPL using all 20 data points.

23.9   If you sing at a level 10,000 times greater than the power of the faintest audible sound, at which sound pressure level are you singing?

23.10   How many times more powerful is a 120-dB sound than a 0-dB sound?

23.11   The OSHA standard for 8-hour exposure to noise is 90 dB(A). EPA suggests that this should be 85 dB(A). Show that the OSHA level is almost 400% louder in terms of sound level than the EPA suggestion.

23.12   If an occupational noise standard were set at 80 dB for an 8-hour day, 5-day-per-week working exposure, what standard would be appropriate for a 4-hour day, 5-day-per-week working exposure? Remember that it is *energy* which causes the damage to the ear.

23.13   Carry a sound level meter with you for one entire day. Measure and record the sound levels as dB(A) in your classes, in your room, during sports events, in the dining hall, or wherever you go during the day. Which of the recorded sound levels surprises you the most and why?

23.14   Seek out and measure the three most obnoxious noises you can think of. Compare these with the noises in Table 24–5.

23.15   In your room measure and plot the sound level in dB(A) of an alarm clock versus distance. At what distance will it still wake you if it requires 70 dB(A) to get you up? Conduct the same experiment out of doors. What is the effect of your room on the sound level?

23.16   Construct a sound level frequency curve for a basketball game. Calculate the noise pollution level.

23.17   A noise is found to give the following responses on a sound level meter: 82 dB(A), 83 dB(B), and 84 dB(C). Is the noise of a high or low frequency?

23.18   A machine produces 80 dB(A) at 100 Hz (almost pure sound).

a. Can a person who has suffered a noise-induced threshold shift of 40 dB at that frequency hear this sound? Explain.
b. What is this noise measure on the dB(C) scale?

23.19   What percent reduction in sound intensity would have been necessary to reduce the takeoff noise of the American SST from 120 dB to 105 dB?

LIST OF SYMBOLS

c	velocity of sound wave, in m/sec
dB	decibel, sound level
EPNdB	effective perceived noise level
Hz	hertz, in cycles/sec
I	sound intensity, in watts
IL	sound intensity level
I_0	intensity of the least audible sound
v	frequency of sound wave, in cycles/sec
NNI	noise and number index

NPL	noise pollution level
OSHA	Occupational Safety and Health Administration
P	pressure, in N/m^2
PNdB	perceived noise level
P_{ref}	reference pressure, in N/m^2
SIL	speech interference level
SL	Sound Level
SPL	sound pressure level, in dB
TNI	traffic noise index
W	power level, in watts
λ	wavelength, in m

Chapter 24

Environmental Impact and Economic Assessment

Ideally, scientists and engineers respond to a given problem in a rational manner. Decision-making in the public sector should follow a definite sequence: (1) problem definition, (2) generation of alternative solutions, (3) evaluation of alternatives, (4) implementation of a selected solution, and (5) review and appropriate revision of the implemented solution.

For projects that may significantly affect the environment, this process is also mandated by federal law. The National Environmental Policy Act (NEPA) requires that alternatives be considered whenever a federal action will have an environmental impact, as well as that the environmental impact be assessed. Many states have enacted similar legislation to apply to state or state-licensed actions. A similar law in Wisconsin (WEPA), for example, requires state agencies to make a conscious effort to develop alternatives before a solution to an environmental problem is implemented. Some state agencies also operate under the Planning, Programming, and Budgeting System (PPBS), which again stresses the need to consider alternative solutions to problems.

Unfortunately, most state systems are now a hybrid of rational planning sequences and disjointed "muddling through," in which limited groups of alternative solutions are proposed and studied successively, with a choice being made by default when all relevant parties cease to disagree. Such a system falls short of rational planning, in which a broad range of feasible alternative solutions are reviewed simultaneously rather than successively. Recent trends in court action have meant delaying or stopping the implementation of many solutions because alternatives were not adequately developed.

Administrative constraints of time, limited information, lack of expertise, unexpected expense, and often conflicting predetermined priorities make realization of rational planning very difficult. However, lack of knowledge of how to perform the steps of rational planning is inexcusable. It is one thing to require that alternatives be reviewed simultaneously; it is another to know how to generate and evaluate truly alternative solutions for any given problem. This chapter discusses how alternative solutions to environmental problems are analyzed in terms of their projected environmental and economic impacts.

ENVIRONMENTAL IMPACT

On January 1, 1970, President Nixon signed into law the National Environmental Policy Act, which declared a national policy to encourage productive and enjoyable harmony between people and their environment. This law established the Council on Environmental Quality (CEQ), which monitors the environmental effects of federal activities and assists the President in evaluating environmental problems and determining their best solutions. However, few people realized that NEPA contained a sleeper: Section 102(2)(C), which requires federal agencies to evaluate the consequences of any proposed action on the environment:

> Congress authorizes and directs that, to the fullest extent possible: (1) the policies, regulations, and public laws of the United States shall be interpreted and administered in accordance with the policies set forth in this chapter, and (2) all agencies of the Federal Government shall include in every recommendation or report on proposals for legislation and other major Federal actions significantly affecting the quality of the human environment, a detailed statement by the responsible official on—
>
> (i) the environmental impact of the proposed action,
> (ii) any adverse environmental effects that cannot be avoided should the proposal be implemented,
> (iii) alternatives to the proposed action,
> (iv) the relationship between local short-term uses of man's environment and the maintenance and enhancement of long-term productivity, and
> (v) any irreversible and irretrievable commitments of resources that would be involved in the proposed action should it be implemented.

In other words, each project funded by the federal government or requiring a federal permit must be accompanied by an environmental impact statement (EIS) that assesses in detail the potential environmental impacts of a proposed action and alternative actions. All federal agencies are required to prepare statements for projects and programs (a programmatic EIS) under their jurisdiction. Additionally, they must generally follow a detailed and often lengthy public review of each EIS before proceeding with the project or permit. In some instances, legislation allows substitution of a slightly less rigidly prescribed environmental impact assessment (EIA) for an EIS. An agency may also publish a Finding of No Significant [Environmental] Impact (FONSI) if it determines by environmental assessment that the impact of the proposed Federal action will be negligible.

The original idea of the EIS was to introduce environmental factors into the decision-making machinery. Its purpose is not to provide justification for a construction project but rather to introduce environmental concerns and discuss them in public before the decision on a project is made. However, this objective is difficult to apply in practice. Interest groups in and out of government articulate plans to their liking, the sum of which provide a set of alternatives to be evaluated. Usually, one or two plans seem eminently more feasible and reasonable than the others. These are sometimes legitimized by just slightly juggling selected time scales or standards of enforcement patterns, for example, and calling them alternatives, as they are in a limited sense. As a result, "nondecisions" are made. Wholly different ways of perceiving the problems and conceiving the solutions may have been overlooked, and the primary objective of the EIS has been circumvented. Over the past few years, court decisions and guidelines by various agencies have helped to mold this procedure for the development of environmental impact statements.

An EIS must be thorough, interdisciplinary, and as quantitative as possible. Its preparation involves three distinct phases: inventory, assessment, and evaluation. The first is a cataloging of environmentally susceptible areas, the second is the process of estimating the impact of the alternatives, and the last is the interpretation of these findings.

Environmental Inventories

The first step in evaluating the environmental impact of a project's alternatives is to inventory factors that may be affected by the proposed action. Existing conditions are measured and described, but no effort is made to assess the importance of a variable. Any number and many kinds of variables may be included, such as

1. The "ologies": hydrology, geology, climatology, anthropology, and archeology
2. Environmental quality: land, surface and subsurface water, air, and noise
3. Plant and animal life
4. Socioeconomic conditions

Environmental Assessment

The process of calculating projected effects of a proposed action or construction project on environmental quality is called *environmental assessment*. A methodical, reproducible, and reasonable method is needed to evaluate both the effect of the proposed project and the effects of alternatives that may achieve the same ends but have different environmental impacts. A number of semiquantitative approaches have been used, among them the checklist, the interaction matrix, and the checklist with weighted rankings.

Checklists are lists of potential environmental impacts, both primary and secondary. Primary effects occur as a direct result of the proposed project, such as the effect of a dam on aquatic life. Secondary effects occur as an indirect result— for example, an interchange for a highway may not directly affect wildlife, but indirectly it will draw such establishments as service stations and fast food stores, thus changing land use patterns.

The checklist for a highway project could be divided into three phases: planning, construction, and operation. During planning, consideration is given to the environmental effects of the highway route and the acquisition and condemnation of property. The construction phase checklist includes displacement of people, noise, soil erosion, air and water pollution, and energy use. Finally, the operation phase lists direct impacts owing to noise, water pollution resulting from runoff, energy use, and so forth, and indirect impacts from regional development, housing, lifestyle, and economic development.

The checklist technique thus lists all of the pertinent factors, and then estimates the magnitude and importance of the impacts. Estimated importance of impact may be quantified by establishing an arbitrary scale, such as

0 = no impact

1 = minimal impact

2 = small impact

3 = moderate impact

4 = significant impact

5 = severe impact

The numbers may then be combined, and a quantitative measure of the severity of the environmental impact for any given alternative estimated.

In the checklist technique most variables must be subjectively valued. Moreover, it is difficult to predict future conditions such as changes in land-use patterns or lifestyle. Even with these drawbacks, however, this method is often used by engineers in governmental agencies and consulting firms, mainly because of its simplicity.

The *interaction matrix* technique is a two-dimensional listing of existing characteristics and conditions of the environment and detailed proposed actions that may affect it. This technique is illustrated in Example 24.1. As an example, the characteristics of water might be subdivided into

- Surface
- Ocean
- Underground
- Quantity
- Temperature
- Groundwater
- Recharge
- Snow, ice, and permafrost

Example 24.1
A landfill is to be placed in the floodplain of a river. Estimate the impact by using the checklist technique.

First the items to be impacted are listed, then a quantitative judgment concerning importance and magnitude of the impact is made. In this example are only a few of the impacts normally considered. The importance and magnitude are then multiplied and the sum obtained. Thus:

Potential Impact	Importance × Magnitude
Groundwater contamination	$5 \times 5 = 25$
Surface water contamination	$4 \times 3 = 12$
Odor	$1 \times 1 = 1$
Noise	$1 \times 2 = 2$
Total	40

This total of 40 may then be compared with totals calculated for alternative courses of action.

Similar characteristics must also be defined for air, land, socioeconomic conditions, and so on. Opposite these listings in the matrix are lists of possible actions. In our example, one such action is labeled resource extraction, which can include the following actions:

- Blasting and drilling
- Surface extraction
- Subsurface extraction
- Well drilling
- Dredging
- Timbering
- Commercial fishing and hunting

The interactions, as in the checklist technique, are measured in terms of magnitude and importance. The magnitudes represented by the extent of the interaction between the environmental characteristics and the proposed actions typically may be measured as well. The importance of the interaction, on the other hand, is often a judgment call.

Example 24.2

Lignite (brown) coal is to be surface-mined in the Appalachian Mountains. Construct an interaction matrix for the water resources (environmental characteristics) versus resource extraction (proposed actions).

	Proposed Action							
Environmental Characteristics	Blasting + drilling	Surface excavation	Subsurface excavation	Well drilling	Dredging	Timbering	Commercial fishing	Total
Surface water	3/2	5/5						8/7
Ocean water								
Underground water			3/3					3/3
Quantity								
Temperature			1/2					1/2
Recharge								
Snow, ice								
Total	3/2	9/10						

FIGURE 24–1
Projected environmental quality index for dissolved oxygen

We see that the proposed action will have a significant effect on surface water quality and that the surface excavation phase will have a large impact. The value of the technique is seen when the matrix is applied to alternative solutions. The individual elements in the matrix, as well as row and column totals, can be compared.

If an interaction is present between underground water and well drilling, for example, a diagonal line is placed in the block. Values may then be assigned to the interaction, with 1 being a small and 5 being a large magnitude or importance, and these are placed in the blocks with the magnitude above and importance below. Appropriate blocks are filled in, using a great deal of judgment and personal bias, and then are summed over a line, thus giving a numerical grade for either the proposed action or environmental characteristics.

Example 24.2 is trivial, and cannot fully illustrate the advantage of the interaction technique. With large projects having many phases and diverse impacts, it is relatively easy to pick out especially damaging aspects of the project as well as the environmental characteristics that will be most severely affected.

The search for a comprehensive, systematic, interdisciplinary, and quantitative method for evaluating environmental impact has led to the *checklist-with-weighted-rankings* technique. The intent here is to use a checklist as before to ensure that all aspects of the environment are covered, as well as to give these items a numerical rating in common units.

The first step is to construct a list of items that could be impacted by the proposed alternative, grouping them into logical sets. One grouping might be:

- Ecology
  - Species and populations
  - Habitats and communities
  - Ecosystems
- Aesthetics
  - Land
  - Air
  - Water
  - Biota
  - Human-made objects
- Environmental Pollution
  - Water
  - Air
  - Land
  - Noise
- Human Interest
  - Educational/scientific
  - Cultural
  - Mood/atmosphere
  - Life patterns
  - Economic impact

Each title might have several specific subtopics to be studied; for example, under Aesthetics, "Air" may include odor, sound, and visual impacts. Numerical ratings may now be assigned to these items by a number of techniques. One procedure is to first estimate the ideal or natural levels of environmental quality (without anthropogenic pollution) and take a ratio of the expected condition to the ideal. For example, if the ideal dissolved oxygen in the stream is 9 mg/L, and the effect of the proposed action is to lower that to 3 mg/L, the ratio would be 0.33. This is sometimes called the *environmental quality index* (EQI). Another option is to make the relationship nonlinear, as shown in Figure 24–2. Lowering the dissolved oxygen by a few milligrams per liter will not affect the EQI nearly as much as lowering it, for example, below 4 mg/L, since a dissolved oxygen below 4 mg/L definitely has a severe adverse effect on the fish population.

EQIs are calculated for all checklist items, and the values are tabulated. Next, the weights are attached to the items, usually by distributing 1000 Parameter Importance Units (PIU) among the items. The product of EQI and PIU, called the *Environmental Impact Unit* (EIU), is thus the magnitude of the impact multiplied by the importance:

$$EIU = PIU \times EQI \qquad (24.1)$$

This method has several advantages. We may calculate the sum of EIUs and evaluate the "worth" of many alternatives, including doing nothing. We may also detect points of severe impact, for which the EIU after the project may be much

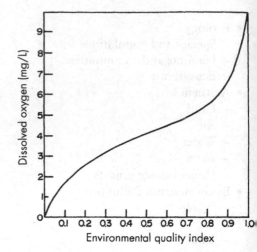

FIGURE 24–2
Projected environmental quality
index curve for dissolved oxygen

lower than before, indicating severe degradation in environmental quality. Its
major advantage, however, is that it makes it possible to input data and evaluate the impact on a much less qualitative and a much more objective basis.

Example 24.3

Evaluate the effect of a proposed lignite strip mine on a local stream. Use 10 PIU
and linear functions for EQI.

The first step is to list the areas of potential environmental impact. These may
be

- Appearance of water
- Suspended solids
- Odor and floating materials
- Aquatic life
- Dissolved oxygen

Other factors can be listed, but these will suffice for this example. Next, we need
to assign EQIs to the factors. Assuming a linear relationship, we can calculate
them as follows:

Item	Condition Before Project	Condition After Project	EQI
Appearance of water	10	3	0.30
Suspended solids	20 mg/L	1000 mg/L	0.02
Odor	10	5	0.50
Aquatic life	10	2	0.20
Dissolved oxygen	9 mg/L	8 mg/L	0.88

Note that we had to put in subjective quantities for three of the items—"Appearance of water," "Odor," and "Aquatic life"—based on an arbitrary scale of decreasing quality from 10 to 1. The actual magnitude is not important since a ratio is calculated. Also note that the sediment ratio had to be inverted to make its EQI indicate environmental degradation—EQI < 1. The EQI indices are weighted by the 10 available PIU, and the EIU are calculated.

Item	Project PIU	After Project $EQI \times PIU = EIU$
Appearance of water	1	$0.3 \times 1 = 0.3$
Suspended solids	2	$0.02 \times 2 = 0.04$
Odor	1	$0.5 \times 1 = 0.5$
Aquatic life	5	$0.2 \times 5 = 1.0$
Dissolved oxygen	1	$0.88 \times 1 = 0.88$
Total	10	2.72

The EIU total of 2.72 for this alternative is then compared with the total EIU for other alternatives.

Evaluation

The final part of the environmental impact assessment is the evaluation of the results of the preceding studies. Typically, the evaluation phase is out of the hands of the engineers and scientists responsible for the inventory and assessment phases. The responsible governmental agency ultimately uses the EIS to justify past decisions or support new alternatives.

SOCIOECONOMIC IMPACT ASSESSMENT

Historically, the President's Council on Environmental Quality has been responsible for overseeing the preparation of EISs, and CEQ regulations list what should be included in all EISs developed by federal agencies. For the proposed projects discussed earlier in this chapter the primary issues are public health dangers and environmental degradation. Under original NEPA and CEQ regulations, both issues must be addressed whenever alternatives are developed and compared.

Recently, federal courts have ruled that consideration of public health and environmental protection alone are not sufficient grounds on which to evaluate a range of alternative programs. Socioeconomic considerations such as population increases, need for public services like schools, and increased or decreased job availability are also included under NEPA considerations. Frequently, public acceptability is also a necessary input to an evaluation process. Although an alternative may protect public health and minimize environmental degradation, it may not be generally acceptable. Factors that influence public acceptability of a given alternative are usually discussed in terms of economics and broad social concerns. Economics includes the costs of an alternative, including the state, regional, local, and private components, the resulting impacts on user charges and prices; and the ability to finance capital expenditures. Social concerns include public preferences in siting (e.g., no local landfills in wealthy neighborhoods) and public rejection of a particular disposal method (e.g., food-chain landspreading of municipal sludge rejected on "general principle"). Consequently, each alternative developed to address the issues of public health and environmental protection must also be analyzed in the context of rigid economic analyses and broad social concerns.

CONCLUSION

Environmental impact assessment requires that a range of solutions to any given environmental pollution problem be developed, analyzed, and compared. This range of alternatives must be viewed in terms of respective environmental impacts and economic assessments. A nagging question exists throughout any such viewing: Can individuals really measure, in the strict "scientific" sense, degradation of the environment? For example, can we place a value on an unspoiled wilderness area? Unfortunately, qualitative judgments are required to assess many impacts of any project. This balancing of values is the foundation of environmental ethics, the topic of the first chapter of this book.

PROBLEMS

24.1 Develop and apply an interaction matrix for the following proposed actions designed to clean municipal wastewater in a community: (a) construct a large activated wastewater treatment plant, (b) require septic tanks for households and small-scale package treatment plants for industries, (c) construct decentralized, small-scale treatment facilities across town, (d) adopt land application technology, (e) continue direct discharge of untreated wastewater into the river. Draw conclusions from the matrix. Which alternatives appear to be superior? Which environmental characteristics appear to be the most important? What should the town do?

24.2 Discuss the advantages and disadvantages of a benefit-cost ratio in deciding whether a town should build a wastewater treatment facility. Focus on the valuation problems associated with analyzing the impacts of such a project.

24.3 Compare the environmental impacts of a coal-fired electric generating plant with those of a nuclear power plant. In your presentation, look at the flow of fuel from its natural state to the facility. Finalize your comparison with waste disposal considerations.

24.4 This problem is an exercise in resource allocation for the entire class. Each member of the class gets a 3×5 index card. Each member of the class also gets *one* of the following "resources": red construction paper to represent housing, green construction paper to represent food, a gold star to represent 10 "money" points, scissors, or tape (to tape the housing and food to the card). Students who do not get red or green paper, or scissors, or tape, each get a gold star. One square inch of housing "costs" 5 points; one square inch of food "costs" 2 points. The amount of red paper should equal exactly 1 square inch (one house) per student; the amount of green paper, 3 square inches per student. There are no other rules!! The class then barters until all housing and food are distributed. There is a bonus (e.g., 10 extra points, a cookie?) for the student who accumulates the most points on his or her card.

Analyze the results—for example: How was wealth distributed? Was each "resource" equally important? Did anyone end up "homeless"? Was all barter legal? Other analyses will become obvious as well. (The authors wish to thank Dr. Robin Matthews for contributing this exercise.)

LIST OF SYMBOLS

CEQ	Council on Environmental Quality
EIA	environmental impact assessment
EIS	environmental impact statement
EIU	Environmental Impact Unit
EPA	Environmental Protection Agency
EQI	environmental quality index
FONSI	Finding of No Significant [Environmental] Impact
NEPA	National Environmental Policy Act
PIU	Parameter Importance Units
POTW	Publicly Owned (Wastewater) Treatment Works
PPBS	Planning, Programming, and Budgeting System
WEPA	Wisconsin Environmental Policy Act

Conversion Factors

Multiply	By	To Obtain
acre	0.404	ha
acre ft	1233	m^3
atmospheres	14.7	lb/in^2
British thermal units (Btu)	252	cal
Btu	1.054×10^3	J
Btu/ft^3	8905	cal/m^3
Btu/lb	2.32	J/g
Btu/lb	0.555	cal/g
Btu/sec	1.05	kW
Btu/ton	278	cal/tonne
calories (cal)	4.18	joules
calories	3.9×10^{-3}	Btu
cal/g	1.80	Btu/lb
cal/m^3	1.12×10^{-4}	Btu/ft^3
cal/tonne	3.60×10^{-3}	Btu/ton
centimeters	0.393	in
feet (ft)	0.305	m
ft/min	0.00508	m/sec
ft/sec	0.305	m/sec
ft^2	0.0929	m^2
ft^3	0.0283	m^3
ft^3	28.3	liters
ft^3/sec	0.0283	m^3/sec
ft^3/sec	449	gal/min
ft lb (force)	1.357	joules
ft lb (force)	1.357	newton meters
gallons (gal)	3.78×10^{-3}	m^3
gallons	3.78	liters
$gal/day/ft^2$	0.0407	$m^3/day/m^2$
gal/min	2.23×10^{-3}	ft^3/sec

Multiply	By	To Obtain
gal/min	0.0631	liter/sec
gal/min	0.227	m^3/hr
gal/min	6.31×10^{-5}	m^3/sec
gal/min/ft^2	2.42	m^3/hr/m^2
million gal/day (mgd)	43.8	liters/sec
mgd	3785	m^3/day
mgd	0.0438	m^3/sec
grams	2.2×10^{-3}	lb
hectares	2.47	acre
horsepower	0.745	kW
inches	2.54	cm
inches of mercury	0.49	lb/in^2
inches of mercury	3.38×10^3	newton/m^2
inches of water	249	newton/m^2
joule (J)	0.239	calorie
joule	9.48×10^{-4}	Btu
joule	0.738	ft lb
joule	2.78×10^{-7}	kWh
joule	1	newton meter
J/g	0.430	Btu/lb
J/sec	1	watt
kilograms (kg)	2.2	lb (mass)
kg	1.1×10^{-3}	tons
kg/ha	0.893	lb/acre
kg/hr	2.2	lb/hr
kg/m^3	0.0624	lb/ft^3
kg/m^3	1.68	lb/yd^3
kilometers (km)	0.622	mi
km/hr	0.622	mph
kilowatts (kW)	1.341	horsepower
kWh	3600	kilojoule
liters (L)	0.0353	ft^3
liters	0.264	gal
liters/sec	15.8	gal/min
liters/sec	0.0288	mgd
meters (m)	3.28	ft
meters	1.094	yd
m/sec	3.28	ft/sec
m/sec	196.8	ft/min
m^2	10.74	ft^2
m^2	1.196	yd^2
m^3	35.3	ft^3

Multiply	By	To Obtain
m^3	264	ga
m^3	1.31	yd^3
m^3/day	264	gal/day
m^3/hr	4.4	gpm
m^3/hr	6.38×10^{-3}	gpm
m^3/sec	35.31	ft^3/sec
m^3/sec	15,850	gpm
m^3/sec	22.8	mgd
miles	1.61	km
mi^2	2.59	km^2
mph	0.447	m/sec
milligrams/liter	0.001	kg/m^3
million gallons (mgd)	3785	m^3
mgd	43.8	liter/sec
mgd	157	m^3/hr
mgd	0.0438	m^3/sec
newton (N)	0.225	lb (force)
newton/m^2	2.94×10^{-4}	inches of mercury
newton/m^2	1.4×10^{-4}	lb/in^2
newton meters	1	joule
newton sec/m^2	10	poise
pounds (force)	4.45	newton
pounds (force)/in^2	6895	N/m^2
pounds (mass)	454	g
pounds (mass)	0.454	kg
pounds (mass)/ft^2/yr	4.89	kg/m^2/yr
pounds (mass)/yr/ft^3	16.0	$kg/yr/m^3$
pounds/acre	1.12	kg/ha
pounds/ft^3	16.04	kg/m^3
pounds/in^2	0.068	atmospheres
pounds/in^2	2.04	inches of mercury
pounds/in^2	7140	newton/m^2
rad	0.01	gray
rem	0.01	sievert
tons (2000 lb)	0.907	tonne (1000 kg)
tons	907	kg
ton/acre	2.24	tonnes/ha
tonne (1000 kg)	1.10	ton (2000 lb)
tonne/ha	0.446	tons/acre
yd	0.914	m
yd^3	0.765	m^3
watt	1	J/sec

Appendix B

Elements and Atomic Weights

	Symbol	Atomic No.	Atomic Weight
Actinium	Ac	89	227*
Aluminum	Al	13	26.98
Americium	Am	95	242*
Antimony	Sb	51	121.75
Argon	Ar	18	39.95
Arsenic	As	33	74.92
Astatine	At	85	215*
Barium	Ba	56	137.34
Berkelium	Bk	97	247*
Beryllium	Be	4	9.01
Bismuth	Bi	83	208.98
Boron	B	5	10.81
Bromine	Br	35	79.91
Cadmium	Cd	48	112.4
Calcium	Ca	20	40.08
Californium	Cf	98	250*
Carbon	C	12	12.01
Cerium	Ce	58	140.12
Cesium	Cs	55	132.91
Chlorine	Cl	17	35.45
Chromium	Cr	24	52.0
Cobalt	Co	27	58.93
Copper	Cu	29	63.54
Curium	Cm	96	244*
Dysprosium	Dy	66	162.5
Einsteinium	Es	99	252*
Erbium	Er	68	167.26
Europium	Eu	63	151.96
Fermium	Fm	100	254*
Fluorine	F	9	19.0
Francium	Fr	87	223*
Gadolinium	Gd	64	157.25
Gallium	Ga	31	69.72

	Symbol	Atomic No.	Atomic Weight
Germanium	Ge	32	72.59
Gold	Au	79	196.97
Hafnium	Hf	72	178.49
Helium	He	2	4.0
Holmium	Ho	67	164.93
Hydrogen	H	1	1.01
Indium	In	49	126.90
Iodine	I	53	138.91
Iridium	Ir	77	192.2
Iron	Fe	26	55.85
Krypton	Kr	36	83.8
Lanthanum	La	57	138.91
Lawrencium	Lw	103	**
Lead	Pb	82	207.19
Lithium	Li	3	6.94
Lutetium	Lu	71	174.97
Magnesium	Mg	12	24.31
Manganese	Mn	25	54.94
Mendelevium	Md	101	**
Mercury	Hg	80	200.59
Molybdenum	Mo	42	95.94
Neodymium	Nd	60	144.24
Neon	Ne	10	20.18
Neptunium	Np	93	237*
Nickel	Ni	28	58.71
Niobium	Nb	41	92.91
Nitrogen	N	7	14.01
Nobelium	No	102	**
Osmium	Os	76	190.2
Oxygen	O	8	16.0
Palladium	Pd	46	106.4
Phosphorus	P	15	30.97
Platinum	Pt	78	195.09
Plutonium	Pu	94	237*
Polonium	Po	84	209*
Potassium	K	19	39.10
Praseodymium	Pr	59	140.91
Promethium	Pm	61	**
Protactinium	Pa	91	234*
Radium	Ra	88	228*
Radon	Rn	86	222*
Rhenium	Re	75	186.2
Rhodium	Rh	45	102.91
Rubidium	Rb	37	85.47
Ruthenium	Ru	44	101.07
Samarium	Sm	62	150.35
Scandium	Sc	21	44.96

	Symbol	*Atomic No.*	*Atomic Weight*
Selenium	Se	34	78.96
Silicon	Si	14	28.09
Silver	Ag	47	107.87
Sodium	Na	11	23.0
Strontium	Sr	38	87.62
Sulfur	S	16	32.06
Tantalum	Ta	73	180.95
Technetium	Tc	43	99*
Tellurium	Te	52	127.60
Terbium	Tb	65	158.92
Thallium	Tl	81	204.37
Thorium	Th	90	232.04
Thulium	Tm	69	168.93
Tin	Sn	50	118.69
Titanium	Ti	22	47.90
Tungsten	W	74	183.85
Uranium	U	92	92
Vanadium	V	23	50.94
Xenon	Xe	54	131.30
Ytterbium	Yb	70	173.04
Yttrium	Y	39	88.91
Zinc	Zn	30	65.37
Zirconium	Zr	40	91.22

*All istopes of these elements are radioactive. The mass given here is that of the prevalent isotope.

**All isotopes of these elements are so short-lived that no mass number is given.

Appendix C
Physical Constants

UNIVERSAL CONSTANTS

Avogadro's Number = 6.02252×10^{23}

Gas constant R = 8.315 kJ/kg-mole-°K = 0.08315 bar-m³/kg-mole-°K
= 0.08205 L-atm/mole-°K = 1.98 cal/mole-°K

Gravitational acceleration at sea level g = 9.806 m/sec² = 32.174 ft/sec²

Properties of water:

Density = 1.0 kg/L
Specific heat = 4.184 kJ/kg-°C = 1000 cal/kg-°C = 1.00 kcal/kg-°C
Heat of fusion = 80 kcal/kg = 334.4 kJ/kg
Heat of vaporization = 560 kcal/kg = 2258 kJ/kg
Viscosity at 20°C = 0.01 poise = 0.001 kg/m-sec

Glossary and Abbreviations

Absorption: process by which one material is captured in another either chemically or by going into solution.

Activated Carbon: material made from coal when hydrocarbons are driven off under intense heat but no oxygen, leaving a tremendous surface area on which many chemicals can adsorb.

Activated Sludge: suspension of microorganisms taken from the bottom of the final clarifier.

Activated Sludge System: an aerated basin in which microorganisms reduce organics to CO_2, H_2O, other stable materials, and more microorganisms; and then a settling tank (final clarifier) in which the microorganisms are separated out and recirculated into the aeration basin.

Acute: severe and short-lived (a disease, for example).

Adiabatic Lapse Rate: change in temperature with elevation as the result of atmospheric pressure. In dry air the adiabatic lapse rate is 1°C/100 m. Adiabatic means that heat is neither added nor removed.

Adsorption: process by which one material is attached to another, such as an organic on activated carbon. It is a surface phenomenon.

Advanced Waste Treatment: wastewater treatment beyond the secondary or biological stage. It may include the removal of nutrients such as phosphorus and nitrogen or any other potential problems.

Aeration: process of being supplied or impregnated with air. Aeration is used in wastewater treatment to foster biological purification.

Aerobic: presence of free oxygen.

Aerosol: suspension of fine particles in a gas.

Algae: one-celled aquatic organisms with chlorophyll that grow in the presence of light, CO_2, and nutrients, and release oxygen.

Algal Bloom: proliferation of living algae on the surface of lakes or ponds.

Alum: aluminum sulfate.

Alveoli: air sacs in the lung where oxygen and carbon dioxide transfer takes place.

Ambient Air: any unconfined portion of the atmosphere; the outside air.

Anaerobic: absence of free oxygen.

Anticyclone: high-pressure cell with winds circulating about a center (clockwise in the northern hemisphere).

Appropriations Doctrine: basis for water law in the western United States.

Aquifer: water-bearing geologic stratum.

Asbestos: mineral fiber with countless industrial uses; a hazardous air pollutant when inhaled.

Assimilation: ability of a body of water to purify itself of organic pollution.

Asthma: difficulty in breathing caused by constriction of bronchial tubes.

Attrition: wearing or grinding down by friction. This is one of the three basic contributing processes of air pollution; the others are vaporization and combustion.

Audiometer: instrument for measuring hearing sensitivity.

BOD: biochemical oxygen demand.

Bag Filter: device for removing particulates in an air stream.

Baling: means of reducing the volume of solid waste by compaction.

Bar: unit of atmospheric pressure measurement equal to 1 dyne/cm^2. 1 bar = 1000 millibars.

Bar Screen: in wastewater treatment, a screen that removes large floating and suspended solids.

Benthic Region: bottom of a body of water.

Beryllium: metal that when airborne has adverse effects on human health. It has been declared a hazardous air pollutant.

Biodegradation: metabolic process by which high-energy organics are converted to low energy, CO_2, and H_2O.

Biochemical Oxygen Demand (BOD): amount of oxygen used by microorganisms (and chemical reactions) in the biodegradation process. BOD is usually measured at 20°C for 5 days.

Biodegradable: having the capacity of decomposing quickly as a result of the action of microorganisms.

Brackish Water: mixture of fresh and salt water.

Bronchiole: small air tube in the lung.

Bronchitis: acute inflammation of air passages.

Bubbler: device for measuring gaseous air pollutants.

COD: chemical oxygen demand.

cfs: cubic feet per second; a measure of the amount of water passing a given point.

Calorie: amount of heat required to raise the temperature of one gram of water one degree centigrade.

Carcinogen: cancer-producing substance.

Catalyst: substance that speeds up a chemical reaction (such as combustion) without entering into the reaction.

Centrifuge: device for dewatering slurries such as wastewater sludge.

Chemical Oxygen Demand (COD): amount of oxygen used in chemically oxidizing a substance.

Chlorinated Hydrocarbons: a class of generally long-lasting, broad-spectrum insecticides, of which the best known is DDT.

Chlorination: application of chlorine to drinking water, sewage, or industrial waste for disinfection or oxidation of undesirable compounds.

Chlorophyll: substance found in plants that absorbs energy from light.

Chlorosis: loss of green color.

Chronic: less severe and longer lived than acute.

Cilia: tiny hair cells lining air passages.

Clarifier: settling tank.

Clear Well: storage tank for finished potable water.

Coagulation: process of chemically treating a turbid waste to reduce the charge on the particles and thus make it possible for them to flocculate.

Coliforms: group of bacteria that produce gas and ferment lactose, some of which are found in the intestinal tracts of warm-blooded animals.

Combined Sewer: carries sanitary wastes as well as stormwater.

Comminutor: grinds up large solids as preparation for further wastewater treatment.

Convection: transmission of energy by movement of fluids.

Cyclone: low-pressure system circulating winds (counterclockwise in the northern hemisphere).

Cyclone: device for removing large dust particles.

dB: decibel.

dB(A): decibels as measured on the A scale.

DDT: first of the modern chlorinated hydrocarbon insecticides, whose chemical name is 1,1,1-tricholoro-2,2-bis (p-chloriphenyl)-ethane.

DO: dissolved oxygen.

DOC: dissolved organic carbon.

Decibel: measure of sound intensity.

Decomposition: reduction of the net energy level and change in chemical composition of organic matter because of the actions of aerobic or anaerobic microorganisms.

Desalinization: salt removal from sea or brackish water.

Detention Time: average time required to flow through a basin.

Diffused Air: method of aerating the microorganisms in the aeration tank by blowing in air through porous diffusers.

Digestion: decomposition of organics, either aerobically or anaerobically, usually at high solids concentrations and, in the case of anaerobic digestion, at elevated temperatures.

Dust: fine solid particles.

Dustfall: gravimetric measurement of dust by settling in a jar.

Dystrophic Lakes: lakes between the eutrophic and swamp stages of aging.

Ecology: interrelationship of living things to one another and to their environment, or the study of such interrelationships.

Edema: swelling of tissues and accumulation of fluid in a body or organ.

Effluent: liquid flowing out.

Electrostatic Precipitator: device for removing fine particulate matter in an air stream.

Emphysema: loss of air exchange capacity by deterioration of the alveoli.

Epidemiology: science of statistically evaluating and dealing with diseases in populations.

Epilimnion: top layer of a lake.

Eutrophication: process of aging of lakes and other still water bodies; characterized by excessive aquatic growth.

Eutrophic Lakes: shallow lakes, weed choked at the edges and very rich in nutrients.

Fecal Coliforms: coliforms specifically originating from warm-blooded intestines.

Feedlots: concentrations of animals for fattening before slaughter.

Final Clarifier: last settling tank in wastewater treatment prior to discharge.

Flocculation: process of forming large clumps from small particles once the particles have coagulated.

Fluorosis: bone disease caused by excessive ingestion of fluorides.

Fly Ash: fine particles generated by noncombustible materials during the burning of coal.

Fume: dust forming from condensation of vapors.

Fungi: small plants without chlorophyll.

Garbage: food waste in refuse.

Genetic: pertaining to origin and development.

Greenhouse Effect: warming due to trapping of heat radiation.

Grit Chamber: used to remove sand and other large, heavy particles in a wastewater treatment plant.

HC: hydrocarbons.

Hz: Hertz.

Heavy Metals: metallic elements with high molecular weights, generally toxic to plant and animal life in low concentrations. Examples include mercury, chromium, cadmium, arsenic, and lead.

Hemoglobin: iron-containing protein pigment in blood that carries oxygen from the lung to other parts of the body.

Heat Island: concentration of warm air over a city, preventing external circulation.

Hertz (Hz): unit of frequency, in cycles/second.

Hi Vol: high-volume sampler.

High-Volume Sampler: device for measuring particulates in air by capture on a filter.

Humus: decomposed organic material.

Hydrocarbons: chemicals containing hydrogen and carbon.

Hypolimnion: bottom layer of a lake.

Incineration: oxidation in the presence of heat and free oxygen, ideally producing CO_2, H_2O, and other stable compounds.

Infiltration: water entering sanitary sewers through broken pipes, illegal connections, and so forth.

Influent: liquid flowing in.

Interceptor Sewer: carries waste from a sewerage system to a treatment plant or to another interceptor.

Inversion: atmospheric condition where a warmer body of air is above a colder air mass.

L_{10}: symbol for indicating 10 percent of data are greater than the stated number.

Lagoon: hole-in-the-ground for temporary disposal of wastes.

Lapse Rate: rate at which temperature varies with altitude.

Lime: calcium hydroxide.

Limnology: study of the physical, chemical, meteorological, and biological aspects of fresh waters, specifically lakes.

MGD: millions of gallons per day, commonly used to express rate of flow.

Masking: covering over of one sound or odor by another.

Mechanical Aeration: method of aerating the microorganisms in an aeration tank by beating and splashing the surface.

Mercaptans: organic compounds containing sulfur.

Micrograms: 1-millionth of a gram.

Microscreening: removal of small particles from water by filtering through a metal screen on a rotating drum.

Mist: small liquid droplets in air.

Morbidity: measure of disease in a population.

Mortality: measure of death in a population.

NPL: noise pollution level.

Necrosis: death of living tissue.

Noise Pollution Level (NPL): calculated value in dB which takes into account the irritation of noise variation.

OSHA: Occupational Safety and Health Administration.

Oligotrophic Lakes: deep lakes that have a low supply of nutrients and contain little organic matter.

Organophosphates: group of pesticide chemicals containing phosphorus, such as malathion and parathion, intended to control insects.

Oxidation Pond: method of wastewater treatment allowing biodegradation to take place in a shallow pond.

Oxygen Sag: drop in DO following pollution of a stream with subsequent recovery.

ppm: parts per million; weight/weight for water and volume/volume for air.

PAN: peroxyacetyl nitrate, a component of photochemical smog.

Particulates: finely divided solid or liquid particles in the air or in an emission.

Pathogen: microorganism that causes disease.

Percolation: movement of water from the surface into the ground.

Photochemical: pertaining to reactions affected by light.

Polyelectrolytes: chemicals used in water and wastewater treatment for coagulation and flocculation.

Potable Water: safe, drinkable, and pleasing water.

Primary Clarifier: first settling tank in a wastewater treatment plant.

Primary Treatment: removal of solids in wastewater.

Pulmonary: pertaining to lungs.

Pyrolysis: combustion in the absence of oxygen.

Rapid Mix: device used for mixing chemicals in water treatment.

Rapid Sand Filter: device for removing turbidity in water by seepage through a bed of sand and washing the sand by flow reversal.

Rainout: removal of air pollution by condensation into rain droplets.

Raw Sewage: untreated domestic or commercial wastewater.

Raw Primary Sludge: slurry from the bottom of the primary clarifier.

Refuse: urban solid waste.

Ringelmann: number used to report density of smoke.

Riparian Doctrine: basis for water law in the eastern United States.

Rubbish: nongarbage fraction of refuse.

SL: sound level.

SPL: sound pressure level.

Salinity: degree of salt in water.

Sanitary Landfill: solid waste disposal in the ground using approved techniques.

Sanitary Sewers: carry only domestic wastewater.

Scrubber: device for removing air contaminants by bringing them into contact with water.

Secondary Treatment: removal of oxygen demand in wastewater.

Sedimentation: settling of solids in water and wastewater.

Septic Tank: underground tank for treating small flows of domestic wastewater.

Settling Tank: in water and wastewater treatment, a tank where solids are allowed to settle.

Sewage: domestic wastewater.

Sewers: pipes used to carry wastewater.

Sewerage System: system of sewers used to carry wastewater.

Sludge: wastewater solids suspended in water.

Smog: originally a combination of smoke and fog, as occurred in London, now synonymous with polluted air.

Smoke: waste from incomplete combustion expelled with the air stream.

Sound Level Meter: device for measuring sound in dB, on an A, B, or C scale.

Sound Level: sound approximately perceived by human ears, expressed as a reading on the sound level meter.

Sound Pressure: fluctuating air pressure as propagated sound waves.

Sound Pressure Level: change in pressure due to a sound, measured in dB.

Spray: large droplets of liquid.

Storm Sewers: carry stormwater only.

Synergism: cooperative action of separate substances so that the total effect is greater than the sum of the effects of the substances acting independently.

Tertiary Treatment: follows secondary wastewater treatment and used to polish the effluent.

Thermocline: inflection point in a lake temperature profile.

Threshold: limit below which no effect is discernible.

Tile Field: pipes laid in the ground with spaces in between so as to promote percolation of wastewater into the ground.

Tone: pure sound uniform in frequency.

Trachea: air duct from the larynx to the bronchial tubes.

Transpiration: process of water transport to the atmosphere through plants.

Trickling Filter: device for removing oxygen demand from wastewater by dribbling the water over rocks covered with a zoological slime.

Turbidity: interference with the passage of light through water caused by suspended matter.

Turnover: mixing of a lake due to thermal variations.

USPHS: United States Public Health Service.

Vacuum Filter: device used for dewatering wastewater sludge.

Venturi Scrubber: high-efficiency device for scrubbing contaminated air.

WHO: World Health Organization.

Washout: removal of air pollutants by rain.

Weir: metal plate over which liquid effluent flows; used in clarifiers.

Wind Rose: graphic representation of wind data.

Zooplankton: tiny aquatic animals.

Index

ISO 14000

Environmental Quality Management Handbook

By Ken Whitelaw

ISO 14000 is the new international environmental quality management standard. This book contains the most recent information about the standard, including all its final modifications. Its purpose is to do for environmental control what ISO9000 has done for quality management. *ISO14000* will outline the scope and purpose of the standard and make it accessible to everyone. It will lead the reader through the theory and implementation of an ISO14000-compliant environmental management system, including the consultant's and auditor's perspective, and give case studies from industries which have actually undergone the process. Finally, there will be input from training organizations and certification and accreditation bodies that will help with trouble-shooting and assessment.

CONTENTS: Preface; Foreword; Acknowledgments; A History of Environmental Systems; Concepts of ISO 14001, Clauses of ISO 14001; Assessment and certification to the Standard by third party Certification Bodies; Integration of Environmental Management Systems with existing or planned: Quality Management Systems, Health and Safety Management, Investors in people/ Case histories of Companies who have successfully implemented ISO 14001 and how they addressed some of the more subjective issues; An insight into the third party certification auditor; EMAS and ISO 14001; Appendix 1: Glossary of Terms; Appendix 2: Further reading

---

January 1998   0 7506 3766 8   320pp   hb   75 line illus

CPSIA information can be obtained
at www.ICGtesting.com
Printed in the USA
LVOW04s0805080118
562161LV00001B/1/P